national electrical code®

blueprint reading

based on the 1996 NEC®

AF700291

AMERICAN TECHNICAL PUBLISHERS, INC.
HOMEWOOD, ILLINOIS 60430

R. T. Miller

© 1995 by American Technical Publishers, Inc.
All rights reserved

1 2 3 4 5 6 7 8 9 – 95 – 9 8 7 6 5 4 3 2 1

Printed in the United States of America

ISBN 0-8269-1558-2

Introduction

PREFACE

This edition of *National Electrical Code® Blueprint Reading* has been revised to reflect many of the important changes that appear in the 1996 National Electrical Code®. The changes are presented as they pertain to One-family Dwellings, Multifamily Dwellings, Commercial Locations, Industrial Locations, and Hazardous Locations. In addition to these appropriate changes to the text and illustrations, the Trade Competency Tests and two Final Tests have been thoroughly updated and expanded.

National Electrical Code® Blueprint Reading is designed to enable the student to learn electrical blueprint reading and become familiar with applicable sections of the 1996 National Electrical Code. Complete references to the 1996 NEC are presented throughout the text. Trade Competency Tests appear at the end of each Chapter to help students check their understanding of the text material and the NEC. Two Final Tests are included in Chapter 6 of the text. Answers, with Code validation for the Trade Competency Tests and Final Tests are included in the *Instructor's Guide for National Electrical Code® Blueprint Reading.*

NATIONAL ELECTRICAL CODE

The National Electrical Code is sponsored and controlled by the National Fire Protection Association (NFPA). The primary function of the NEC is to safeguard people and property against electrical hazards. It is mandatory that the *National Electrical Code® Blueprint Reading* be used only in conjunction with the 1996 NEC. Copies of the 1996 NEC (NFPA No. 70) can be ordered directly from its publisher

National Fire Protection Association, Inc.
Batterymarch Park
Quincy, MA 02269

WATTS and VOLT-AMPERES

In general, within the NEC, the term *watts* (W) has been superseded by the term *volt-amperes* (VA) for the computation of loads. However, references to nameplate ratings still reflect the term watts on certain loads.

CALCULATIONS

When total wattage or VA is to be divided by phase-to-phase (3ϕ) voltage times 1.732, the following values may be substituted

for 208 V × 1.732, use 360
for 230 V × 1.732, use 398
for 240 V × 1.732, use 416
for 440 V × 1.732, use 762
for 460 V × 1.732, use 797
for 480 V × 1.732, use 831

MANDATORY USE OF SHALL

Section 110-1 states that mandatory rules use the word *shall.* Always refer to the NEC for mandatory rules.

FRACTIONS OF AN AMP

Chapter 9, B. Examples states that fractions less than .5 may be dropped.

THW Cu

Unless otherwise specified, copper conductors are sized on THW per Table 310-16.

TRADE COMPETENCY TESTS and FINAL TESTS

National Electrical Code® Blueprint Reading contains 19 Trade Competency Tests that should be completed after studying the corresponding chapter. The two Final Tests include questions that cover the content of the entire text. The Final Tests should be taken after completing the text and all of the Trade Competency Tests.

QUESTIONS

Question formats include True-False, Multiple Choice, and Completion. Always record the answer in the space provided in the Answer column and provide Code validation in the Section column.

RECORD CODE VALIDATION HERE

RECORD ANSWER HERE

Section	Answer	
230-74	(T) F	7. The disconnecting means for a lighting service shall disconnect all ungrounded conductors simultaneously.
370-15(a)	metal	8. Nonmetallic boxes may have nonmetallic or__________covers.
250-79(f)	A	9. An equipment bonding jumper is permitted outside a raceway when not over __________′ long. A. 6 C. 10 B. 8 D. none of the above

Contents

NEC Changes

Key changes of the 1996 NEC® include:

- **90-2(a)(5)** Only those installations that are an integral part of the electrical utilities generation, transmission, or control of power are excluded from the scope of the NEC®.
- **90-2(b)(5)** Commercial parking lot lighting is covered by the NEC®, while associated lighting, under control of the utility, is not.
- **100** The definition of *bathroom*, which appears in more than two articles, has been relocated to 100.
- **100** The definition of *energized* has been added to 100. The definition is taken from the NFPA, NEC® Manual of Style.
- **100** The definition of *in sight from* has been revised to clarify that when equipment is required by the NEC® to be in sight from other equipment, it must be within 50′ and visible.
- **100** The definition of *live parts* has been added to 100. The definition is taken from NFPA, NEC® Manual of Style.
- **100** A definition of *nonlinear load* now appears in 100.
- **100** The definition of *service conductors* has been revised and relocated to 100.
- **110-4** All equipment shall have a nominal voltage rating which is not less than the circuit nominal voltage rating to which it is connected.
- **110-14(b)** Devices intended for connecting or splicing conductors underground shall be listed for direct burial use.
- **110-14(c)** Conductors with temperature ratings higher than the terminations are permitted provided conductor ampacity is based on the lower rating.
- **110-16(a)** Equipment is permitted to be stacked within height requirements of the section provided it is of equal depth.
- **110-16(a), Ex. 1** Electrical equipment, for which rear access is required to work on de-energized components, shall have a minimum of 30″ horizontal work space clearance.
- **110-16(e)** The minimum height required is 6½′ about electrical equipment.
- **110-31** A fence around 600 V, nominal equipment shall be at least 7′ in height of barbed wire or equivalent.
- **110-40** Temperature limitation requirements are provided for medium voltage conductors and their terminations. See 110-14(c).
- **200-10(b)** The termination point on receptacles, attachment plugs, and cord connectors for the grounded conductor may be marked with a letter "W" instead of the word "white" or the color white.
- **210-6(c)(1)** Electric-discharge lighting fixtures of all types may be supplied by 277 V circuits provided the fixtures are listed.
- **210-6(c)(2)** Incandescent lighting fixtures of all types may be supplied by 277 V circuits provided the fixtures are listed.
- **210-8(a)(2)** GFCI protection for garages includes other buildings or structures such as workshops or storage sheds with grade-level access.
- **210-8(a)(3)** GFCI protection includes all outdoor receptacles except those not readily accessible and supplied from a dedicated branch circuit for snow-melting equipment.
- **210-8(a)(4)** All receptacles installed in crawl spaces of dwelling units, that are at or below grade, shall have GFCI protection.
- **210-8(a)(6)** All receptacles intended to serve countertop surfaces shall be provided with GFCI protection.
- **210-8(a)(7)** For GFCI protection for receptacles installed within 6′ of a wet bar sink, the 6′ measurement shall be taken from the outside edge of the sink.
- **210-52(b)(1)** Only receptacle outlets covered by 210-52(a) and 210-52(c), and the outlet for refrigeration equipment are required to be on the small appliance branch circuit.
- **210-52(b)(1), Ex. 2** The refrigerator receptacle outlet may be installed on a 15 A, or greater, individual branch circuit.
- **210-52(d)** Receptacle outlets in dwelling bathrooms shall be supplied by a dedicated 20 A branch circuit. No other outlets are permitted on the circuit.
- **210-52(d)** Receptacle outlets installed in bathroom basin countertops or work surfaces shall not be installed in the face-up position.
- **210-52(e)** The required outdoor receptacle(s) for a dwelling shall be installed not more than 6′-6″ above grade.
- **210-63** A receptacle outlet shall be installed within 25′ of HACR equipment on the roof, in an attic, or in a crawl space.
- **210-70(a)** The lighting outlet required for outdoor entrances and exits shall be installed outside of the entrance or exit.
- **220-2** 347 V and 347/600 V have been added to the list of nominal voltages.
- **220-10(b)** Feeders and service conductors supplying continuous loads shall have both the overcurrent devices and conductors sized at 125%.
- **225-1** Branch circuits or feeders may be run on, not just between, buildings or structures.
- **225-8(b)** A disconnecting means required at each building shall be installed at a readily accessible location where the conductors enter the building.
- **225-8(d)** A directory shall be permanently installed at each disconnecting means location specifying all of the other circuits supplying the building.
- **225-19(d)** A 3′ clearance is required only from windows that open.
- **230-7** Service conductors installed in FMC require an equipment bonding jumper.
- **230-9** Balconies, ladders, and stairs require 3′ clearance from service conductors.
- **230-24(b)** The lowest point of the drip loop shall have 10′ clearance.
- **230-28** Only power service-drop conductors shall be attached to the service mast.

- **230-54(b)** Taping or sealing is not required if service cables are continuous from the pole to the service equipment.
- **230-54(e), Ex.** Conductors of different potential shall emerge from service heads through separate openings.
- **230-70(a)** Service equipment shall not be installed in bathrooms.
- **240-3(b)** Three conditions shall be met if the next higher standard overcurrent device is to be used.
- **240-20(b)(1)** Individual SPCBs are permitted with or without handle ties for protection of ungrounded conductors on certain AC circuits.
- **240-20(b)(2)** Individual SPCBs are permitted with or without handle ties for protection of ungrounded conductors on certain DC circuits.
- **240-20(b)(3)** Individual SPCBs are permitted with or without handle ties for protection of ungrounded conductors on line-to-line 3ϕ, 4-wire or 2ϕ, 5-wire grounded systems.
- **240-21(b)(3)** 10′ tap conductors are permitted to terminate in a single safety switch.
- **240-24(e)** All OCPDs, with the exception of supplementary devices, shall not be installed in bathrooms.
- **240-83(d)** CBs used as switches shall be listed and contain the marking "SWD."
- **250-32** Raceways and enclosures both shall be grounded.
- **250-43(g)** The requirement for grounding electric signs has been expanded to include outline lighting.
- **250-80(a)** When separately derived systems are installed, interior metal water piping nearest the area served shall be bonded to the grounded conductor of the separately derived system.
- **250-80(c)** Steel that is not intentionally grounded shall be bonded to the service equipment enclosure, grounded conductor, or GEC at the service or to the grounding electrodes themselves.
- **250-81** Piping beyond the first 5′ in a building shall not be used as a conductor to interconnect GES components or as a part of the GES.
- **250-95** The same rule for sizing EGCs applies whether conductors are run in parallel raceways or parallel cables.
- **250-113** Sheet-metal screws shall not be used to connect grounding conductors to enclosures and equipment.
- **250-119** The symbol may be used in lieu of "green," "ground," "G," or "GR" for identifying the EGC terminal on wiring devices.
- **280-4(a)** All surge arresters used in circuits of less than 1000 V shall be listed.
- **300-4(e)** Cables or raceways installed in shallow grooves and covered with paneling shall be protected with 1/16″ steel plate.
- **300-4(f)** The requirements for insulated fittings apply throughout the NEC®.
- **300-5(j)** Means shall be taken to protect against damage to conductors as a result of ground movement.
- **300-11(a)** Ceiling grids, in either fire-rated or nonfire-rated ceiling assemblies, are not permitted to support any cables or raceways.
- **300-11(a)(1)** For wiring located above fire-rated roof/ceiling/floor assemblies, an independent means of support shall be provided unless the ceiling support system has been tested as part of the fire-rated assembly.
- **300-11(a)(2)** For nonfire-rated roof/ceiling/floor assemblies, wiring shall not be supported by the ceiling support wires unless the ceiling system manufacturer's instructions specifically permit the ceiling wires to be used for support.
- **305-2** All temporary wiring methods shall be approved for the conditions of use.
- **305-3(b)** A new 525 contains provisions for electrical installations in carnivals, circuses, fairs, and similar events.
- **305-6** All 15 A and 20 A, 125 V receptacles for temporary power for construction, remodeling, etc., shall be GFCI protected.
- **305-6(a)** Any receptacle, whether existing or part of permanent wiring, shall have GFCI protection if used to supply temporary power.
- **Note 3 of Notes to Ampacity Tables of 0 to 2000 Volts** The size of feeder conductors to a building is limited to the size of service-entrance conductors.
- **331-4(9)** ENT, installed where exposed to rays of the sun, shall be identified as being sunlight resistant.
- **333-11** Armored cable may be run on underside of joists in all locations if not subject to physical damage.
- **334-1** The definition of metal-clad cable includes a listed metal-clad cable that contains optical-members along with circuit conductors.
- **338-2** USE, used for service laterals, may emerge from the ground if it is suitably protected.
- **343-5(b)** The maximum size permitted for nonmetallic underground conduit with conductors has been increased from 2″ to 4″.
- **350-10(b)** The maximum size permitted for FMC is 4″ electrical trade size.
- **Table 350-12** Many values in the Table have been adjusted downward, allowing fewer conductors in the conduit.
- **352-40** A new Part C has been added to 352 covering Strut-type Channel Raceways.
- **364-10** The OCPD protecting busways may be the next higher standard rating if the current rating does not correspond to a standard OCPD rating.
- **365-1** *Cablebus* is an assembly of insulated conductors installed in a completely enclosed, ventilated, metal enclosure.
- **370-1** The term "conduit body" has replaced "fitting" where appropriate.
- **370-16(a)(2)** 100 cu in. or less nonmetallic and metallic boxes shall be legibly marked by the manufacturer with their capacity.
- **370-16(b)** Small fittings, such as locknuts and bushings, do not need to be considered when calculating box fill.
- **370-16(b)(2)** All internal cable clamps are to be considered when calculating box fill.
- **370-23(b)(2)** Polymeric braces are permissible for enclosures supported directly from grade.
- **380-8(a)** The maximum height of an operating handle for a switch or CB has been increased from 6′-6″ to 6′-7″.
- **380-14(d)** Listed snap switches rated 347 V AC are permitted for controlling noninductive and inductive loads.
- **410-1** The definition of *luminaire* has been added to 410.
- **410-4(d)** Ceiling fans are not permitted to be installed above bathtubs 3′ horizontally and 8′ vertically.

- **410-4(d)** Lighting track is not permitted to be installed above bathtubs 3′ horizontally and 8′ vertically.
- **410-56(c)** IG receptacles installed in nonmetallic boxes shall be covered with a nonmetallic faceplate.
- **410-57(c)** Receptacles are not permitted to be installed in a shower space or bathtub area.
- **410-58 (e)** Grounding-type attachment plugs shall only be used with cords equipped with an EGC.
- **410-102** Lighting track loads are calculated at 150 VA for each 2′ of track.
- **411** New 411 covers requirements for low voltage landscape and interior lighting systems
- **422-8(d)(1)** The installation requirements for electrically-operated kitchen waste disposers now apply to occupancies other than just dwelling units.
- **422-8(d)(2)** The installation requirements for built-in dishwashers and trash compactors now apply to occupancies other than just dwelling units.
- **424-44** The 16½ W/per linear foot limitation only applies to constant wattage heating cables.
- **430** Requirements for sizing Design E, Branch-Circuit, Short-Circuit, and Ground-Fault devices have been added to Table 430-150.
- **430-2** Overload protection is required for power conversion equipment unless equipment is marked otherwise.
- **430-7(a)(9)** Design letters shall be marked on motors.
- **430-34** Increased values for overload relays are based on motor nameplate FLC rating.
- **430-52(c)(3)** ITCBs are permitted only if they are adjustable and part of a listed combination motor controller.
- **430-52(c)(5)** The term "adjustable speed drive system" has been replaced with "power electronic devices."
- **Table 430-150** Values for less than 2 HP motors have been increased. FLC values have been added for 250–500 HP motors.
- **Table 430-152** Motor design types have been added. The column for NTDFs has been expanded.
- **450-23** Installation requirements for LFLI transformers have been completely revised.
- **455-2** *Manufactured Phase* is the derived phase with no electrical connection to 1ϕ input at the phase converter.
- **455-2** *Rotary Phase Converter* is a device with a rotary transformer and capacitor panel(s) for operating 3ϕ loads from a 1ϕ supply.
- **455-2** *Static Phase Converter* is a device which converts 1ϕ input to 3ϕ output without rotating parts.
- **501-18** Multiwire branch circuits are not permitted in Class I, Division 1 locations.
- **502-4** Listed MC cable may be installed in Class II, Division 1 locations with provisions.
- **502-8** TE water-air cooled motors may be used in Class II, Division 2 locations with provisions.
- **511-16** All metal raceways, metal-jacketed cables, and noncurrent-carrying metal parts of equipment in commercial garages shall be grounded per 250.
- **517-18(a)** Two branch circuits (one each from normal and emergency systems) shall be provided in patient bed locations in general care areas.
- **520-2** Twelve definitions for theaters and similar locations have been added.
- **520-10** Portable indoor equipment may be used outdoors temporarily with provisions.
- **520-47** All lamps behind stages which may come into contact with scenery shall be guarded.
- **520-53(k)** Paralleled conductors feeding equipment shall be so marked
- **525** Installation requirements for electrical installations in carnivals, fairs, etc. have been added.
- **530-6** Portable equipment can be used outdoors if attended and properly guarded.
- **530-71(d)** The maximum voltage drop on any branch circuit supplied from a 60 V derived system is 1.5%.
- **547-10** Wiring in agricultural buildings shall be protected from physical damage.
- **550-1** Mobile home parks are now included in 550-1.
- **552-1** New 552-1 covers wiring and equipment installed on or within park trailers.
- **555-1** Boathouses are now included in 555-1.
- **600** The entire article has been rewritten for better clarity and understanding.
- **604-4, Ex. 2** Manufactured wiring systems can be installed outdoors if listed for the purpose.
- **610-11** AC cable is a permitted wiring method for cranes and hoists.
- **620** The entire article has been rewritten for better clarity and understanding.
- **625** New 625 covers electric vehicle charging equipment.
- **645-1** The title and the scope of the article have been changed to Information Technology Equipment.
- **680-4** A definition of *packaged spa or hot tub equipment assembly* has been added.
- **680-6(a)(3)** All receptacles within 20′ of a fountain shall have GFCI protection.
- **680-25(b)(1), Ex. 2** ENT is permitted for use with pool fixtures when installed within buildings.
- **690-2** A new definition of *invertor* has been added to replace *power conditioning unit.*
- **695** New 695 contains installation requirements for fire pumps.
- **700-9(b), Ex. 1** Transfer switches shall supply only emergency loads.
- **710-4(b)(1)** Underground conductors over 600 V shall be protected to a point underground.
- **725** The entire article has been rewritten for better clarity and understanding.
- **760-1** The title of the article and scope have been changed to Fire Alarm Systems.
- **800-7** Communication circuits in hazardous locations shall comply with Chapter 5.
- **Chapter 9** Wire dimension and area tables have been completely revised.
- **Appendix C** New raceway fill tables have been added for conduit fill calculations.

One-family Dwellings 1

WIRING PLANS

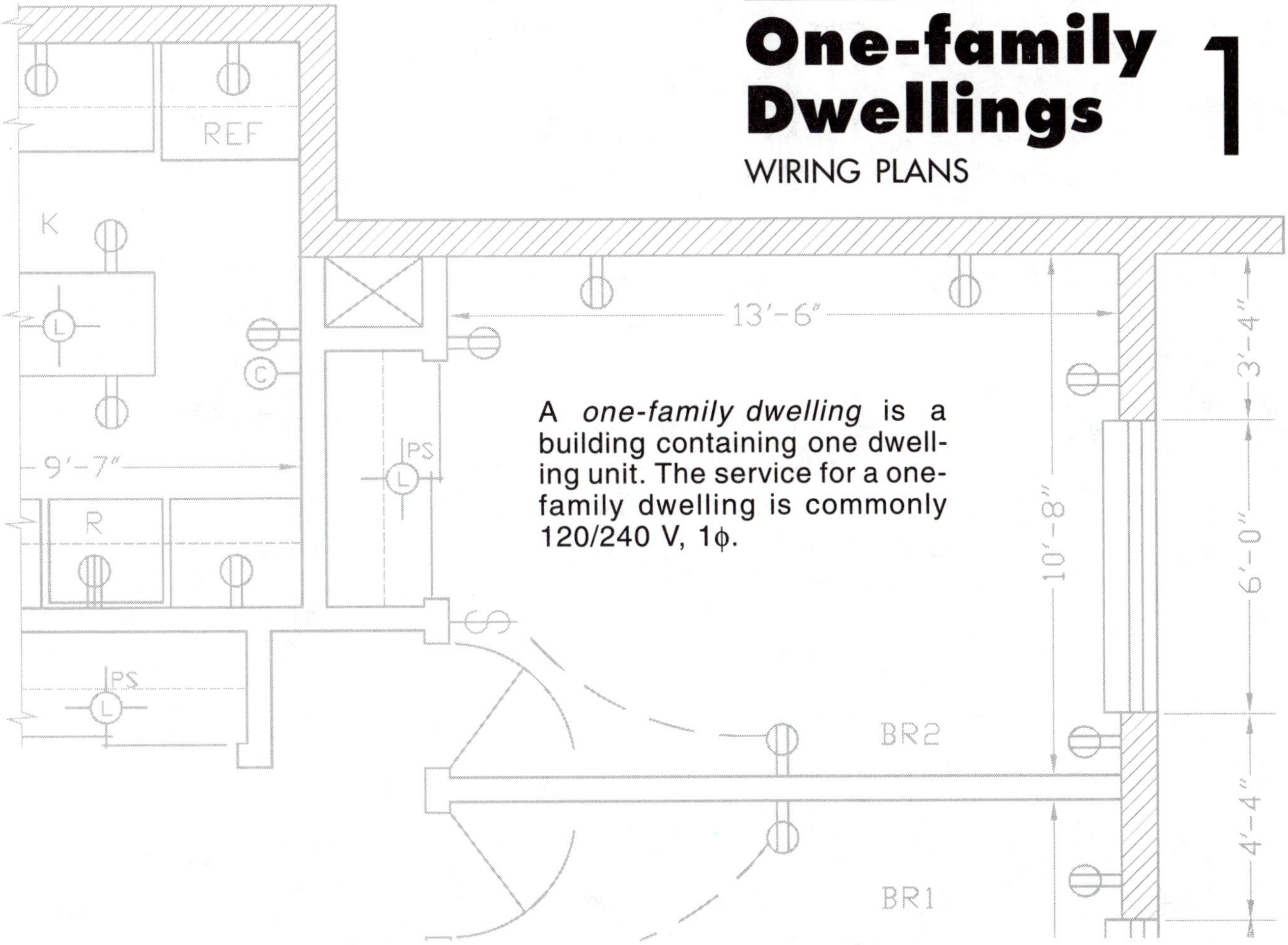

A *one-family dwelling* is a building containing one dwelling unit. The service for a one-family dwelling is commonly 120/240 V, 1ϕ.

GENERAL NOTES TO PRINTS

The first drawing shows a one-family residence, the dwelling area being confined to one floor. Lighting outlets, plug receptacles, and switches are marked. It is not customary to furnish a complete wiring plan for such an installation. The common practice is to indicate switching arrangements by means of broken lines connecting controlled outlets with their switch or switches.

Starting at the entry, two three-way switches, one at the front door, the other at the far side of the living-room arch control the entry light and one in the front hall. Two three-way switches are connected to a pair of lighting outlets in the central hall. Bedroom, laundry, and bath lights operate by means of single-pole switches. Two three-way switches are used with a four-way in the kitchen, and two three-ways are in the dining room. In the living room, one single-pole switch is connected with two bracket lights at the fireplace. Another single-pole switch controls two plug receptacles.

Paragraphs dealing with applicable NEC® Sections are placed around the figure. Observe that certain plug receptacle outlets are marked with one or more of the small letters, from *a* to *f* inclusive, and a certain point on the wall of Bedroom 2 is marked with a small *e*. These indications are explained in the paragraphs. Calculations used to determine feeder and service requirements under the general rules of NEC 220 and 230 are below the figure.

The second illustration, which is somewhat more involved than the first, represents a dwelling that occupies a floor and a half. Switching arrangements are indicated, as before, with broken lines. The optional method of load calculation, permitted by 220-30 is followed here. All necessary calculations are shown on the following drawing.

BRANCH CIRCUIT SERVICE AND FEEDER CALCULATIONS -- ARTICLE 220, PARTS A & B

NEC 220-3(b) Unit lighting load for dwelling occupancies shall not be less than 3 VA sq ft. See Table 220-3(b). Use outside dimensions.

NEC 220-4(d) For general illumination, one 15 A, 120 V branch circuit would be required for every 600 sq ft of area. 3 VA per sq ft × 600 = 1800 VA ÷ 120 V = 15 A.

NEC TABLE 220-3(b) **Receptacles other than those of the two small appliance circuits are considered part of general illumination and require no allowance for additional load.**

NEC 210-23 Branch circuits may be installed for specific loads (heaters marked "(d)" here in bath).

Symbols Used on Wiring Diagrams

- **Ceiling Outlet**
- **Wall Bracket**
- **Duplex Receptacle**
- **Floor Outlet**
- **Special Outlet**
- S-S_3-S_4 **Switches Single-Pole, 3-Way, and 4-Way**
- WP **Duplex Receptacle —Weatherproof**
- **Duplex Receptacle —One Half of Receptacle Controlled by a Wall Switch**
- **Door Bell or Chime Control Button**

NEC 422-28(e) Branch ckt. to single non-motor appliance not to exceed O.C. rating on device. If no appliance rating, 13.3 amps permitted on 20 amp ckt. If over 13.3 amps O.C. the next size OCPD above 150% can be used, where it does not correspond to standard OCPD.

210-52(a) All general occupancy rooms in dwellings shall have receptacle outlets installed so that no point along the floor wall line is more than 6′, measured horizontally, from an outlet.

NEC 230-42(b-3) In general, service-entrance conductors shall have an ampacity of at least 60 amps (see exceptions).

220-4(d) "VA per sq ft" load shall be apportioned evenly in branch circuits, within panelboard.

210-8(a)(3) Outdoor receptacles shall have GFCI protection. Outdoor receptacles are not permitted on small appliance branch circuit per 210-52(b)(1).

210-8(a)(1-6) Ground-fault protection for personnel in dwellings shall be provided for all 15 A and 20 A, 125 V receptacles in all (1) bathrooms, (2) garages and other buildings used for storage or as workshops, (3) outdoor locations, (4) unfinished basements and crawl spaces at or below grade, (5) kitchen countertops, (6) within 6′ of a wet bar sink intended to serve countertops. See Exceptions.

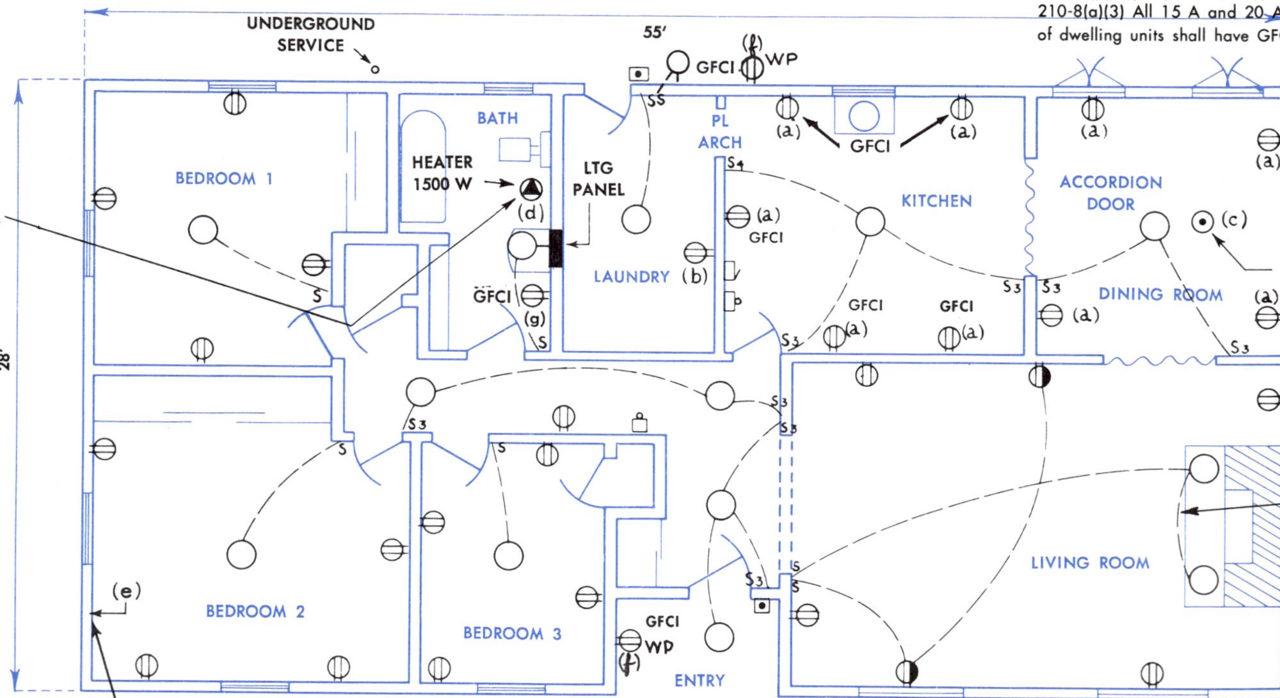

210-52(b) For small appliance load, including all receptacles in kitchen, pantry, breakfast room, dining room, and similar areas, two or more 20 A appliance circuits shall be provided. Such circuits shall have no other outlets (marked (a) on drawing). Countertop outlets to be at least two 20 A circuits. Clock outlets and receptacles for supplemental equipment on ranges, ovens, and counter-mounted cooking units are permitted on these circuits. See 210-52(b)(2), Ex. 1 & 2. Refrigerator receptacle is permitted on an individual 15 A circuit. See 210-52(b)(1), Ex. 2. 210-8(a)(3) All 15 A and 20 A, 125 V receptacles installed outdoors of dwelling units shall have GFCI protection (marked (f) on drawing).

NEC 220-4(c) At least one 20-amp circuit must be provided for laundry receptacle (NEC) 210-52(f) outlet marked (b) on drawing.

NEC 210-7(a) All receptacles installed on 15 A and 20 A circuits shall be of the grounding type. See Exception.

210-52(h) Hallways in dwellings, more than 10′ in length, measured from the centerline of the hall, without passing through a doorway, shall be provided with at least one receptacle outlet.

NOTE—Dashed lines on drawing do not show exact paths of the wires joining outlets and switches but merely represent plan for switching the various outlets.

NEC 220-16(a) Feeder load for the two small appliance circuits of 220-4(b) is to be taken at 3000 VA (1500 VA for each two-wire circuit).

CALCULATIONS BASED ON CODE SECTIONS

Area = 28′ × 55′ = 1540 sq ft . . . NEC 220-3(b)		
MINIMUM VOLT-AMPS, LIGHTING = 1540 × 3 VA =	4620 VA	NEC 220-3(b)
ALLOWANCE FOR TWO APPLIANCE CIRCUITS	3000 VA	NEC 220-16(a)
ALLOWANCE FOR LAUNDRY CIRCUIT	1500 VA	NEC 220-16(b)
TOTAL LTG, SMALL APP, AND LAUNDRY CKTS	9120 VA	NEC 220-16(a,b)
FIRST 3000 VA AT 100%	3000 VA	NEC TABLE 220-11
REMAINING 6120 VA AT 35%	2142 VA	NEC TABLE 220-11
LOAD REQUIRED FOR LTG, APP, AND LAUNDRY CKTS	5142 VA	NEC TABLE 220-11
ALLOWANCE FOR BATHROOM HEATER	1500 VA	NEC 220-15
TOTAL SERVICE LOAD	**6642 VA**	

NEC 220-16(b) Feeder load at 1500 VA for each laundry circuit as required by NEC 220-4(c).

← **NEC 220-16 (a),(b) Small appliance and laundry loads may be included with general lighting load in feeder calculations as per Table 220-11.**

If supply is 3-wire, 120/240 V, then current = 6642 VA divided by 240 V = 28 A (approx). Table 310-16 shows No. 10 Type THW has sufficient carrying capacity, but NEC 230-42(b) would require not less than 100-A, 3-wire because there are more than six two-wire branch circuits.

210-70(a) Dwelling units shall have at least one wall switched-controlled lighting outlet in every habitable room including bathrooms, hallways, stairways, and at outdoor entrances or exits. In rooms other than kitchen and bathroom, switch-controlled receptacles are permitted in lieu of lighting outlet. See Ex. 1.

BRANCH CIRCUITS REQUIRED

TWO 20-AMP APPLIANCE CIRCUITS	**2**
ONE 20-AMP LAUNDRY CIRCUIT	**1**
ONE SPACE HEATER CIRCUIT	**1**
THREE ILLUMINATION CIRCUITS =	
4620 VA ÷ 120 = 39 A or 3 15 A circuits. . . . NEC 220-4(a)	**3**
SEVEN CIRCUITS TOTAL	**7**

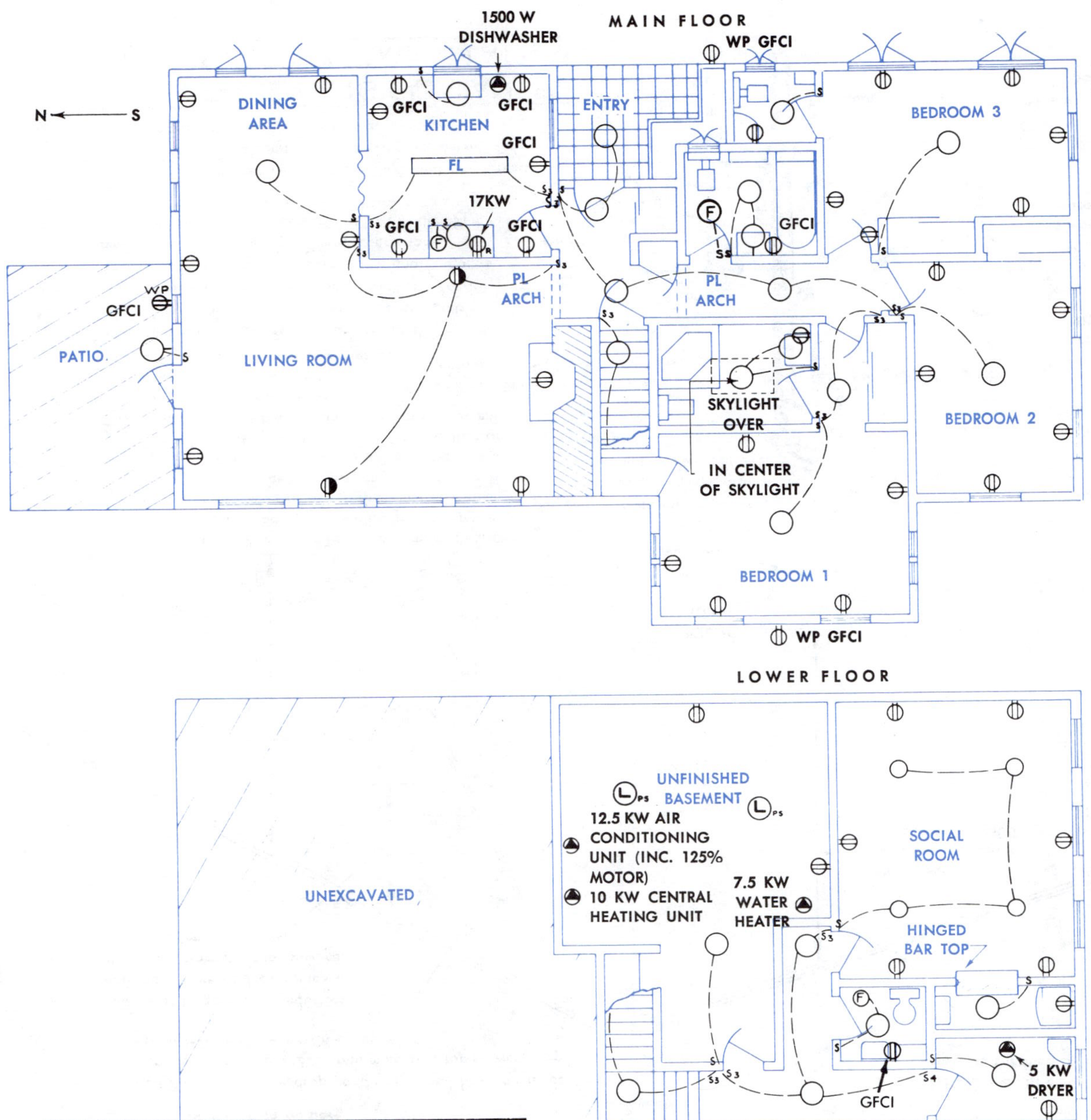

OPTIONAL CALCULATION FOR DWELLING UNIT (NEC 220-30)

Air conditioning, Central heating - 10 kVA - omitted as per Table 220-30(2)(3)(4)			= 12,500 VA
Lighting, small appliance and laundry	NEC 220-30 (b-1&2)	11,450 VA	
Water heater	NEC 220-30 (b-3)	7500 VA	
Dryer	NEC 220-30 (b-3)	5000 VA	
Dishwasher	NEC 220-30(b-3)	1500 VA	
Range	NEC 220-30(b-3)	17,000 VA	
Total Load		42,450 VA	
First 10 kVA of other load		10,000 VA	= 10,000 VA
Over 10 kVA at 40% =	40% ×	32,450 VA	= 12,980 VA
Total Load			**35,480 VA**

Service Calculations
(3-wire, 120/240 V)
35,480 VA ÷ 240 V = 147.8 A
Note 3 of Notes to Ampacity Tables of 0 to 2000 Volts shows No. 1 THW copper conductor is suitable for ungrounded conductors.

OPTIONAL CALCULATION - FOR DWELLING UNIT NEC 220-30

Area, excluding unfinished portions per NEC 220-3(b) = 2316 sq ft	
Lighting load, per Table 220-3(b) = 2316 × 3 VA	= 6950 VA
Allowance for two appliance circuits per NEC 220-16(a)	= 3000 VA
Allowance for one laundry circuit per NEC 220-16(b)	= 1500 VA
Total load, lighting and small appliances	**= 11,450 VA**

Note: Number of illumination circuits = 2316 sq ft × 3 VA = 6950 VA ÷ 120 V = 58 A
NEC 220-4(a) requires four 15 A or three 20 A circuits.

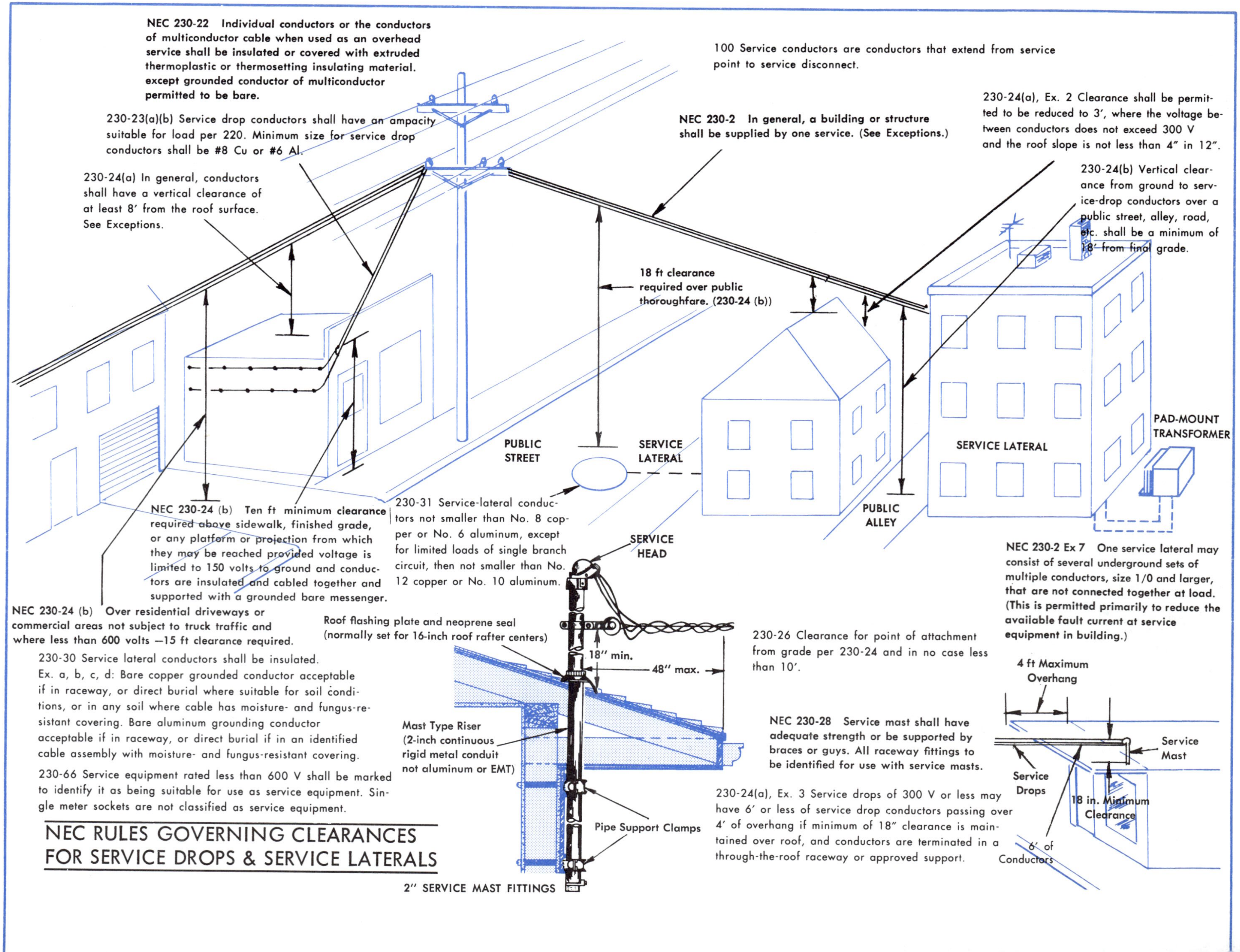

NEC RULES GOVERNING CLEARANCES FOR SERVICE DROPS & SERVICE LATERALS

NEC RULES CONCERNING CLEARANCES FOR SERVICE DROPS FROM BUILDING OPENINGS

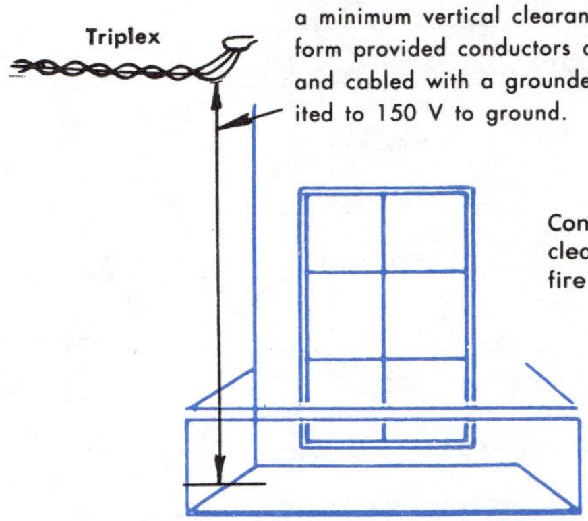

230-24(b) Service drops, 600 V nominal or less, shall have a minimum vertical clearance of not less than 10′ from a platform provided conductors are insulated and supported on and cabled with a grounded bare messenger and voltage limited to 150 V to ground.

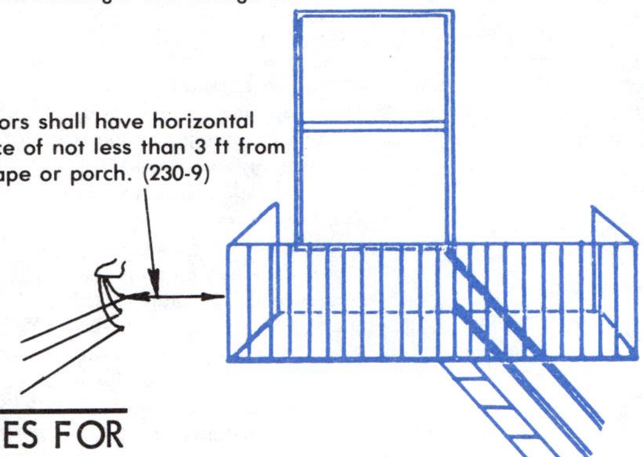

Conductors shall have horizontal clearance of not less than 3 ft from fire escape or porch. (230-9)

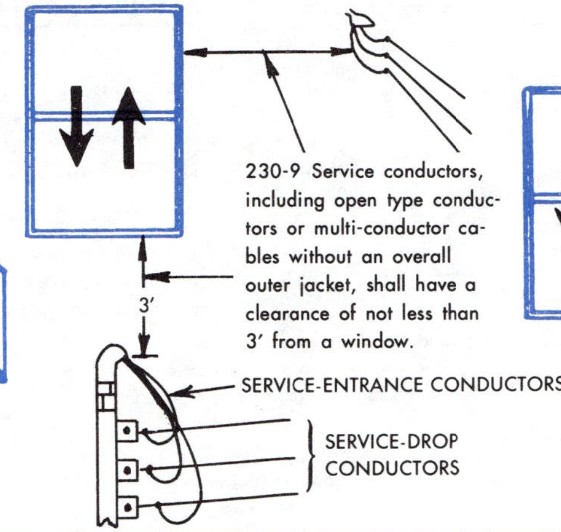

230-9 Service conductors, including open type conductors or multi-conductor cables without an overall outer jacket, shall have a clearance of not less than 3′ from a window.

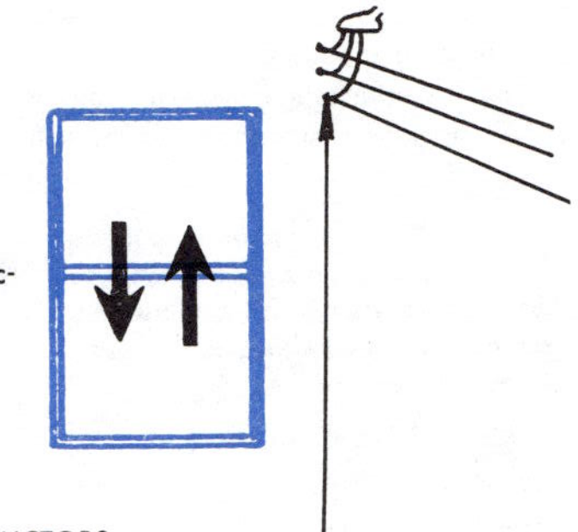

Conductors run above the window are considered as out of reach from the window. (230-9 Ex.)

NEC RULES GOVERNING SERVICE HEADS AND SERVICE-ENTRANCE CONDUCTORS

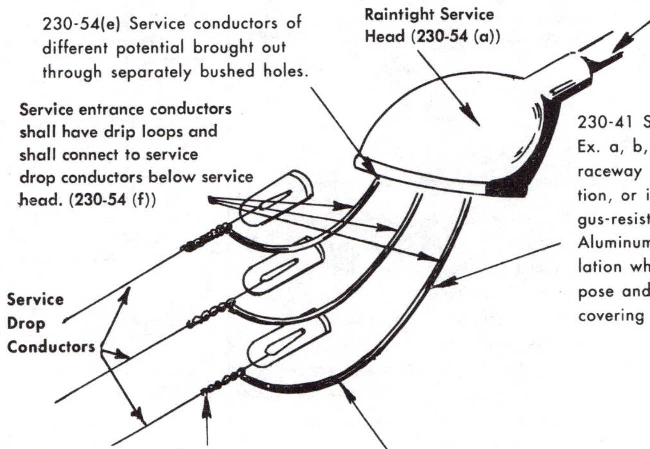

Raintight Service Head (230-54 (a))

Service Raceway

230-54(e) Service conductors of different potential brought out through separately bushed holes.

Service entrance conductors shall have drip loops and shall connect to service drop conductors below service head. (230-54 (f))

230-54(e) Service heads shall be designed so service conductors of different potentials shall be brought out through separate openings. See Ex. for multi-conductor jacketed cable that is continuous from pole to meter socket.

230-23(b) Service drops shall not be smaller than No. 8 copper or No. 6 aluminum, except for limited loads of a single-branch circuit; then they shall not be smaller than No. 12 hard-drawn copper or equivalent

NEC 230-42 (a) Service entrance conductors shall have ampacity to carry loads as per Article 220.

230-42(b)(1)(2) In single family dwelling with six or more 2-wire branch circuits or with initial computed load of 10kW or more, ungrounded conductors shall have ampacity of not less than 100 A.

NEC 230-42 (b-3) Not smaller than 60 amps for other loads.
Exception 1: Not smaller than No. 8 copper or No. 6 aluminum for not more than two 2-wire branch circuits.
Exception 2: Not smaller than No. 8 copper or No. 6 aluminum if by special permission for loads limited by demand or by source of supply.
Exception 3: Not smaller than No. 12 copper or No. 10 aluminum for limited loads of a single-branch circuit but never smaller than the branch-circuit conductors.

230-41 Service-entrance conductors shall be insulated.
Ex. a, b, c, d: Bare copper conductor acceptable if in raceway or direct burial where suitable for soil condition, or in any soil when cable has moisture- and fungus-resistant covering.
Aluminum grounded conductor acceptable without insulation when part of cable assembly approved for purpose and has a moisture- and fungus-resistant outer covering when in a raceway or direct burial.

230-54(c) Service-drop conductors connected to service entrance conductors at a point below the level of the service head or gooseneck. Exception: Where impracticable, service head shall be permitted to be located not over 24″ from the point of attachment.

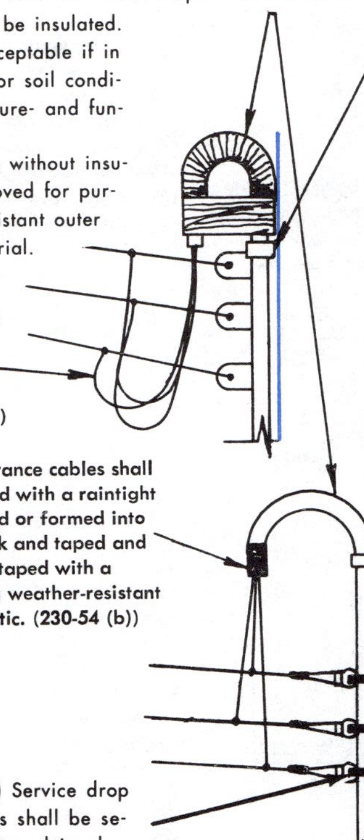

Cable securely held in place.

Drip loops formed to exclude moisture. (230-54 (f))

Service-entrance cables shall be equipped with a raintight service head or formed into a gooseneck and taped and painted or taped with a self-sealing weather-resistant thermoplastic. (230-54 (b))

230-54(d) Service drop conductors shall be securely fastened in place.

Watertight Service—Entrance Cable Connector

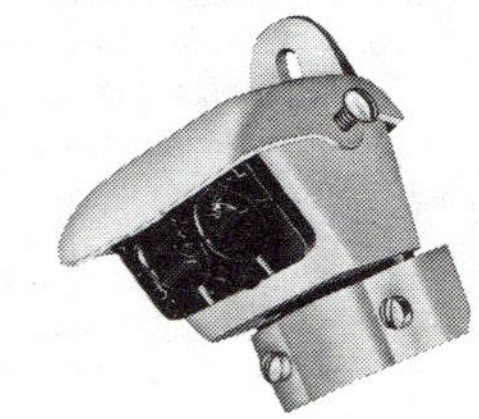

(RACO INC., subsidiary of Harvey Hubbell Inc.)
Raintight Service Head

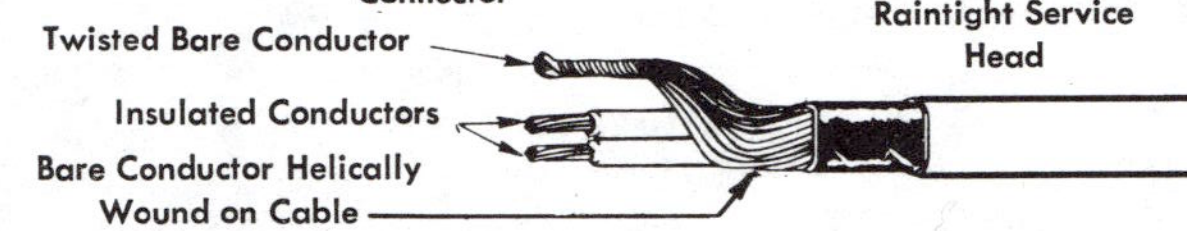

Type SE Service-Entrance Cable. Cable has two insulated conductors and one spirally-wound conductor which is covered but not insulated. Uninsulated conductor is twisted at ends and used as a grounded neutral conductor. 338-1(a) requires cable to have an outer covering that is flame-retardant and moisture-resistant.

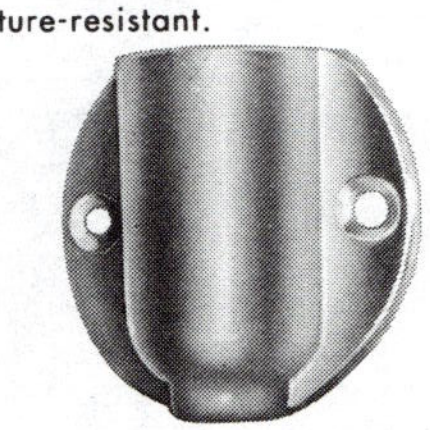

(RACO INC., subsidiary of Harvey Hubbell Inc.)
Sill Plate

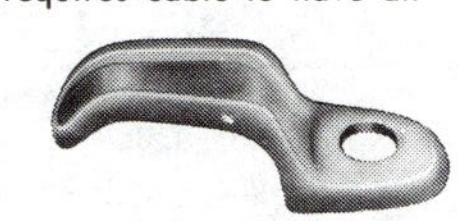

(RACO INC., subsidiary of Harvey Hubbell Inc.)
Cable Strap. Cable shall be supported within 12 inches of service head, gooseneck, or connection to raceway or enclosure and at intervals not exceeding 30 inches. (230-51(a))

SERVICES–OVERHEAD, UNDERGROUND, & DISCONNECTING MEANS

Utility supply system—overhead or underground.

230-2(b) Where more than one service, or any combination of branch circuits, feeders, and services supplies a building or structure, a directory or plaque shall be installed denoting all other services, branch circuits, and feeders.

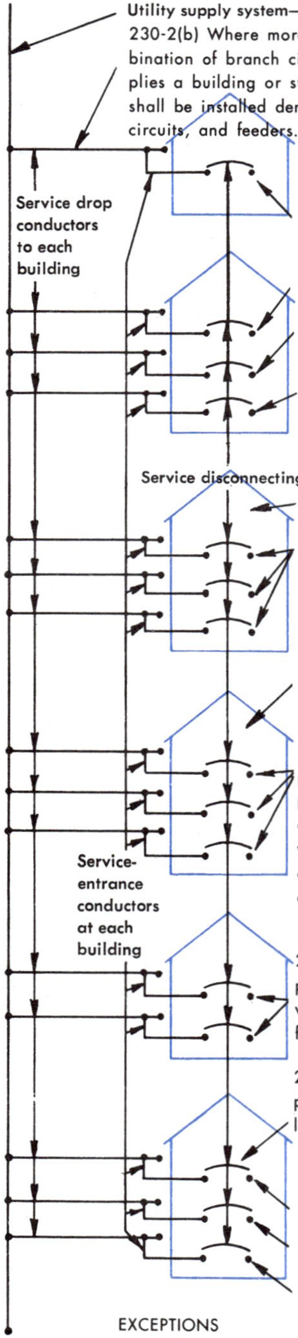

230-2(a), Ex. 7 Service lateral may have several sets of multiple conductors, #1/0 and larger, tied together at source but not at load.

230-2 Normal service for light and power load.

230-2(a) Normal service.

230-2(a), Ex. 2 Additional services for emergency or parallel power production systems (solar photovoltaic).

230-2(a), Ex. 2 Additional service for emergency systems as permitted.

Multiple-occupancy building

230-2(a), Ex. 3 More than one service permitted in multiple-occupancy building where no space available to locate service equipment that is accessible to all tenants and then permitted only by special permission.

Building requiring a large amount of power

230-2(a), Ex. 4 Additional services permitted in buildings with large capacity requirements: (a) in excess of 2000 A at 600 V or less; or (b) where single-phase load is in excess of that supplied by serving agency; or (c) by special permission.

230-2(a), Ex. 6 Additional service permitted for loads having different voltages, frequencies, phases, or different rate schedules.

230-2(a), Ex. 5 Additional services permitted by special permission for large buildings.

Lighting load

Power load

Electric heat load

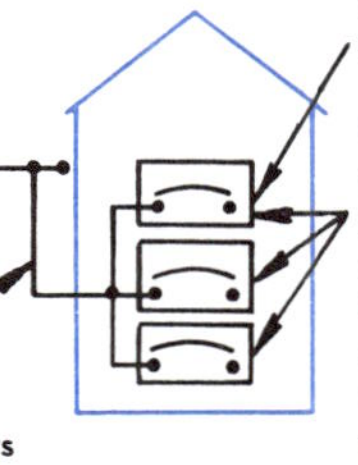

Disconnects in separate enclosures.

One set of service-entrance conductors permitted to feed up to 6 disconnects in separate enclosures. (230-40 Ex. 2) (Disconnects not required to be all of the same size.)

SEPARATE ENCLOSURES
(230-40 Ex. 2)

MAXIMUM NUMBER OF
DISCONNECTS (NEC 230-71)

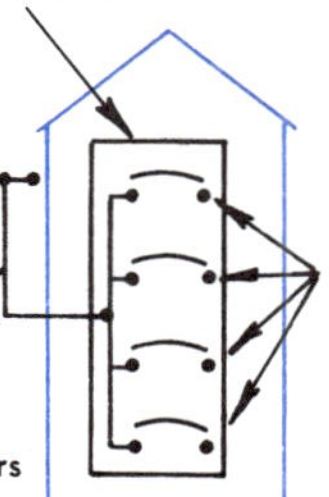

Not more than 6 disconnects in a single enclosure, a group of separate enclosures, or on a switchboard. (230-71 (a)) Two or three single-pole switches or circuit breakers with "handle ties" or a "master handle" are counted as one disconnect. (230-71 (b)) (Not more than six operations of the hand permitted.

**RATING OF DISCONNECT
(NEC 230-79)**

The service disconnecting means shall have a rating not less than the load as determined by Article 220 and not smaller than:
(a) for limited loads of one branch circuit it may have a rating of not less than 15 amps;
(b) where not more than two 2-wire branch circuits, it shall have rating of not less than 30 amps;
(c) for single-family dwelling where initial computed load is 10 kW or more, or where initial installation has 6 or more 2-wire branch circuits, the service disconnecting means shall have rating of 100 amps, 3-wire;
(d) all other installations require rating of not less than 60 amps. (230-79)

NEC RULES FOR INSTALLATION OF SERVICE ENTRANCE CONDUCTORS

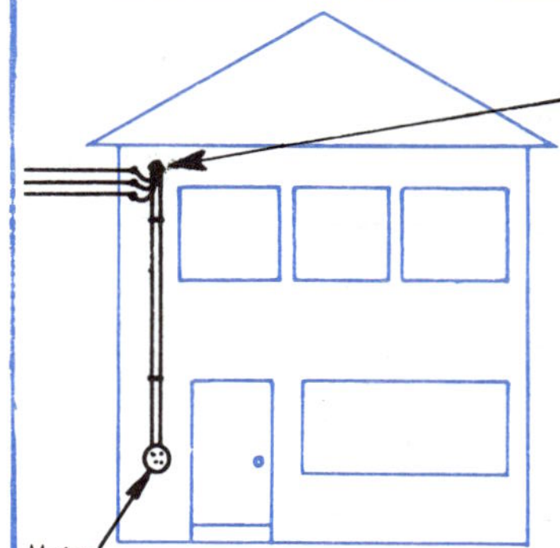

WIRING METHODS–
NOT OVER 600 V (230-43)
Service-entrance conductors permitted to be installed as: (1) open wiring on insulators, (2) rigid metal conduit, (3) intermediate metal conduit, (4) electrical metallic tubing, (5) service-entrance cables, (6) wireways, (7) busways, (8) auxiliary gutters, (9) rigid nonmetallic conduit, (10) cablebus, (11) type MC cable, (12) MI cable, (13) flexible metal conduit or liquidtight flexible metal conduit, with equipment bonding jumper per 250-79(a)(c)(f) and not over 6′ in length, and (14) liquid-tight flexible nonmetallic conduit.

Service raceway or cable shall terminate in a box, cabinet or equivalent, except at switchboard it may terminate in a conduit bushing. (230-55)

NEC 230-70(a)(b)(c) Disconnecting means shall be provided for all service-entrance conductors. Each shall be suitable for use as service equipment and shall be permanently marked to identify it as a service disconnecting means. Shall be suitable for prevailing conditions.

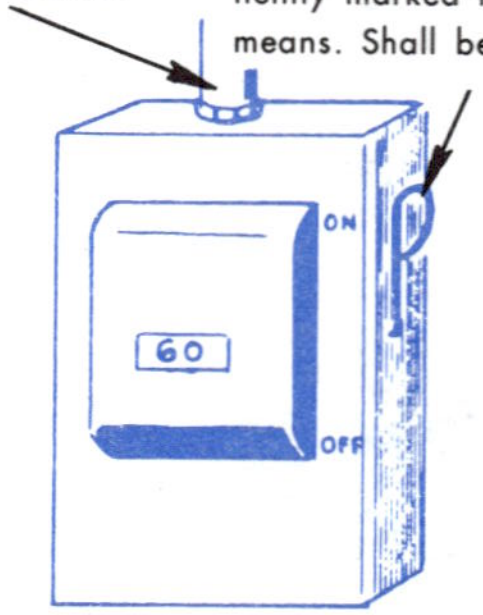

NEC 230-64 Sufficient working space shall be provided—never less than specified by 110-16.

NEC 230-74 Disconnect means shall simultaneously disconnect all ungrounded service conductors from premises wiring system.

230-75 Means must be provided to disconnect the grounded conductor from the premises wiring. A terminal or bus used in conjunction with pressure connectors is permissible.

NEC 230-76 (1), (2) Shall be a switch or circuit breaker and be manual or power operated. If power operated, it must be able to be opened by hand.

230-77 Service disconnecting means shall plainly indicate if it is in open or closed position.

NEC 230-78 An enclosed disconnect shall be externally operable with no exposed live parts, except, when power operated, it need only be able to be opened by hand.

384-16(a), Ex. 2 Individual protection for lighting and appliance panelboards not required for existing installation where used as service equipment for one-family residence.

Lighting and appliance panelboard required to be prctected by not more than two main circuit breakers or two sets of fuses. (384-16 (a))

NEC 384-16 (a) Ex 1 permits feeder overcurrent to also protect panelboard if not larger than rating of panelboard.

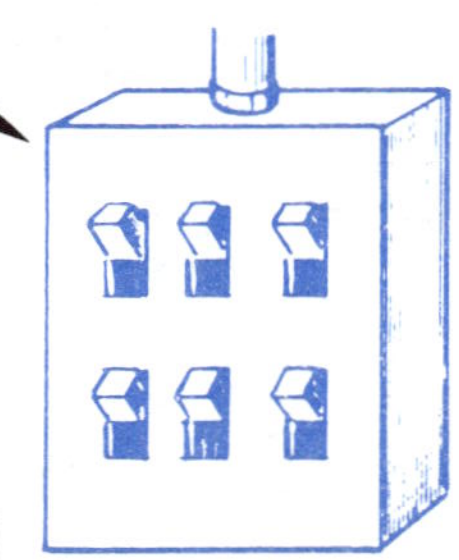

230-80 If more than one disconnect is used per 230-71, total rating of switches or CBs shall not be less than required by 230-79.

SERVICE DISCONNECTING MEANS
NEC 230-70 It shall be installed at a readily accessible place nearest entrance of conductors, either inside or outside of building.

Service switch may be placed on supply side of service overcurrent devices. (230-94 Ex 1)
Note: This is usual method.

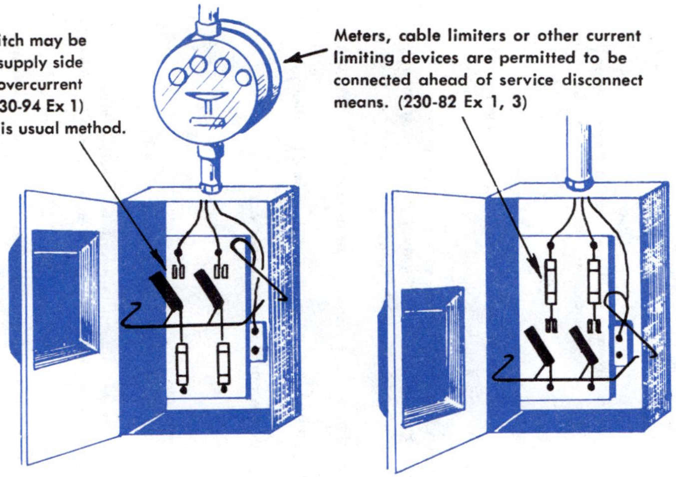

Meters, cable limiters or other current limiting devices are permitted to be connected ahead of service disconnect means. (230-82 Ex 1, 3)

230-82 In general, equipment shall not be connected to line or supply side of service disconnect. See Exceptions.

240-61 Cartridge fuses and fuseholders are required to be classified according to voltage and current ranges.

NEC RULES GOVERNING SEQUENCE OF SERVICE SWITCH AND FUSES

230-81 Service conductors shall be attached to the disconnect by clamps, pressure connectors, or other approved means. Soldered connections are not permitted.

380-6(a) Single-throw knife switches shall be installed so they will not tend to close by gravity.

230-90(a) Each ungrounded service conductor shall have overcurrent device in series with and not exceeding ampacity of conductor. See Exceptions.

Service overcurrent device shall be integral part of disconnect or located immediately adjacent thereto. (230-91(a))

Where a feeder carries the total service load, the ampacity of the feeder conductors shall not be less than the ampacity of the service conductors for loads of 55 amps or less. (215-2 (b))

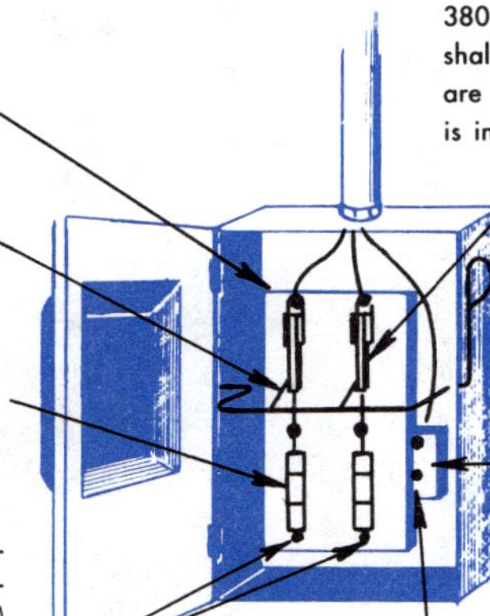

380-6(c) Motor-circuit switches shall be installed so blades are not energized when switch is in open position.

Neutral Bar

Means for disconnecting the grounded service conductor shall be provided. (230-75)

NEC RULES COVERING SERVICE DETAILS

① Ground clamp listed for the purpose shall be used to connect grounding electrode conductor to grounding electrode. (250-115)

② Water piping grounding electrode (250-80(a), 250-81(a))

③ Metal grounding electrode conductor enclosures shall be electrically continuous from cabinet to grounding electrode. (250-92(b))

④ If metal enclosure is not electrically continuous it must be bonded at each end to grounding conductor. (250-92(b))

⑤ Grounding electrode conductor

⑥ Service-entrance equipment

⑦ Neutral connection bar

⑧ Grounding electrode conductor to also ground all metal service-entrance equipment, service raceways and system grounded conductor (neutral). (250-23 (a), 250-53 (a), 250-32)

⑨ Grounding electrode conductor bonded to metal enclosure so as to be electrically continuous. (250-92(b))

⑩ 250-112 Grounding electrode connection shall be made in a permanent manner and shall be accessible. Bonding around water meters, valves, unions, etc. is required.

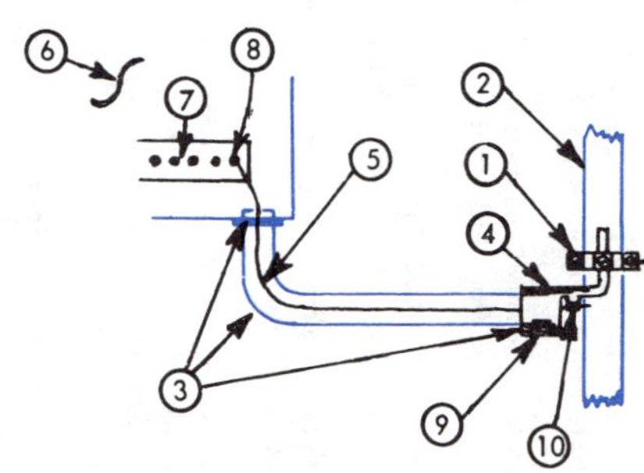

(Electrical Fittings Corporation)

NEC RULES CONCERNING SERVICE GROUNDING

SIZE OF EQUIPMENT BONDING CONDUCTOR ON SUPPLY SIDE SERVICE AND MAIN BONDING SERVICE (NEC 250-79(d)) NOTES TO PAGE 9:

(1) Service Raceways & Enclosures: Metal service equip. and enclosures to be grounded. (250-32)

(2) THREADED BOSS OR HUB—acceptable means of bonding service conduit to meter socket (250-72 (b))

(3) METER SOCKET

(4) Grounded circuit conductor can be used to ground meter enclosures, service raceways, etc., on supply side of service disconnecting means. (250-61 (a))

(5) NEUTRAL CONDUCTOR (250-25 (2))

(6) Service equipment, raceways, cable armor, cable sheaths, etc., and any service conductor that is required to be grounded by 250-5 shall be grounded as per Article 250. (230-63)

(7) For a single-phase 3-wire system the neutral conductor shall be grounded. (250-25 (2))

(8) Grounded system conductor (neutral) shall be connected to grounding electrode conductor. (250-53 (a))
Means must be provided in the service equipment to disconnect the service grounded conductor (neutral) from the premises wiring. In this case the terminal or bus to which all grounded conductors are attached by means of pressure connectors is permitted for this purpose. (230-75)

(9) To grounding electrode

(10) Grounding conductor connected to grounding fitting by exothermic welding, or by listed lugs, pressure connectors, clamps, or other listed means. Solder connections not permitted (250-115)

(11) When in conduit, the equipment grounding conductor must be included the same as any other conductor when determining conduit size by using Tables in Chapter 9. See Note 2 to Tables in Chapter 9.

(12) Aluminum or copper-clad aluminum conductors not permitted where subject to corrosive conditions or in contact with masonry or earth and when outside shall not be within 18 inches of earth. (250-92 (a))

(13) GROUNDING ELECTRODE CONDUCTOR – used to connect the equipment grounding conductors, the service-equipment enclosures and the grounded service conductor (when present) to the grounding electrode [250-53 (a) (Definitions)].

(14) Size of grounding electrode conductor determined by the size of the largest service-entrance conductor or equivalent for parallel conductors as per Table 250-94. Need not be larger than No. 6 copper or its equivalent in ampacity when it is the sole connection to made electrodes as permitted in 250-83.

(15) Grounding electrode conductor shall be copper, aluminum, copper-clad aluminum, or other corrosion-resistant material. Conductor shall be solid or stranded, insulated, covered, or bare and shall be continuous without splice or joint except splices in busbars permitted. (250-91 (a) Ex 1)
Taps are permitted to GEC where more than one service-entrance enclosure is installed if such tap extends inside of such enclosure and is sized per 250-94 for largest conductors entering enclosure. (250-91(a), Ex. 2.)

(16) No. 4 or larger grounding electrode conductor needs no protection if it is not exposed to *severe physical damage.* A No. 6 needs no protection if it follows surface of building and is not exposed to physical damage. No. 8 shall be in rigid metal conduit, IMC, rigid nonmetallic conduit, EMT, or cable armor. (250-92 (a))

(17) Path to ground from circuits, equipment, and enclosures shall: (a) be permanent and electrically continuous, (b) have capacity to safely conduct available fault currents, (c) have low impedance so as to limit voltage to ground and to facilitate the circuit overcurrent devices. (250-51(1)(2)(3))

(18) When underground metal water pipe used as the grounding electrode, a supplemental electrode is required. (250-81 (a))

(19) Secondary AC systems to be grounded shall have the grounding electrode conductor connected at each service. Grounding electrode conductor connected to AC system grounded conductor (neutral) on supply side of disconnecting means on premises, preferably within service enclosure. This connection not to be made on load side of disconnecting means. (250-23 (a)—see Exceptions)

(20) Equipment grounding conductor connections at service equipment shall be made: (a) for grounded systems, this connection made by bonding the equipment grounding conductor, grounded service conductor and the grounding electrode conductor, (b) for ungrounded systems, this connection made by bonding the equipment grounding conductor to the grounding electrode conductor. 250-50(a)(b). See Exceptions and Article 100.

(21) MAIN BONDING JUMPER—is the connection between grounded circuit conductor and equipment grounding conductor at service. (Definitions)
Main bonding jumper shall be used to connect equipment grounding conductor and service equipment enclosure to system-grounded conductor (neutral) within Service Enclosure. (250-53(b)). Shall be a wire, bus, screw or similar suitable conductor and without splice. (250-79(a))

(22) EQUIPMENT BONDING JUMPER—Main and equipment bonding jumper shall be: (a) copper or other corrosion-resistant material; (b) attached as per 250-113 and 250-115. (250-79 (a), (c))
Equipment bonding jumper on supply side of service and main bonding jumper. Sized as per Table 250-94 but not less than 12½ percent of largest phase conductor (250-79(d))
Where service-entrance conductors are parallel in separate raceways as permitted in 310-4 the size of the bonding jumper for each raceway shall be based on size of conductors in each raceway or cable (250-79(d))

(23) SERVICE EQUIPMENT ENCLOSURE—contains service disconnecting means.

(24) GROUNDING BUSHING – Must be approved for purpose. 250-72.

(25) All metal service equipment enclosures containing service-entrance conductors, meter fittings, boxes, or the like, interposed in service raceway or armor and any conduit or armor that forms part of the grounding conductor to the service raceway shall be effectively bonded together, except as permitted by 250-55 for underground service cables. (250-71 (a), (1), (2), (3))

Grounding continuity of service equipment shall be assured by: (a) bonding equipment to neutral as per 250-113; (b) threaded bosses, hubs or couplings; (c) threadless couplings for rigid conduit or EMT; (d) bonding jumpers installed as per Article 250. Bonding jumpers required around concentric or eccentric knockouts; (e) other approved devices such as bonding-type locknuts and bushings. (Standard locknuts and bushings not accepted.) Joints shall be made up wrenchtight where rigid metal conduit or IMC is involved. 250-72

(26) NEUTRAL SERVICE ENTRANCE CONDUCTOR—In this drawing only the identified neutral conductor is shown, the two "ungrounded" or hot conductors are not shown.

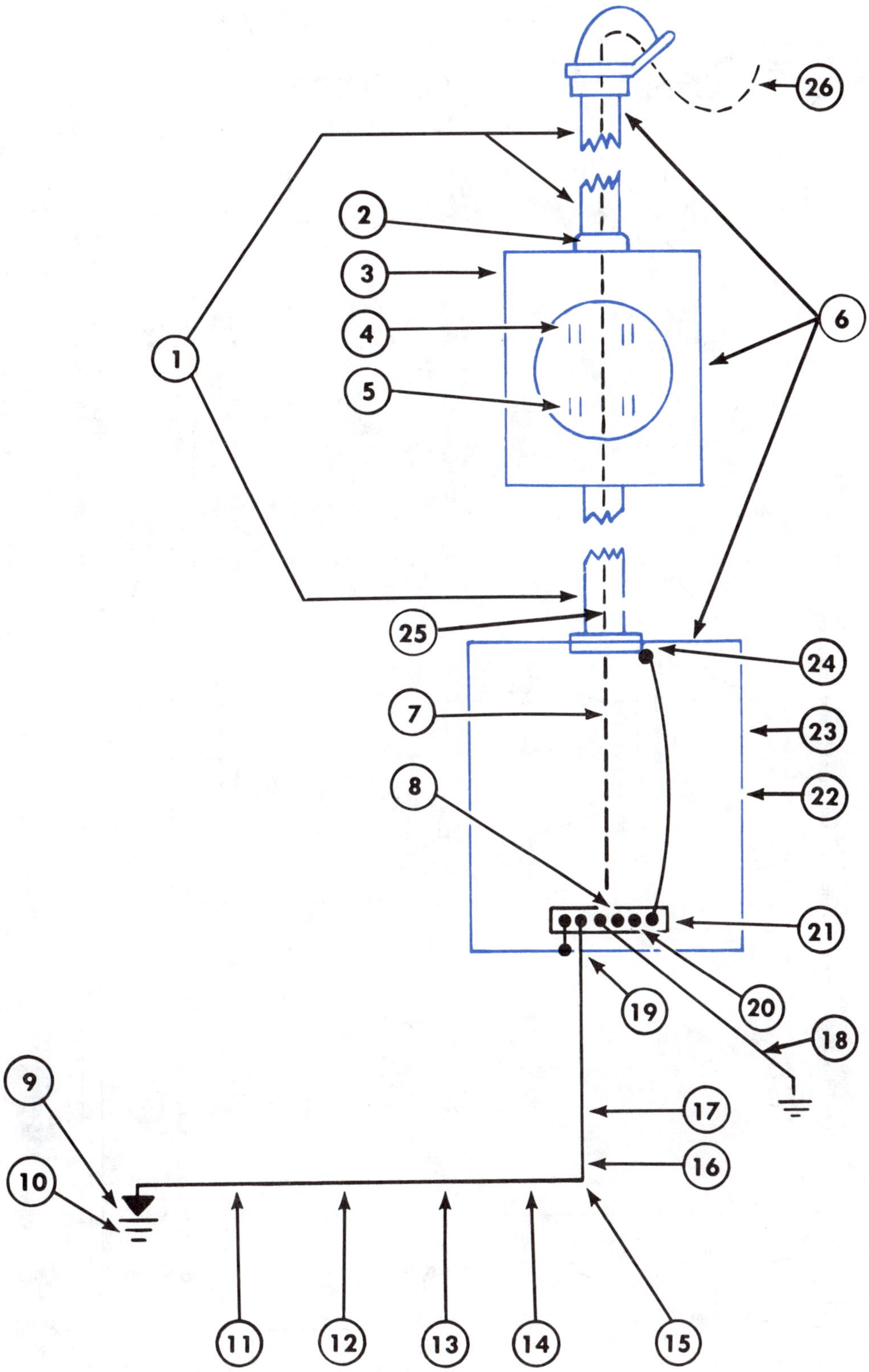

SERVICE GROUNDING AND BONDING

MAIN AND EQUIPMENT BONDING JUMPERS (NEC 250-79)

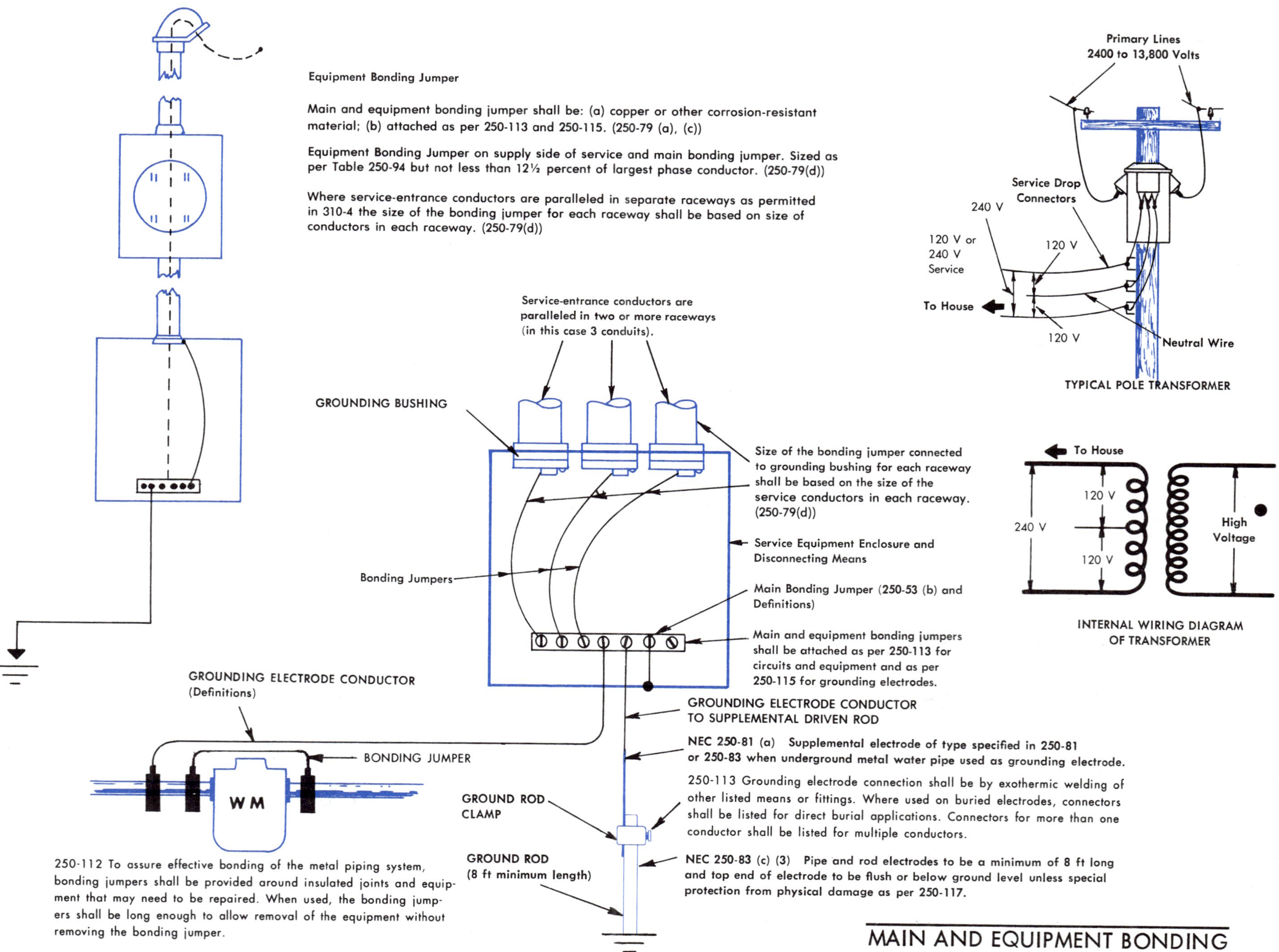

Equipment Bonding Jumper

Main and equipment bonding jumper shall be: (a) copper or other corrosion-resistant material; (b) attached as per 250-113 and 250-115. (250-79 (a), (c))

Equipment Bonding Jumper on supply side of service and main bonding jumper. Sized as per Table 250-94 but not less than 12½ percent of largest phase conductor. (250-79(d))

Where service-entrance conductors are paralleled in separate raceways as permitted in 310-4 the size of the bonding jumper for each raceway shall be based on size of conductors in each raceway. (250-79(d))

NEC 250-81 (a) Supplemental electrode of type specified in 250-81 or 250-83 when underground metal water pipe used as grounding electrode.

250-113 Grounding electrode connection shall be by exothermic welding of other listed means or fittings. Where used on buried electrodes, connectors shall be listed for direct burial applications. Connectors for more than one conductor shall be listed for multiple conductors.

NEC 250-83 (c) (3) Pipe and rod electrodes to be a minimum of 8 ft long and top end of electrode to be flush or below ground level unless special protection from physical damage as per 250-117.

250-112 To assure effective bonding of the metal piping system, bonding jumpers shall be provided around insulated joints and equipment that may need to be repaired. When used, the bonding jumpers shall be long enough to allow removal of the equipment without removing the bonding jumper.

GROUNDING A SEPARATELY DERIVED SYSTEM (NEC 250-26)

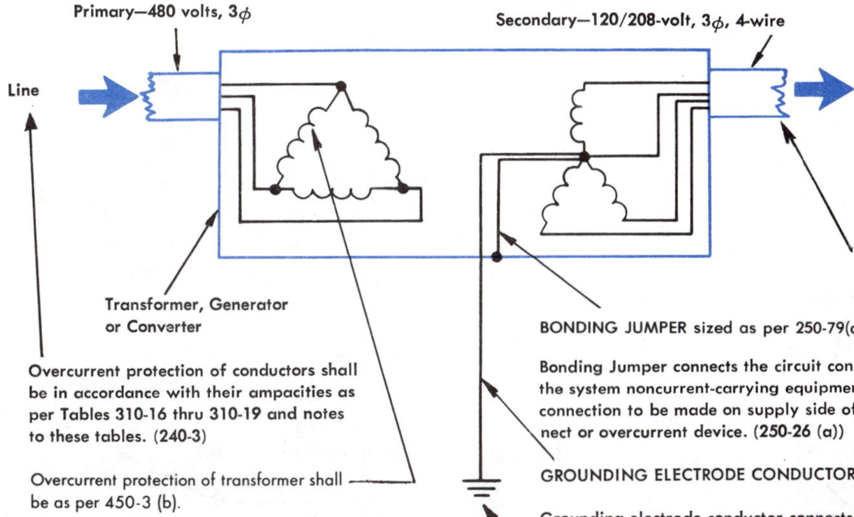

See NEC 250-5 (b-4) Ex 3 for control circuits and 250-5 (d) where generator is used and where premises wiring neutral connects to generator neutral.

Overcurrent device shall be installed where conductor to be protected receives its supply except as permitted in 240-21(a–o). See 240-21(d) for transformer taps not over 25′ long.

Overcurrent protection of conductors shall be in accordance with their ampacities as per Tables 310-16 thru 310-19 and notes to these tables. (240-3)

Overcurrent protection of transformer shall be as per 450-3 (b).

BONDING JUMPER sized as per 250-79(d) for derived phase conductors. (250-26(a))

Bonding Jumper connects the circuit conductor of the system that is required to be grounded to the system noncurrent-carrying equipment enclosures (except as permitted in 250-23 (a) Ex 4). This connection to be made on supply side of separately derived system and ahead of any system disconnect or overcurrent device. (250-26 (a))

GROUNDING ELECTRODE CONDUCTOR sized as per 250-94 for derived phase conductors. (250-26 (b))

Grounding electrode conductor connects the circuit conductor of the system that is required to be grounded to the grounding electrode. Except as permitted by 250-23 (a) Ex 4, this connection to be made at source of separately derived system and ahead of any disconnect means or overcurrent device. (250-26 (b))

GROUNDING ELECTRODE shall be as near as practicable to system and shall be: (1) nearest effectively grounded structural metal member of structure or, (2) nearest effectively grounded metal water pipe or, (3) other electrodes as per 250-81 and 250-83 where electrodes specified by (1) or (2) above are not available. (250-26 (c))

GROUNDED CONDUCTOR BROUGHT TO SERVICE EQUIPMENT FOR SERVICES OPERATING AT 1000 VOLTS OR LESS (250-23 (b))

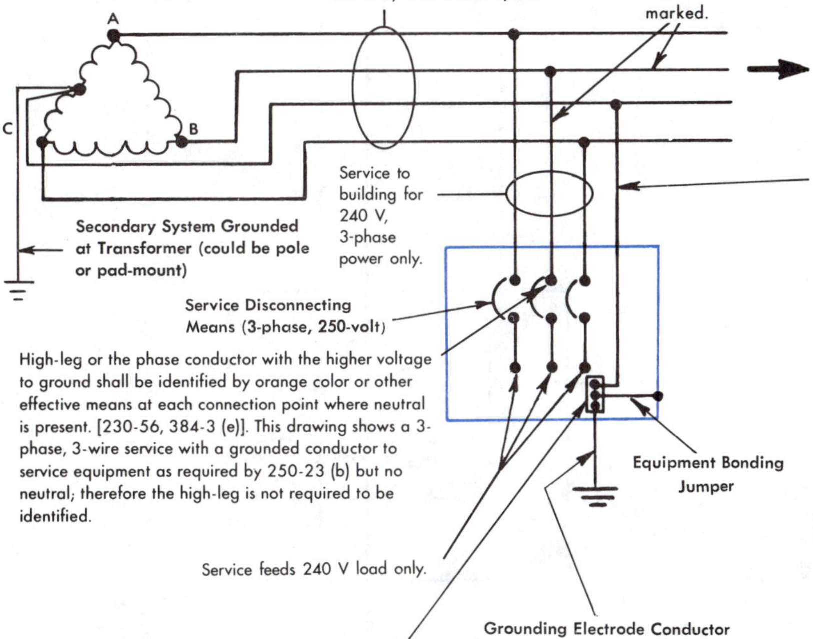

384-3(f) On 3ϕ, 4-wire delta-connected systems, phase arrangement shall be such that B phase shall have the higher voltage to ground. Where additions to existing installations are made, other arrangements shall be permitted but shall be marked.

High-leg or the phase conductor with the higher voltage to ground shall be identified by orange color or other effective means at each connection point where neutral is present. [230-56, 384-3 (e)]. This drawing shows a 3-phase, 3-wire service with a grounded conductor to service equipment as required by 250-23 (b) but no neutral; therefore the high-leg is not required to be identified.

Service feeds 240 V load only.

In this case there is only 3-phase, 240 V power load and no 120/240 V load on service; therefore there is no identified neutral conductor installed on load side of service.

GROUNDED CONDUCTORS—Where an AC system operating at 1000 volts or less is grounded at any point, the grounded conductor shall be run to each service. This conductor sized as per 250-94 and, in addition, for service-phase conductors larger than 1100 kcmil copper or 1750 kcmil aluminum the grounded conductor shall not be smaller than 12½ percent of largest phase conductor except the grounded conductor need not be larger than the largest phase conductor. (250-23 (b))

Example: Each service phase consists of three 600,000 cir mil copper conductors in parallel. Total area of phase conductors is 3 x 600,000 cir mil, or 1,800,000 cir mil. The minimum size of the grounded conductor run to the service equipment would be 1,800,000 x 12½ percent, or 225,000 cir mil copper. 225 cir mil requires a 250 cir mil conductor.

Explanation: The purpose of this requirement is to insure a low impedance path for any line to ground fault currents that could develop in the premises served. Section 250-51 (3) requires the path to ground shall have sufficiently low impedance to facilitate the operation of circuit protective devices in the circuit.

High-leg to be orange or otherwise identified. [384-3 (e)]

NEC 384-3 (f) High-leg to be B phase except that it may be same phase as metering equipment.

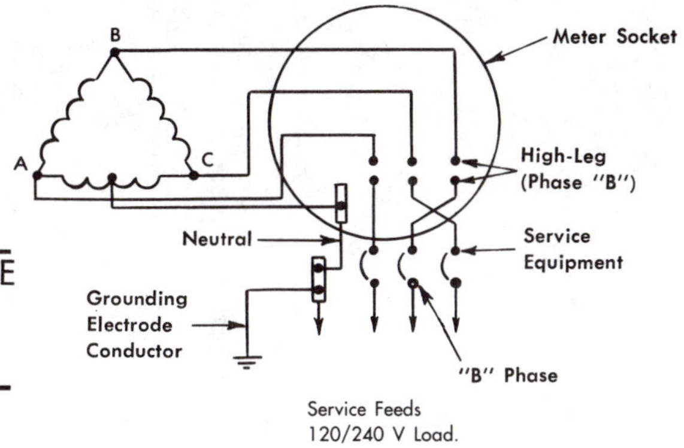

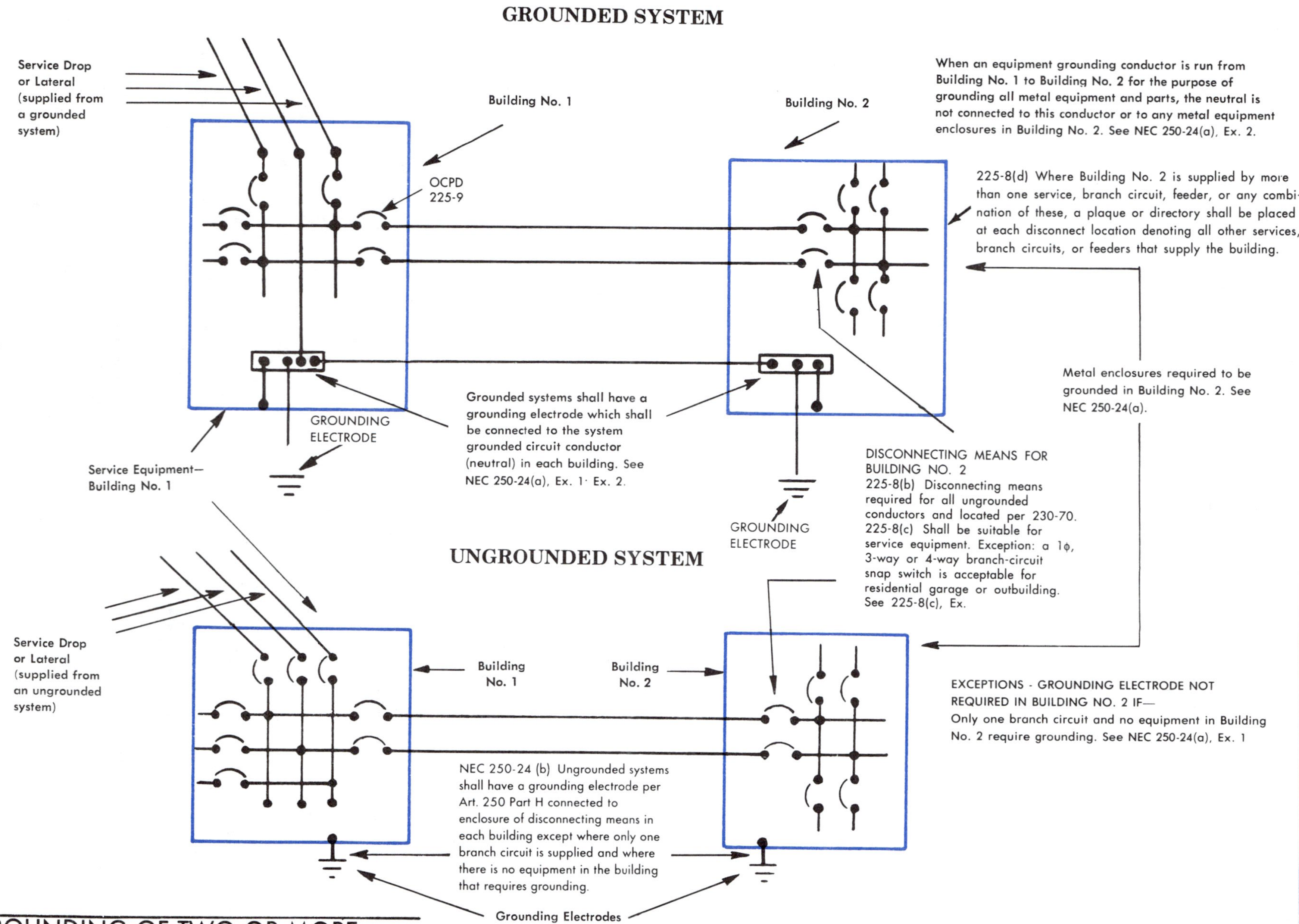

GROUNDING OF TWO OR MORE BUILDINGS SUPPLIED FROM SINGLE SERVICE EQUIPMENT (250-24)

Alternating-current circuits and systems that are required to be grounded (250-5):

Less than 50 volts: [250-5 (a)]

1. Where supplied by a transformer whose primary exceeds 150 volts to ground;
2. Where supplied by a transformer whose primary is ungrounded;
3. Where installed as overhead conductors outside of buildings.

From 50 V to 100 V 250-5(b)

1. Where system can be grounded so that maximum voltage-to-ground does not exceed 150 V.
2. Where the neutral is used as a circuit conductor on 3ϕ, 4-wire wye systems.
3. Where the midpoint of one phase winding is used as a circuit conductor on 3ϕ, 4-wire delta systems.
4. Where a service conductor is uninsulated as permitted in 230-22, 230-30, and 230-41, see Exceptions.

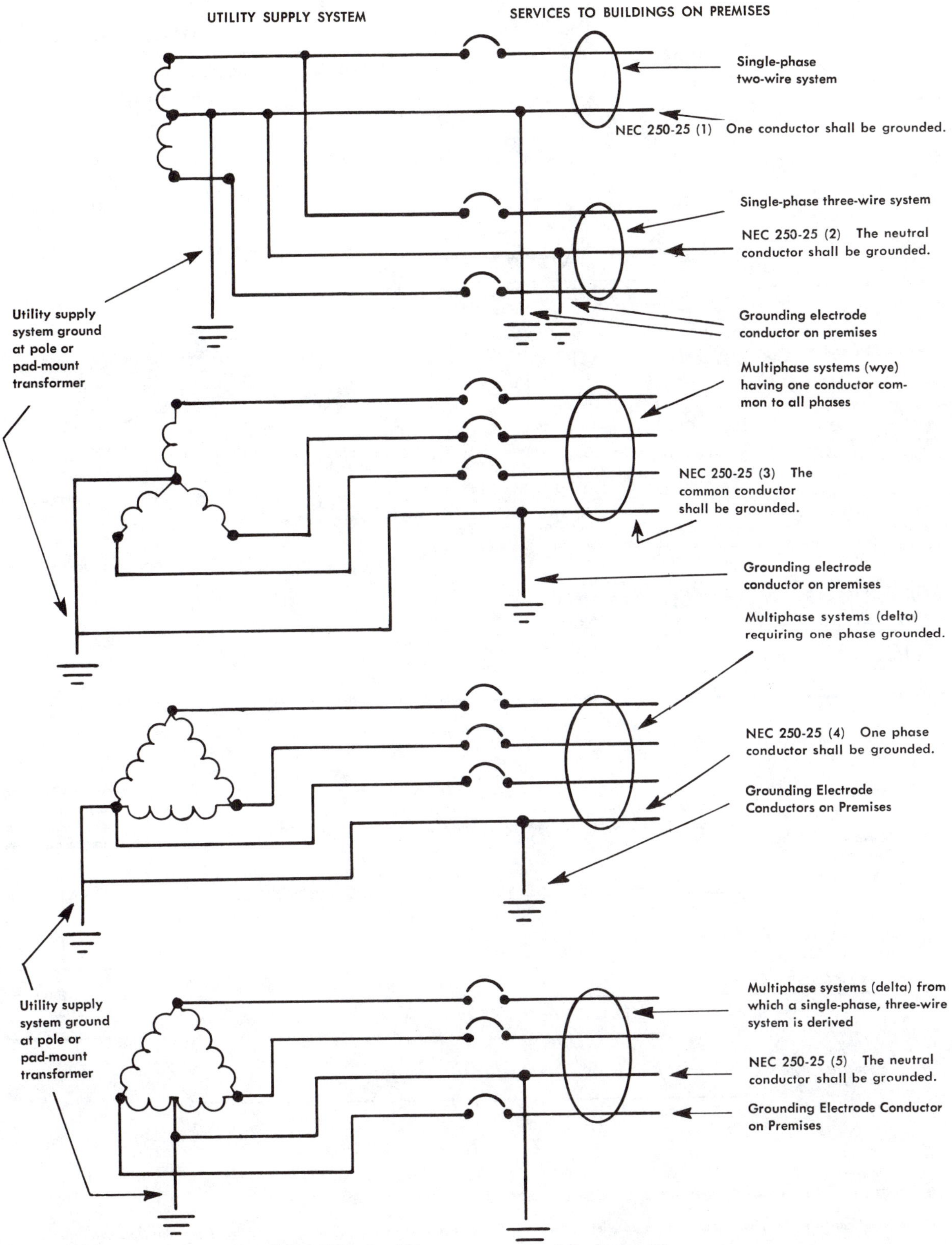

Exceptions to grounding AC systems of 50 to 1000 volts which supply premises wiring and premises wiring systems (250-5 (b) Ex 1, 2, 3 & 4)

Exception 1. Used only to supply industrial electric furnaces for melting, refining, tempering, and the like.

Exception 2. Separately derived systems used only for rectifiers which supply only adjustable speed industrial drives.

Exception 3. Separately derived systems fed by transformers with primary of less than 1000 volts and where all of the following are met:

a. System used for control circuits only.
b. Only qualified persons will service the installation.
c. Require continuity of control power.
d. Control system has ground detectors.

Exception 4. Isolated systems as permitted in Article 517 – Health Care Facilities and 668 – Electrolytic cells.

Exception 5. For high impedance grounded neutral systems see NEC 250-5(b), Ex. 5.

THE CONDUCTOR TO BE GROUNDED FOR ALTERNATING CURRENT SYSTEMS (250–25)

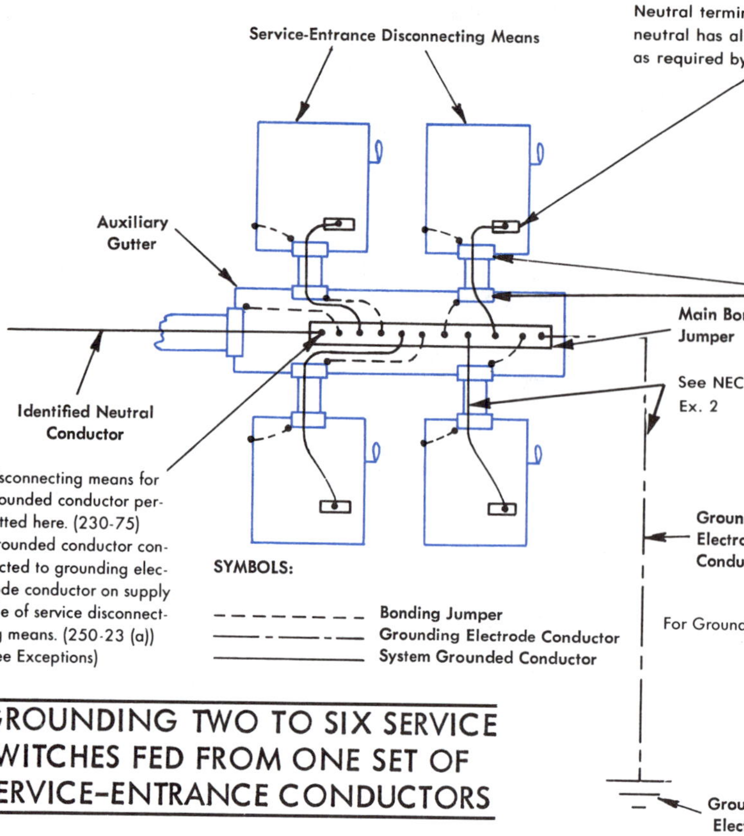

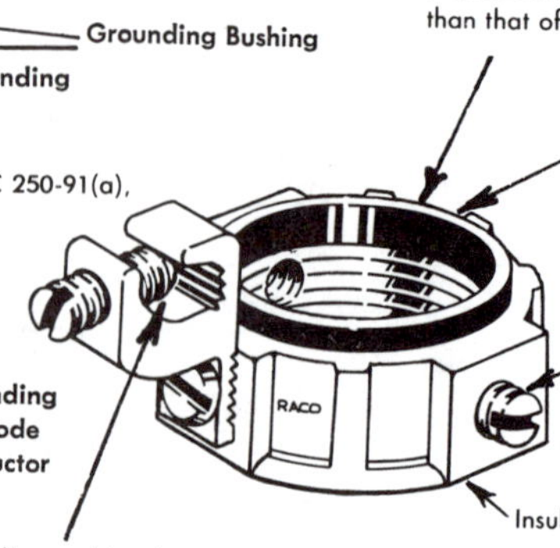

(RACO INC., subsidiary of Harvey Hubbell Inc.)

GROUNDING TWO TO SIX SERVICE SWITCHES FED FROM ONE SET OF SERVICE-ENTRANCE CONDUCTORS

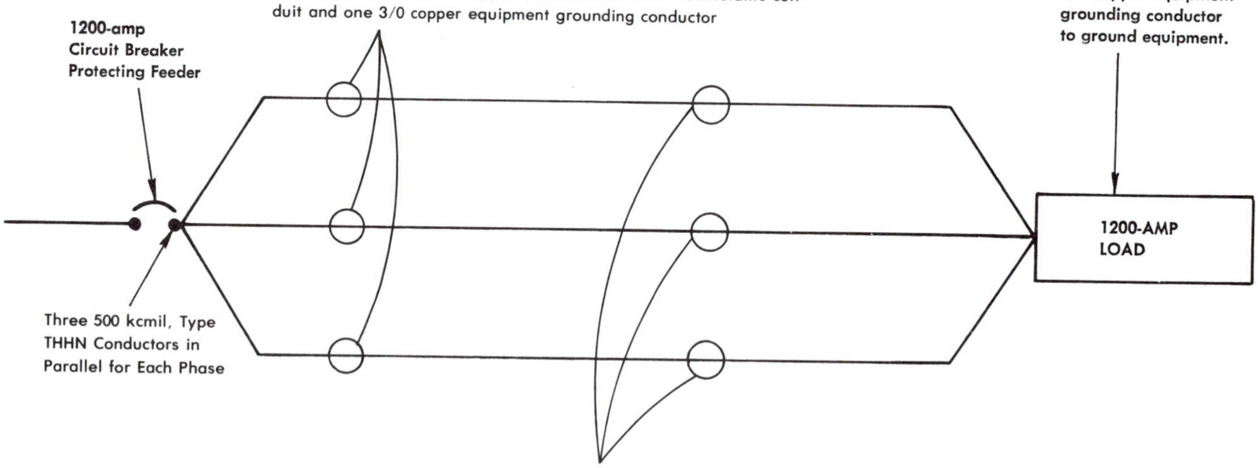

EACH NONMETALLIC CONDUIT MUST CONTAIN A 3/0 COPPER EQUIPMENT GROUNDING CONDUCTOR.

250-95 requires that where conductors are run in parallel, either in multiple raceways or cables, as permitted by 310-4 the EGC shall also be run in parallel. Each parallel EGC shall be sized in relation to size of overcurrent device protecting circuit conductor.

Example: In the above diagram each of the 3-inch,nonmetallic conduits contain three 500,000 cir mil, THHN copper conductors. The three conductors are connected in parallel and are protected by a 1200-amp circuit breaker. Table 250-95 requires a 3/0 copper equipment grounding conductor and 250-95 requires that these three 3/0 conductors must be installed in each 3-inch conduit and connected in parallel. If all nine 500 kcmil conductors were installed in one raceway, one 3/0 copper equipment grounding conductor would satisfy the requirements of 250-95.

EQUIPMENT GROUNDING CONDUCTORS ON LOAD SIDE OF SERVICE (NEC 250-95)

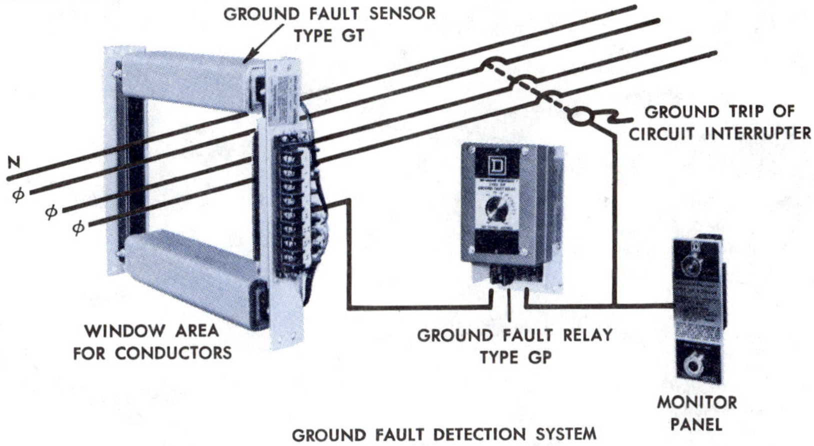

GROUND FAULT DETECTION SYSTEM

APPLICATION — CIRCUIT BREAKER

The ground current powered ground-fault detection system is designed for use on power systems which include a grounded conductor (neutral or grounded phase). When circuit conditions are normal, currents from all the phase conductors and neutral (if used) add up to zero, and the sensor produces no signal. If a ground fault occurs, the currents add up to equal the ground-fault current, and the sensor provides a signal proportional to the ground-fault current. This signal powers the ground-fault relay which activates the ground trip mechanism on the circuit breaker. See NEC 215-10 and 240-13.

APPLICATION – SWITCH AND FUSE

A typical ground-fault protective system as applied to a manually operated shunt trip BOLT-LOC bolted pressure contact switch is shown below.

Ground-fault protection of equipment required for grounded wye services of more than 150 volts to ground but not more than 600 volts phase-to-phase for each service disconnect rated 1000 amperes or more. (230-95)

Operation of ground-fault protection shall cause service disconnect to open all ungrounded wires of faulted circuit. Ground-fault protection shall have maximum setting of 1200 amps and a maximum time delay of one second for ground-fault currents of 3000 amps or more. (230-95 (a))

For switch and fuse combination the fuses shall be capable of interrupting any current higher than the interrupting capacity of the switch during a time when the ground-fault protective system will not cause the switch to open. (230-95 (b))

Overcurrent devices shall be selected and coordinated so as to clear a fault without extensive damage to electrical components of the circuit. (110-10)

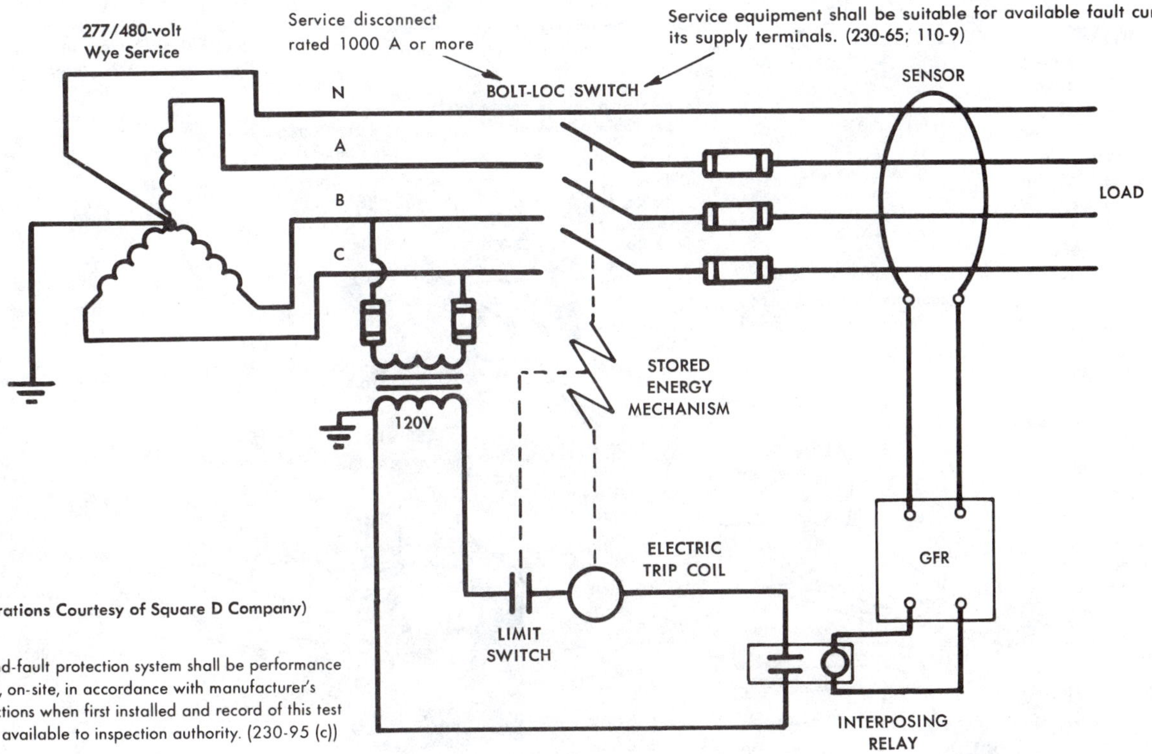

(Illustrations Courtesy of Square D Company)

Ground-fault protection system shall be performance tested, on-site, in accordance with manufacturer's instructions when first installed and record of this test made available to inspection authority. (230-95 (c))

DESCRIPTION OF OPERATION

Under normal operation there is no output from the ground-fault sensor and therefore the GFR contact is open. If a ground-fault current of sufficient magnitude occurs the GFR contact will close to energize the interposing relay which in turn will energize the electric trip coil of the BOLT-LOC switch to open the circuit.

GROUND FAULT PROTECTION OF EQUIPMENT (NEC 230-95)

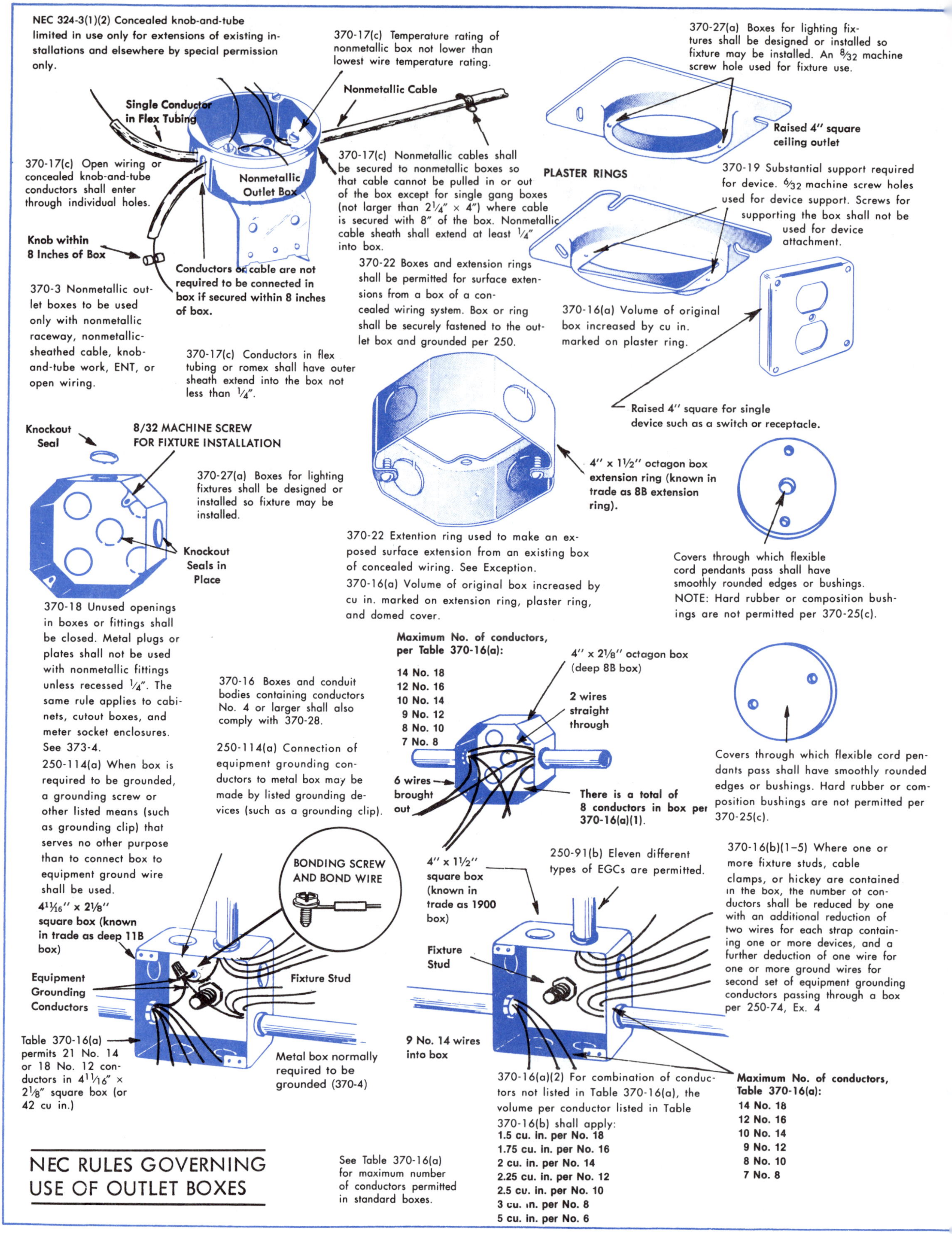

NEC RULES GOVERNING USE OF OUTLET BOXES

NEC 370 INSTALLATION RULES FOR OUTLET, SWITCH, AND JUNCTION BOXES

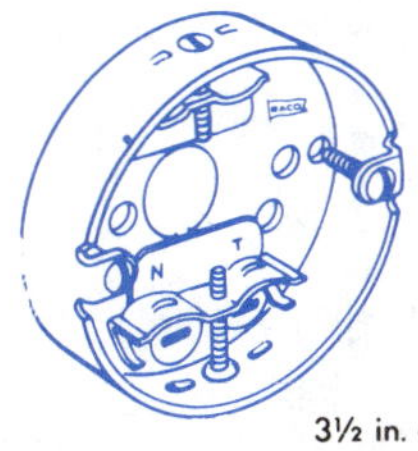

3½ in. ceiling pan, ¾ in. deep with cable clamps – normally 5 cu. in. capacity

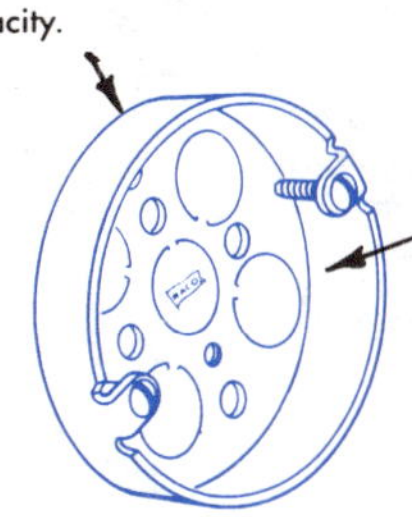

4 in. ceiling pan, ½ in. deep with ½ in. knockouts in bottom of box – normally has 6 cu. in. capacity.

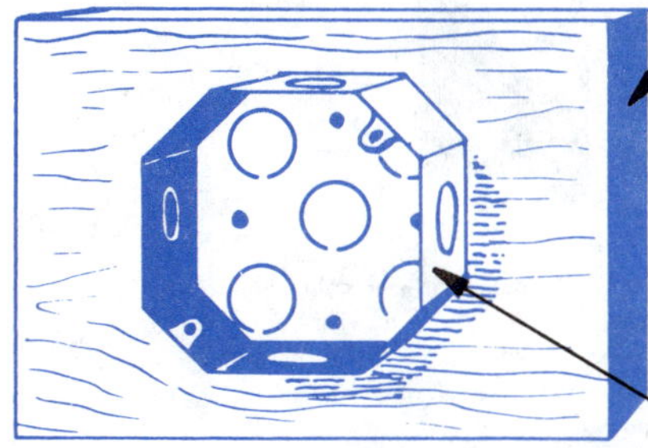

Wood brace shall have a nominal 1″ × 2″ thickness per 370-23(b)(2).

Boxes shall be supported from a structural member either directly or by using an approved metallic or wooden brace per 370-23(b).

No box shall be less than ½″ deep. Where box is intended to contain flush device, it shall be at least $1\frac{5}{16}$″ deep per 370-24. The intent of this rule is to permit ½″ deep box for use with a fixture with a canopy, which gives added depth.

370-2 Where conduits or connectors utilizing locknuts or bushings are installed, round boxes shall not be permitted.

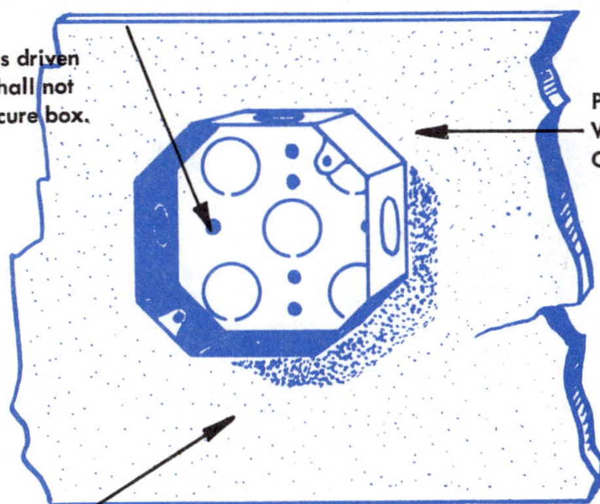

Wooden plugs driven into plaster shall not be used to secure box. (110-13 (a))

Boxes shall be securely and rigidly fastened to the surface on which they are mounted per 370-23(a).

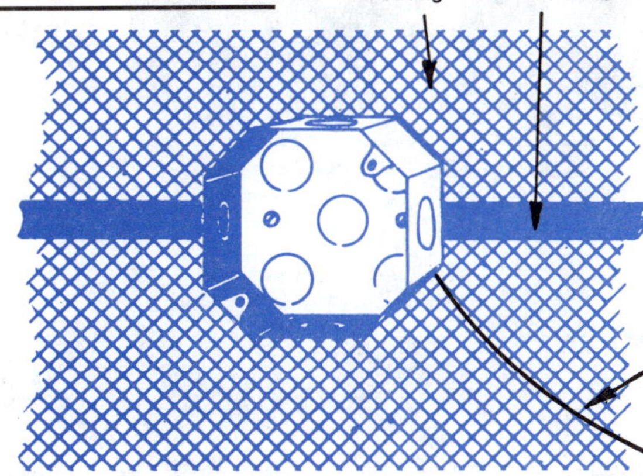

When metal boxes are used with knob-and-tube or nonmetallic-sheathed cable and are mounted on or in contact with metal or metal lath they shall be grounded. Art. 250 (370-4)

Metal boxes located within 8 feet vertically or 5 feet horizontally of ground or grounded metal objects and subject to contact by persons, shall be grounded. (250-42 (a))

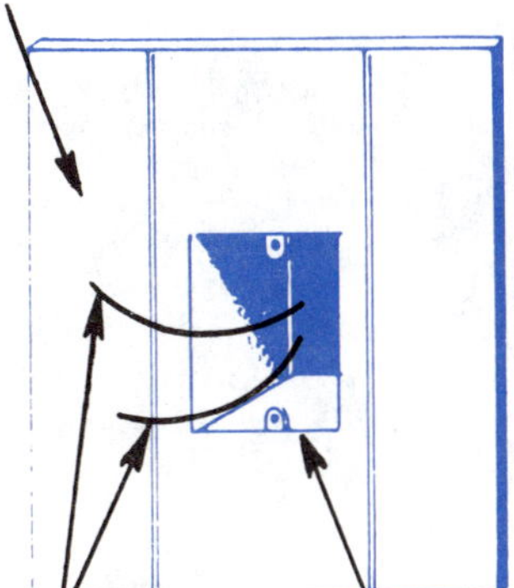

At least 6 inches of free conductor shall be left at each outlet and switch point, except for conductors pulled straight through box. (300-14)

Box in wall of combustible material must be flush with surface or project therefrom per 370-20.

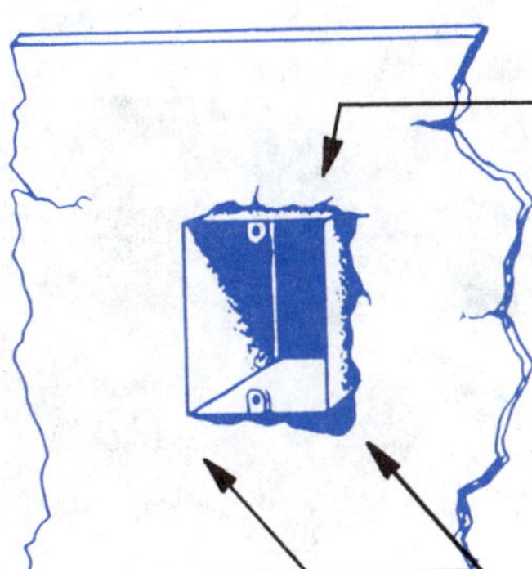

370-20 Boxes and fittings, when installed in walls or ceilings constructed of noncombustible material, shall be permitted to be recessed no more than ¼″.

A broken plaster surface shall be repaired per 370-21.

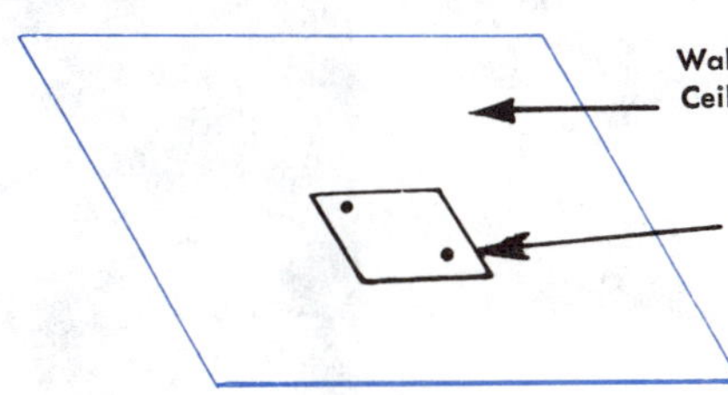

370-29 In general, boxes and conduit bodies shall be installed so that wiring within them can be rendered accessible without damage to the building. Listed boxes shall be permitted to be covered by gravel, soil, and light aggregate under some conditions. See Exception.

Boxes installed in areas above ceiling lift-out panels are considered accessible and exposed. See Definitions of Accessible, Concealed, and Exposed, as applied to wiring methods.

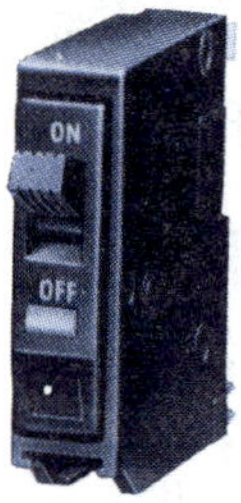

Single-pole Breaker

NEC 240-81 Shall clearly indicate when open (off) or closed (on). On switchboards or panelboards where mounted vertically, the up position of handle shall be "on" position.

(Photos courtesy Square D Company)

Two-pole Breaker

Three-pole Breaker

NEC 380-8(a) CBs used as switches shall be readily accessible and operating handle not over 6′-7″ from floor or platform. See Ex. 1, 2, 3.

NEC 380-3 Circuit breakers shall be externally operable and installed in an enclosure listed for the intended use.

NEC 380-2 (b) Circuit breakers shall not open grounded conductor except: (1) where all conductors of circuit are disconnected simultaneously; (2) where so arranged that grounded conductors cannot be disconnected until all ungrounded conductors of circuit have been disconnected.

NEC 110-9, 230-65 Shall have interrupting capacity sufficient for available fault current.

NEC 110-10 Shall be selected and coordinated to clear without damage to electrical components of circuit.

NEC 240-82 Shall be designed so that any alteration of the trip point or time for operation will require dismantling or breaking of seal.

NEC 240-83 MARKING – (a) Shall be marked with rating and be visible after installation. Shall be visible after removal of trim or cover. (b) If 100 amps or less and 600 volts or less, rating shall be marked into handles or escutcheon areas. (c) If other than 5000 amps rating shall be on breaker. Exception: rating not required if used for supplementary protection. (d) If used to switch 120 V or 277 V, fluorescent fixtures shall be listed for such switching duty and marked "SWD".

240-6 Standard ratings for ITCBs shall be: 15, 20, 25, 30, 35, 40, 45, 50, 60, 70, 80, 90, 100, 110, 125, 150, 175, 200, 225, 250, 300, 350, 400, 450, 500, 600, 700, 800, 1000, 1200, 1600, 2000, 2500, 3000, 4000, 5000, and 6000 A. Additional standard ratings for fuses are: 1, 3, 6, 10, and 601 A.

NEC 240-8 Shall not be connected in parallel. Exception: permitted if factory-assembled in parallel and listed as a unit.

NEC 110-9, 230-65. When checking interrupting rating of a circuit breaker, also check the voltage rating because generally the I.C. is lower as the voltage rating is higher.

i.e. at 240 volts—65,000 A.I.C.
at 480 volts—25,000 A.I.C.
at 600 volts—18,000 A.I.C.

If the foregoing is not evaluated, this can result in a dangerous misapplication of the breaker insofar as interrupting rating is concerned.

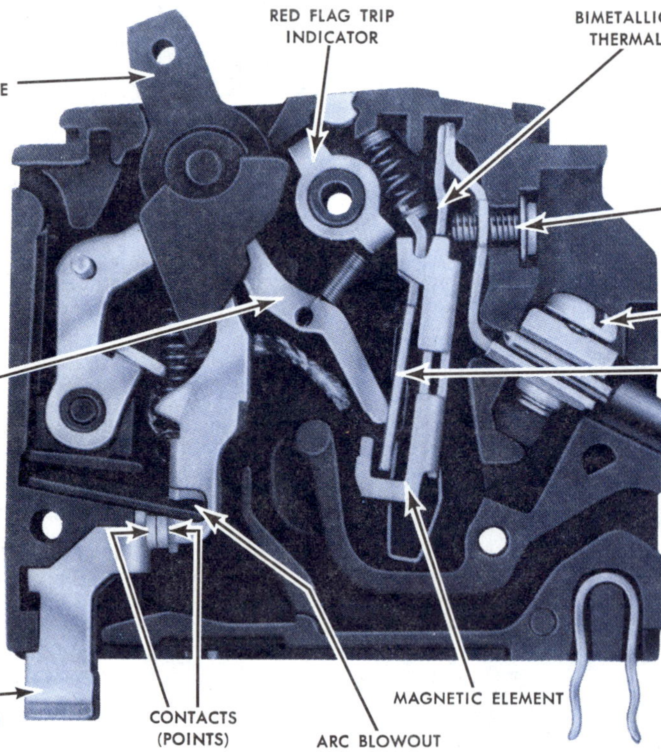

NEC 240-85 CB with straight marking can be used in neutral grounded, grounded wye, and grounded or ungrounded delta systems. CBs with slash markings can be used only in grounded neutral systems.

CUTAWAY VIEW OF QO CIRCUIT BREAKER

(Square D Company)

CIRCUIT BREAKERS

NEC ARTICLE 100—DEFINITION OF CIRCUIT BREAKER:
A device designed to open and close a circuit by nonautomatic means and to open the circuit automatically on a predetermined overcurrent without injury to itself when properly applied within its rating. (See Article 100 for definition of adjustable, instantaneous trip, inverse time, nonadjustable and setting.)

1-Pole
Qwik-Gard

2-pole
Qwik-Gard

(Square D Company)

Some of the areas where GFCI (Ground-Fault Circuit-Interrupters) are used in the NEC for the Protection of Personnel are listed below:

NEC 210-8(a) Dwelling Unit receptacles
- (1) Bathrooms
- (2) Garages and accessory buildings
- (3) Outdoors
- (4) Crawl spaces at or below grade
- (5) Unfinished basements
- (6) Kitchens
- (7) Wet bar sinks

210-8(b) Other than dwelling units
- (1) Bathrooms
- (2) Rooftops

215-9 Feeders

NEC 305-6 Temporary Wiring Installations

NEC 422-8(d)(3) Portable high-pressure spray washing machines.

NEC 426-31 EX. De-icing and Snow Melting Equip.

NEC 511-10 Commercial Repair Garages

NEC 517-20(a) Patient Care Areas

NEC 550 Mobile Homes
- 550-8(b) Outdoors and bathrooms
- 550-8(e) Outside receptacles

NEC 551 Recreational Vehicles
- 551-40(c) Power supply
- 551-41(c) Single phase receptacles inside unit
- 551-53(b) Shower fixtures
- 551-71 Site receptacles

NEC 555-3 Marinas and Boatyards

NEC 600-10(c)(2) Outdoor Portable Signs

NEC 680 Swimming Pools

NEC 680-6(a)(3) Receptacles
NEC 680-20(a)(1) Underwater Lighting Fixtures
NEC 680-26(b) Electric Pool Covers
NEC 680-31 Storable Pool Pumps
NEC 680-40, Ex. 2 Spas and Hot Tubs—Outdoors
NEC 680-41(a)(2) Spas and Hot Tubs—Indoors
NEC 680-62(a) Therapeutic Tubs in Health Care
NEC 680-70 Hydromassage Bathtubs

DEFINITION OF GROUND-FAULT CIRCUIT-INTERRUPTER (GFCI)
(See Definitions in NEC).
A device whose function is to de-energize a circuit within established time when current to ground exceeds a predetermined value which is less than that required to operate protective device of supply circuit.

GROUND FAULT CIRCUIT INTERRUPTERS (GFCI) FOR PERSONAL PROTECTION

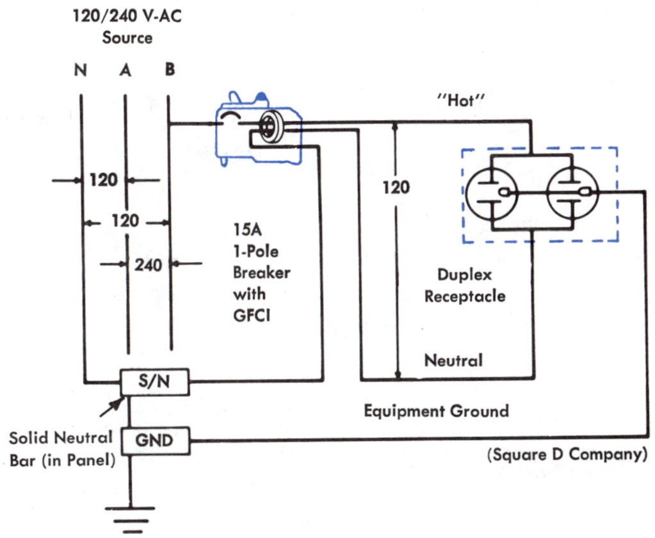

(Square D Company)

WIRING OF 1-POLE QWIK-GARD

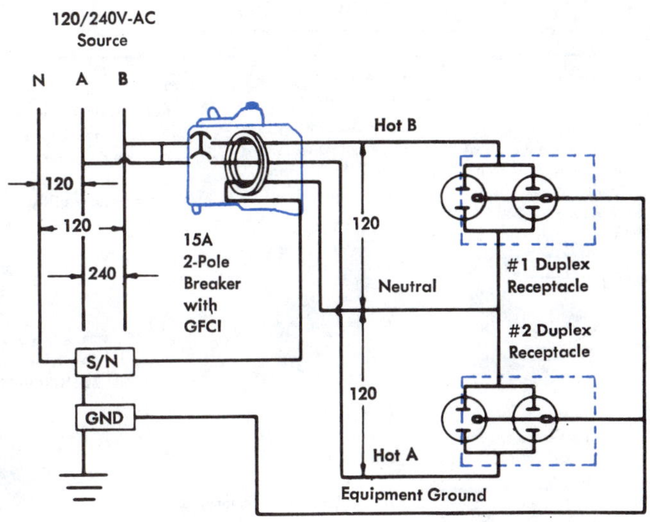

(Square D Company)

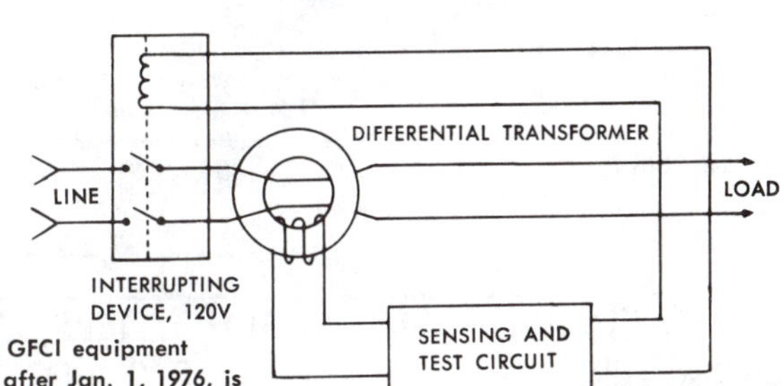

(Pass & Seymour)

All U.L. listed GFCI equipment manufactured after Jan. 1, 1976, is required to have a 5 ma (±1) fault current to ground trip value.

For Motors of 125 Volts or Less

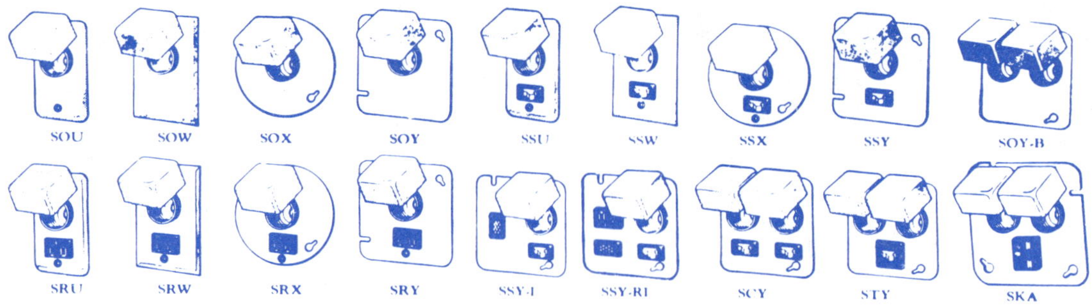

(Bussman-Cooper Industries)

Covers fit standard outlet or switch boxes. Units are UL listed and are made with Edison-base fuseholder. Units without switch for motor 3/4 hp or smaller. Units with switch for AC motors 1/2 hp or smaller. Do not use on DC.

Shall be no exposed live parts after fuses or fuses and adapters are installed. (240-50 (d))

Edison-base fuseholders installed only where made to accept Type S fuses by use of adapters. (240-52)

240-24(a) Overcurrent devices required to be readily accessible. See Exceptions.

FUSETRON BOX COVER UNITS

Plug fuses are tested by UL on circuits that deliver 10,000 amps AC short-circuit current.

Plug fuses of Edison-base type to be used only as replacements where no evidence of overfusing or tampering. (240-51 (b))

Edison-base plug fuses shall not be classified over 125 volts and 0-30 amps. (240-51 (a))

Ordinary Type W, Edison-base (no time delay)

Fusetron, Dual-Element Edison-Base Plug Fuse (time delay)

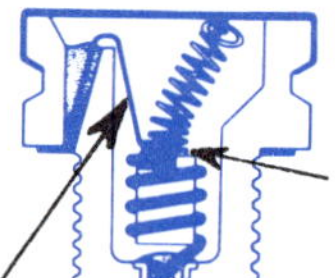

On overload, heat generated is transmitted to solder in thermal cutout. Solder softens from heat and is pulled out of contact with thermal circuit and opens circuit.

Dual-element Type S Fuse—Cut-Away View

On short circuit the fuse link blows and opens circuit.

Type S fuseholders and adapters designed to make tampering or shunting difficult. (240-54 (d))

Type S fuses shall be plug type and: (a) classified not over 125 volts and 0-15, 16-20, and 21-30 amps; (b) shall not be interchangeable with lower amp classification and designed to be used only with Type S fuseholder or Type S adapter inserted. (240-53 (a), (b))

Type S fuseholders and adapters to be designed only for Type S fuses. (240-54 (b))

Dual-Element Type S Fuse

When once inserted Type S adapters to be nonremovable. (240-54 (c))

Type S adapters to fit Edison-base fuseholders. (240-54 (a))

Adapter

(Illustrations, courtesy of Bussmann Mfg. Div., McGraw-Edison Co.)

PLUG FUSES, FUSES, FUSEHOLDERS, AND ADAPTERS (PART E, ARTICLE 240 NEC)

Type S fuses, fuseholders, and adapters to be standardized by manufacturers to permit interchangeability. (240-54 (e))

240-50(a) Plug fuses shall not be used in circuits over 125 V between conductors except in grounded circuits at not more than 150 V to ground. See Exception.

Each fuse, fuseholder, and adapter shall be marked with amp rating. (240-50 (b))
15 amps or less shall have hexagonal top or cap. (240-50 (c))
Screw-shell of plug-type fuseholder shall be connected to load side of circuit. (240-50 (e))

70-600 amp, knife-blade type fuseholder, pull-out type. This type of fusible load center normally has 100 to 200 amp rating.

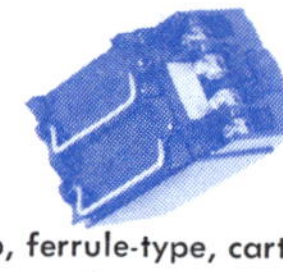

0-60 amp, ferrule-type, cartridge fuseholder, pull-out type.

0-30 amp plug-type, single pole, twin plug fuseholder.

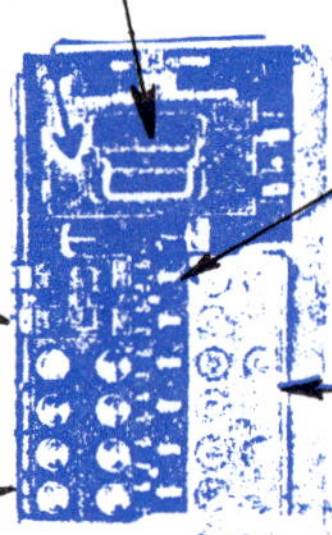

Connection bus for component cartridge or plug fuseholder blocks.

Provisions to mount terminal bar in panelboard.

LIGHTING AND APPLIANCE PANELBOARD
(Enclosure and trim not shown)

Equipment Grounding Conductor Terminal Bar

Terminal Bar Mounting Screws

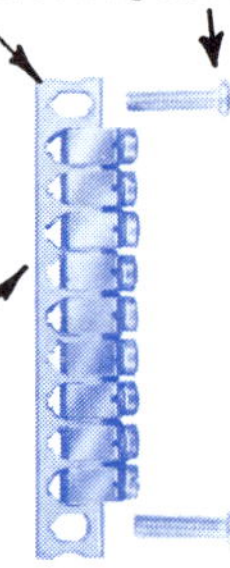

Equipment Grounding Terminal Bar

COMPONENTS FOR SERVICE PANELBOARD

(Square D Company)

240-30 In general, overcurrent devices shall be enclosed in cutout boxes or cabinets. See Exceptions.

240-32 Enclosures for overcurrent devices in damp or wet locations shall be mounted so there is a minimum of 1/4″ air space between enclosure and wall. Cabinets or cutout boxes installed in wet locations shall be weatherproof.

Enclosures shall be mounted in vertical position except where impracticable and in compliance with 240-81. (240-33)

384-14 Classified as a lighting and appliance panelboard if 10% of its circuits are 30 amps or less and have provisions for neutral connections.

384-16(a) General rule: Each lighting and appliance panelboard shall be protected at not more than the rating of the panelboard.

384-15 Lighting and appliance branch-circuit panelboards shall not contain more than 42 overcurrent devices in any one cabinet or cutout box. A 2-pole device counts as 2; a 3-pole device counts as 3.

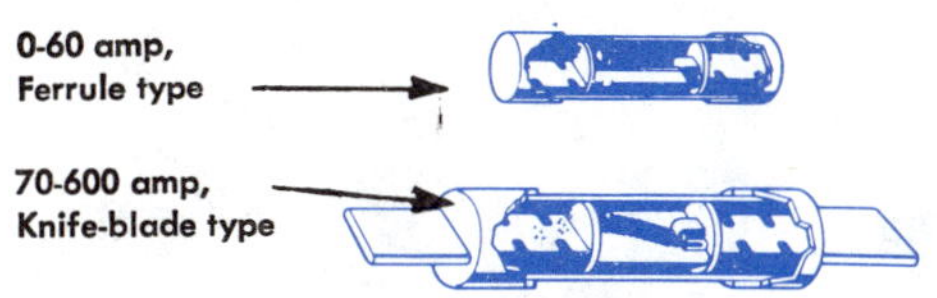

DUAL ELEMENT, TIME-DELAY CARTRIDGE FUSE

The dual element, time-delay fuse is designed to provide a time delay in the harmless low overload range and also to be fast-opening when exposed to dangerous short circuits. This fuse contains two fusible elements in series and the magnitude of a potentially damaging current flow determines which one of these elements opens. The thermal cutout element will open at currents of up to about 500% of fuse rating (overload range), whereas the short-circuit element will open at 500% or more of fuse rating.

DUAL ELEMENT, TIME-DELAY CURRENT-LIMITING FUSE

Dual-element current-limiting fuses function similar to regular dual-element, time-delay fuses except that they are faster operating in short-circuit range, thus are more current-limiting. The short-circuit element is normally made of silver, with a quartz sand arc quenching filler.

CURRENT-LIMITING FUSE

Single-element current-limiting fuses function similar to regular dual-element, time-delay low-overload range. Has lowest let-thru values in comparision with other fuses. Used to protect circuit components having inadequate interrupting, bracing, or withstand rating, such as low I.C. breakers.

NEC 240-11 Current-limiting fuse when interrupting currents in its range, will reduce current in faulted circuit to magnitude substantially less than obtainable in same circuit if replaced with solid wire if same impedance.

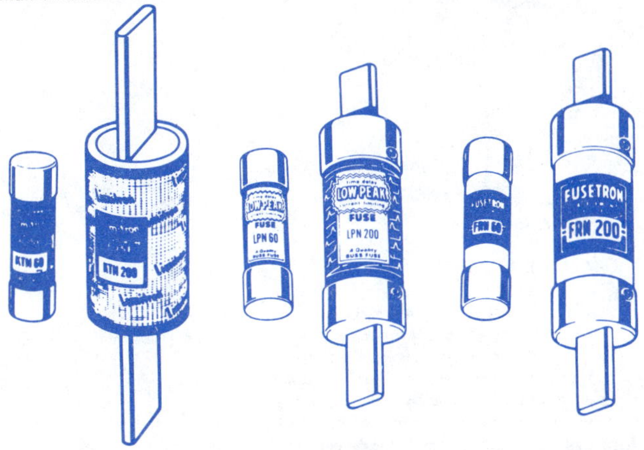

LIMITRON FAST-ACTING FUSE, CLASS K1 | LOW-LEAK DUAL ELEMENT FUSE, CLASS K5 | FUSETRON DUAL ELEMENT FUSE, CLASS K5

Class K fuses have ratings of 0-600 amps in both 250 and 600 volts. Have dimensions same as Class H fuses. Grouped into three categories: K1, K5, and K9 and may have UL listing of 50,000, 100,000 or 200,000 amps symmetrical rms amps. UL has assigned a maximum level of peak-let-thru current (Ip) and amp-squared seconds (I^2t) for each K rating. K rating determines degree of current limiting. K9 has least current-limiting ability and K1 the greatest. May also be listed by UL as time delay; then are required to have minimum time delay of 10 seconds at 500% of rated current.

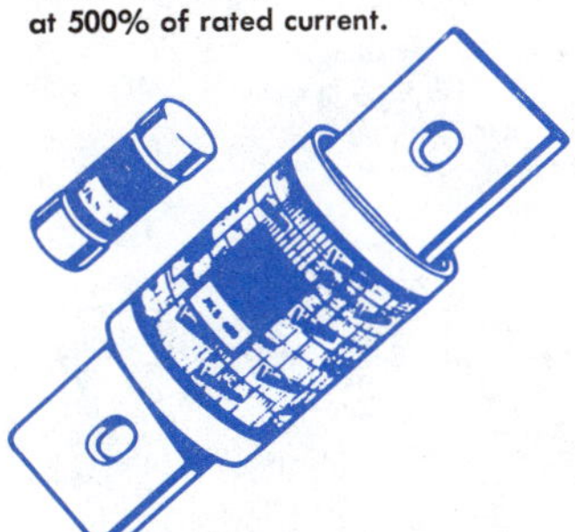

LIMITRON FAST-ACTING FUSE, CLASS J

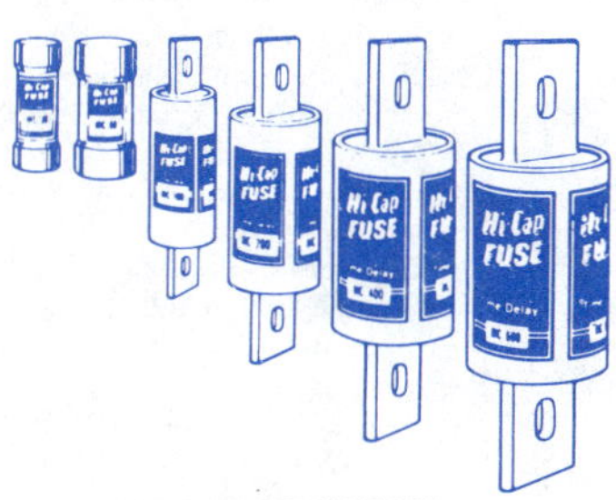

HI-CAP TIME-DELAY FUSE, CLASS J DIMENSION

Class J fuses are 600-volt AC, 0-600 amps, current limiting and listed by UL with interrupting rating of 200,000 symmetrical rms amps. Fuseholders are smaller than Class H, thus are noninterchangeable.

OVERCURRENT PROTECTION--CARTRIDGE FUSES

240-1, FPN Overcurrent protection opens circuit when current reaches a value that can damage either equipment or conductors. Overcurrent devices shall be selected based on circuit impedance, short circuit, and with-stand ratings. Equipment intended to interrupt current at fault levels shall have sufficient interrupting rating. See 110-9 and 110-10.

NEC 110-9 Fuses shall have sufficient interrupting capacity for the voltage employed and the current that must be interrupted. (Also see 230-65.)

NEC 230-65 Service equipment shall have interrupting capacity suitable for available fault current.

NEC 110-10 Fuses shall be selected and coordinated to clear fault without extensive damage to electrical components of circuit.

240-6 Standard ratings for ITCBs shall be: 15, 20, 25, 30, 35, 40, 45, 50, 60, 70, 80, 90, 100, 110, 125, 150, 175, 200, 225, 250, 300, 350, 400, 450, 500, 600, 700, 800, 1000, 1200, 1600, 2000, 2500, 3000, 4000, 5000, and 6000 A. Additional standard ratings for fuses are: 1, 3, 6, 10, and 601 A.

NEC 240-8 Fuses shall not be connected in parallel. Exception: permitted if factory assembled and listed as a unit.

NEC 240-20 Shall be connected in series with each ungrounded conductor.

NEC 240-21 Overcurrent device connected where conductor to be protected receives supply per 240-21(a–o).

240-22 In general, overcurrent devices shall not be, placed in series with any grounded conductor. See Exceptions for use in (1) motor overload protection and (2) simultaneous disconnection.

NEC 240-24 Fuses shall be readily accessible except: (1) busways (364-12); (2) supplemental (240-10); (3) services (230-92).

NEC 240-30 Shall be enclosed in cabinets or cutout boxes. (See Exceptions 1-3).

NEC 240-32 In damp or wet locations shall be in enclosures identified for such use and 1/4" air space between enclosure and mounting surface.

NEC 240-33 Enclosures shall be mounted in vertical position except where impracticable and in compliance with 240-81.

NEC 240-40 Disconnect required on supply side of cartridge fuses where accessible to other than qualified persons except: (1) service as per 230-82; (2) group operation of motors as per 430-112 and fixed space heating as per 424-22.

(Illustrations courtesy of Bussman Mfg. Division, McGraw-Edison Co.)

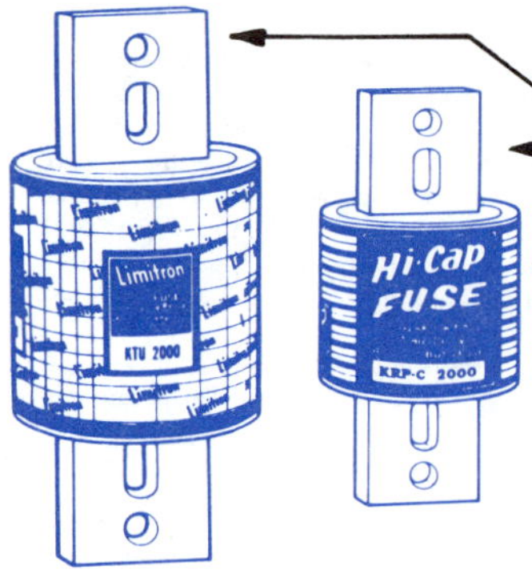

LIMITRON FAST-ACTING FUSE, CLASS L

HI-CAP TIME-DELAY FUSE, CLASS L

Class L listed by UL in sizes 601-6000 amps at 600 volts AC. They bolt into place, have a current-limiting characteristic and interrupting rating of 200,000 symmetrical rms amps and are used in bolted pressure contact switches.

240-60 Cartridge Fuses and Fuseholders:
(a) In general, cartridge fuses and fuseholders of the 300 V type are not permitted to be installed in circuits of over 300 V between conductors. In 1ϕ, line-to-neutral circuits supplied from a solidly grounded neutral system. However, fuses and fuseholders are permitted when the line-to-neutral voltage does not exceed 300 V.

(b) Cartridge fuseholders (0-6000 amps) shall be so designed to make it difficult to put fuse of given class into fuseholder for current lower or voltage higher than intended. Current-limiting fuseholders shall permit insertion only of current-limiting fuses;
(c) Fuses shall be marked to show: (1) amp rating, (2) volt rating, (3) interrupting rating if other than 10,000 amps, (4) current-limiting where applicable except where used as supplementary protection, (5) trademark of manufacturer.

240-100 Feeders over 600 V shall have a short-circuit protective device that complies with 710, Part C. It must be capable of detecting and interrupting currents that can occur in excess of fuse setting or melting point. Fuse rating in continuous amps shall not exceed 3 times amp rating of conductor. For CBs, long-time trip setting or minimum-trip setting of an electronically-activated fuse shall not exceed 6 times ampacity of conductor.

240-101 Branch-circuits shall have a short-circuit protective device that complies with 710, Part C. It must be capable of detecting and interrupting currents that can occur at their location in excess of fuse setting or melting point.

240-61 Classification of cartridge fuses and fuseholders shall be on basis of amperage and voltage. Fuses of a higher rating are permitted to be used on lower rated circuits, but fuses of a lower rating are not permitted to be used with higher voltages. 600 V rated fuses could be used on a 277 V circuit but 250 V rated fuses could not.
Some common classifications for not over 600 volts—

Amperes		
0-30	401-600	2001-2500
31-60	601-800	2501-3000
61-100	801-1200	3001-4000
101-200	1201-1600	4001-5000
201-400	1601-2000	5001-6000

NEC 240-3 Conductors shall be protected at ampacities per Tables 310-16 through 310-19 except where ampacity of conductor and standard rating of fuse do not correspond, then next larger fuse may be used only up to 800 A per 240-3(a–m).

NEC 240-2 Equipment shall be protected against overcurrent. (See 240-2 for list of equipment.)

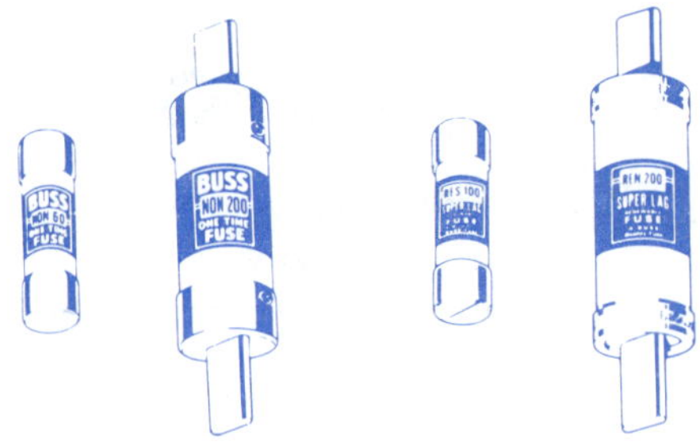

One-time Fuse Renewable Fuse

CLASS H DIMENSION FUSES

Class H fuses are probably oldest and best known. In the past referred to as "NEC" or "Code" fuses. Interrupting rating not marked on label. Tested by UL on circuits of 10,000 amps. Rated 0-600 amps in both 250- and 600-volt range.

61-600 Ampere Knife-Blade Type

0-60 Ampere Ferrule Type

CLASS R FUSES

Class R fuses are of the nonrenewable, cartridge type and are the most recently listed fuse in the standards of UL. This fuse is the same size as Class H and K fuse except in 0-60 amps one end-cap ferrule will have an annular grooved ring and in 61-6000 amps one knife blade will have a slot in its side. Fuseholders to be designed to accept only Class R fuses, but rejection features in fuse would not prohibit the use of Class R fuse in a Class H or K fuseholder. Fuse rated to interrupt fault current up to 200,000 rms amps. If Class R to have UL label as time-delay, it would basically have a minimum of 10 seconds delay at 5 times fuse rating.

Class G are available in sizes 0-60 amps for circuits of 300 volts or less to ground (i.e., 277/480-volt systems). They are a small dimension cartridge fuse (commonly referred to as Type SC fuses) and basically Type S cartridge fuses because they are noninterchangeable in that a 15-amp fuse will fit only into a 15-amp fuseholder, and a 20-amp fuse will fit only into a 20-amp fuseholder. They are available in sizes 15, 20, 30 and 60 amps. They are listed by UL at 100,000 symmetrical rms amps interrupting capacity.

OVERCURRENT PROTECTION--CARTRIDGE FUSES

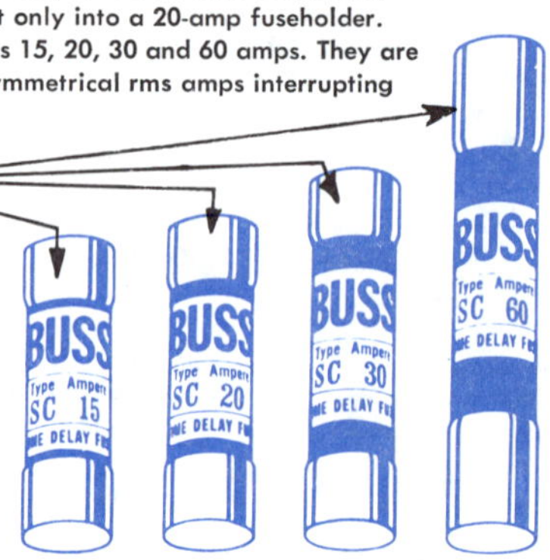

BUSS TYPE SC FUSES, CLASS G

(Illustrations, courtesy of Bussman, Cooper Industries)

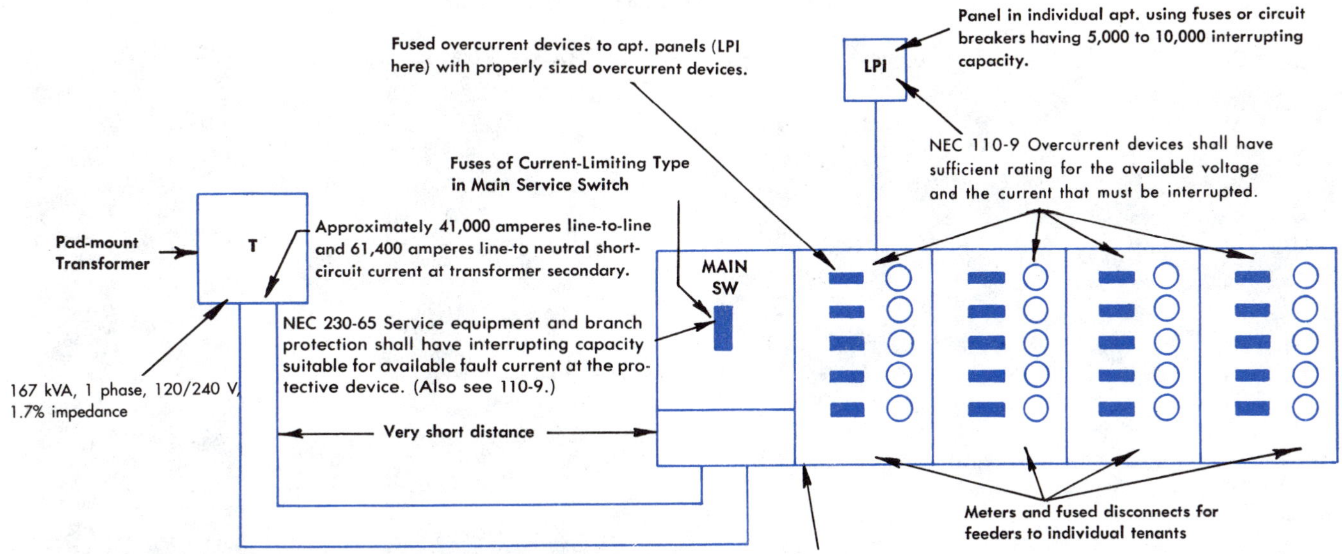

NEC 110-10 The overcurrent protective devices, the total impedance, the component short-circuit withstand ratings, and other characteristics of circuit to be protected shall be selected and coordinated to permit protective devices to clear fault without extensive damage to electrical components of circuit. In this diagram, the properly selected current-limiting fuses will protect the individual breakers in the panels rated 5,000 or 10,000 A.I.C.

THE ABOVE INSTALLATION MEETS REQUIREMENTS OF NEC 110-9, 110-10 AND 230-65.

Simple formula for calculating the short-circuit currents for above transformer:

Step 1. Calculate transformer full load current (single-phase):

$$I = \frac{\text{kVA} \times 1000}{E} = \frac{167 \times 1000}{240} = 696 \text{ amperes (FLA)}$$

Step 2. Find multiplier:

$$\frac{100}{\text{Trans \% Impedance}} = \frac{100}{1.7} = 58.8$$

Step 3. Find short-circuit current at transformer secondary:
Trans FLA $\times$ multiplier $= 696 \times 58.8 = 40{,}941$ amperes short-circuit L-L (line-to-line).

Note 1. The L-N (line-to-neutral) fault current at transformer (single-phase center-topped) secondary is approximately 1.5 $\times$ L-L value, i.e., 40,941 $\times$ 1.5 = 61,411 amperes L-N. (Multiplier 1.5 varies from 1.33 to 1.67.)

Note 2. For three-phase systems, use the following formula in Step 1, then proceed with Steps 2 and 3.

$$I = \frac{\text{kVA} \times 1000}{E \times 1.73}$$

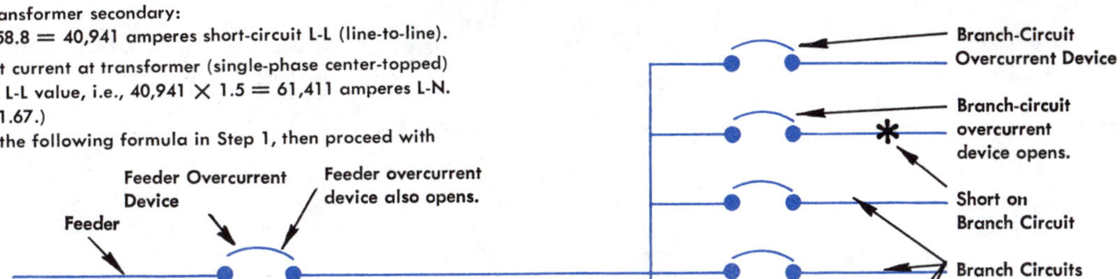

NON-SELECTIVE SYSTEM—A fault in one branch circuit not only opens the branch-circuit device but also opens the feeder overcurrent device.

NEC 517-17(a)(b) In Health Care Facilities where ground-fault protection is provided for services and feeders, they shall be selective, coordinated, and performance tested when first installed.

NEC 240-12 (FPN) Coordination is properly localizing a fault condition to restrict outages to affected equipment by using proper selective fault-protective devices.

When a fault on an individual branch circuit not only opens the branch-circuit overcurrent device but also opens the feeder overcurrent device, it is referred to as "non-selective coordination." See 110-10, and 240-12, FPN Recommends that an internal failure in one branch will not interrupt the supply to other branches. A selectively coordinated system is basically one where a fault on a branch circuit or feeder occurs and only the overcurrent device immediately ahead (upstream) from the fault opens. A fault on a branch circuit would open only the branch-circuit device and not affect the feeder or service overcurrent device.

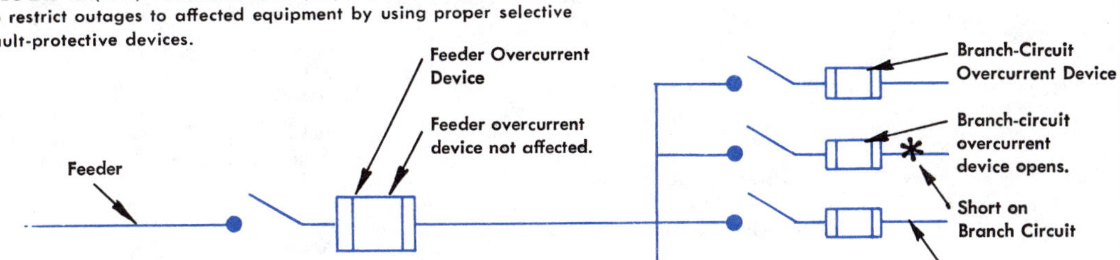

SELECTIVE SYSTEM—A fault in one branch circuit opens only branch-circuit device and does not affect feeder overcurrent device.

OVERCURRENT DEVICES AND AVAILABLE FAULT CURRENT

(Illustrations, Courtesy of Bussman-Cooper-Industries)

The 30A, 300V T-Tron fuse is no bigger than a nickel (interrupting rating is 200,000 Amps).

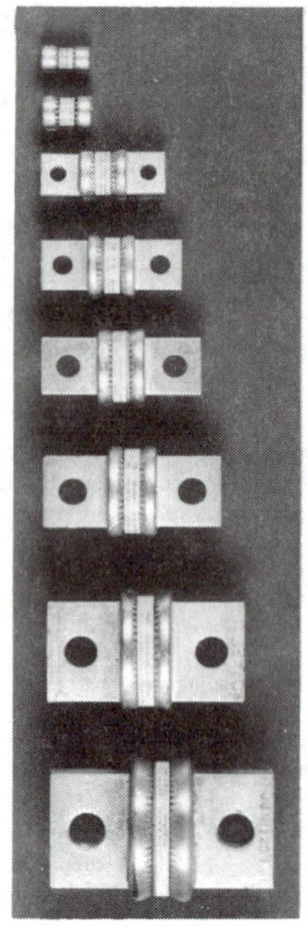

JJN (300 Volt) T-Tron fuses (all case sizes).

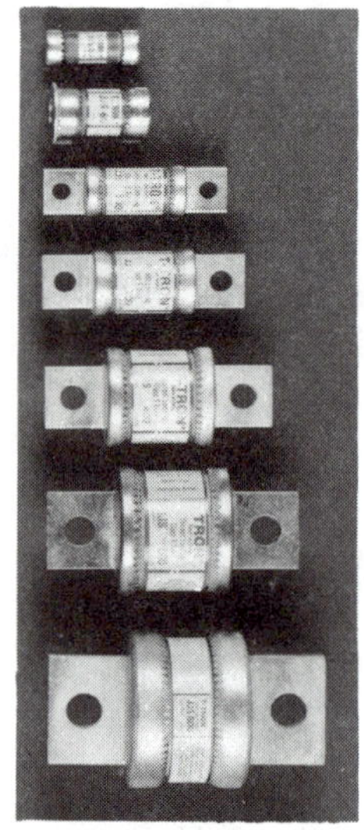

JJS (600 Volt) T-Tron fuses (all case sizes).

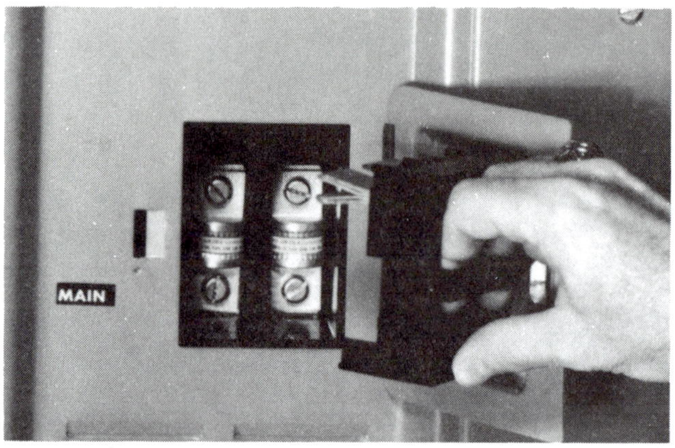

200 ampere Load Center with pull-out disconnect; Class T fuses bolted into panel board.

CLASS T FUSES

Class T fuses are listed by UL in sizes 1-600 amps in both 300 and 600 volt AC range. They are fast-acting with a high degree of current limitation and a high interrupting rating (200,000 amps). They provide overload and high performance short-circuit protection of main, feeder, and branch circuits, as well as general-purpose equipment. Class T fuses are the smallest and most compact of all UL-listed power distribution fuse devices. They comply with NEC 110-9 "Interrupting Rating" and 230-65 "available short-circuit current," which requires service equipment to be suitable for the short-circuit current available at its supply terminals. In addition, Class T fuses, when properly selected, can limit the short-circuit current for the protection of low interrupting capacity (IC) circuit breakers. (In a 10,000 amp, IC branch-circuit, the circuit breakers are protected by Class T fuses in the mains.)

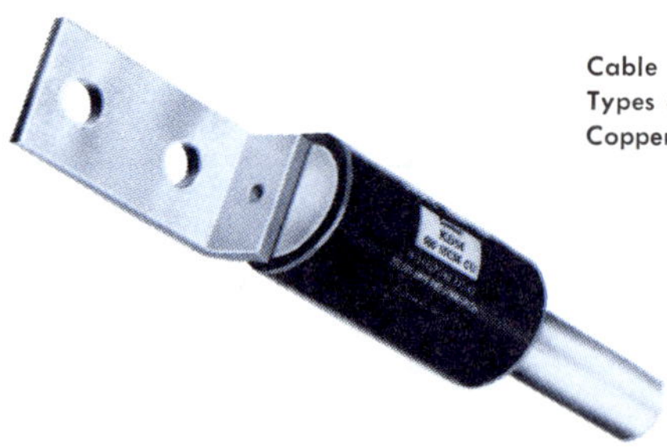

Cable Limiters Available in Many Types of Terminals to Take Either Copper or Aluminum Cable.

Typical Application for Cable Limiters

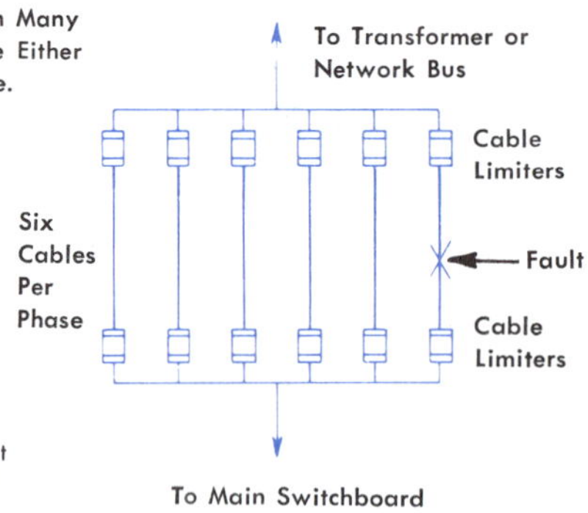

Multiple Cables Per Phase Cable Protection and Fault Isolation

Isolating faulted cables in multiple cable runs require cable limiters at each end of each cable.

Faulted cable would be isolated by the two cable limiters on each end of faulted cable.

Continuity of service would be maintained by the five remaining cables.

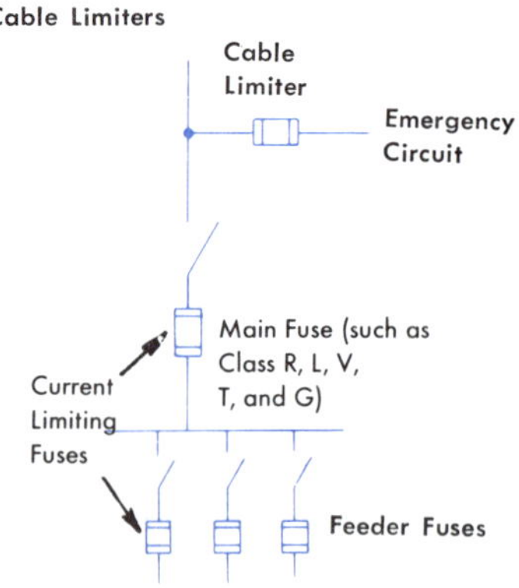

Emergency Circuit Taps Ahead of Mains

Unprotected tapped conductors can result in violent conductor vaporization under short-circuit conditions initiating further switchboard short-circuit hazards. The protection of emergency circuit taps ahead of mains can be protected against short circuits by current-limiting cable limiters. The various types of cable limiter terminations makes adaptation to service switchboard construction relatively simple.

NEC 240-11 Basic definition of current limiting device: an overcurrent device that when interrupting currents in current-limiting range, will reduce current in faulted circuit to a value less than obtainable in same circuit if replaced with a solid conductor having comparable impedance.

240-40 In circuits over 150 V to ground, a disconnecting means shall be installed on line side of all fuses, where accessible to non-qualified personnel. See Exceptions for (1) cable limiters and (2) group operations in 430-112 and 424-22.

230-82 In general, equipment shall not be installed on line or supply side of service disconnecting means. See Exceptions 1-8.

230-91(a) Service overcurrent device permitted to be located immediately adjacent to the service disconnecting means or be an integral part of it.

240-60(b) Current-limiting fuseholders shall be designed such that they only accept fuses that are of current limiting type.

CABLE LIMITERS

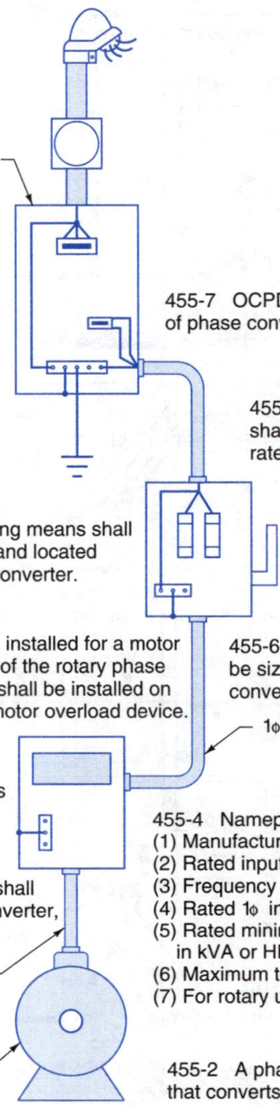

455-7 OCPD shall be sized at 125% of phase converter's nameplate 1ϕ input FLA.

455-8(c) Disconnecting means shall be sized at 115% of maximum rated 1ϕ input FLA. See Exceptions.

455-8(a) Disconnecting means shall be readily accessible and located within sight of phase converter.

455-23 Capacitors, installed for a motor load, shall be a part of the rotary phase conversion system, shall be installed on the line side of the motor overload device.

455-6 Conductors of 1ϕ side shall be sized at 125% of the phase converter input FLA.

455-5 EGC means shall be provided.

455-4 Nameplate shall indicate:
(1) Manufacturer's name
(2) Rated input and output voltage
(3) Frequency
(4) Rated 1ϕ input FLA
(5) Rated minimum and maximum single load in kVA or HP
(6) Maximum total load in kVA or HP
(7) For rotary units, 3ϕ amps at full load.

455-10 Terminal housing shall be provided on a phase converter, per 430-12.

455-2 A phase converter is an electrical device that converts 1ϕ power to 3ϕ power.

PHASE CONVERTERS—455

504-20 Any wiring method suitable for unclassified locations, including Ch 7 and 8 methods, are permitted. Seals shall be provided per 504-70.

504-30(a)(1) Open wiring methods, including conductors and cables, shall be separated and secured from other conductors of power and lighting circuits by at least 2″.

504-30(a)(2) Conductors shall not be placed in any raceway, cable tray, or cable with conductors which are not intrinsically safe.

504-30(a)(3) Intrinsically safe sytems installed within enclosures shall have conductors separated at least 2″ from other conductors.

504-50(a) Associated metal apparatus such as cable shields, enclosures, and raceways shall be grounded.

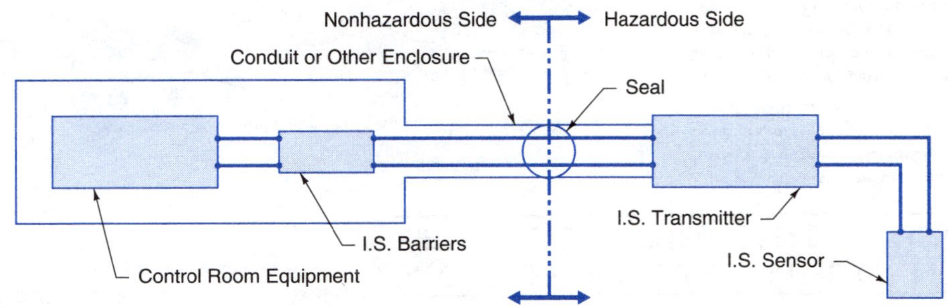

504-50(b) Intrinsically safe systems shall be grounded to the GES per 250-81 where available.

504-50(c) Shields, for shielded conductors or cables, shall be grounded unless they are part of an intrinsically safe circuit. See Exception.

504-60(a) Bonding, in hazardous locations, shall comply with 250-78.

504-70 Conduits and cables shall be sealed to minimize passage of gases, vapors, or dust per 501-5 and 502-5.

504-80(a) Intrinsically safe circuits, shall be suitably identified at all terminals and junction locations to avoid unintentional interference.

504-80(b) Labels, worded "Intrinsic Safety Wiring", shall be permanently affixed to all wiring so as to be visible after installation and at intervals not exceeding 25′ in length.

504-80(c) Color coding of intrinsically safe conductors and raceways, boxes and cable trays that contain only intrinsically safe wiring shall be permitted to be light blue provided no other light blue conductors are used.

504-2 An intrinsically safe system is an assembly of interconnected apparatus and cables that may be used in hazardous locations because a spark is incapable of causing ignition in circuits.

INTRINSICALLY SAFE SYSTEMS—504

NEC RULES GOVERNING USE OF NON-METALLIC-SHEATHED CABLE (ROMEX)

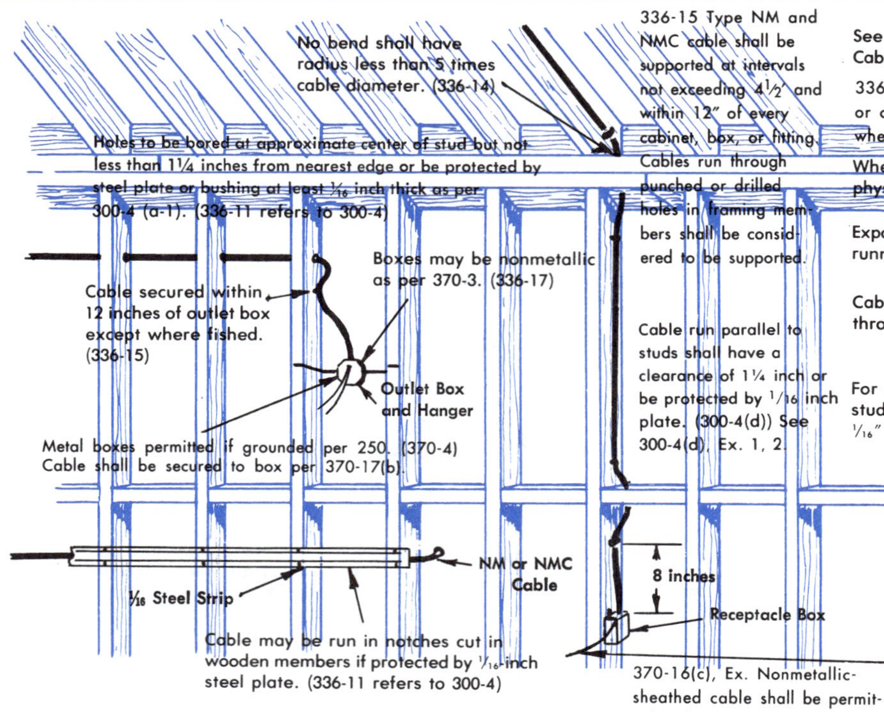

See 336-3 for permitted or not permitted uses. Cable may be fished between boxes. (336-15 Ex. 1)

336-10(b) 6" minimum conduit, tubing, guard strips, or other means required for mechanical protection where cable passes through floor.

Where necessary, exposed cable shall be protected from physical damage by conduit, pipe, guard strips, etc. (336-10(b))

Exposed cable shall follow surface of building finish or of running boards. (336-10(b))

Cables smaller than No. 6 or three No. 8 shall be drilled through basement joists or on running boards. 336-12

For exposed or concealed work where run thru metal studs or members and nail or screw penetration likely, 1/16" steel plate or clip required. (336-11 refers to 300-4(b))

Where cable is run through metal studs or members, an approved bushing or grommet shall be used except where slots or holes are formed or rounded to protect cable. 336-11 refers to 300-4(b).

EGC shall have same ampacity as circuit wires for 15 A, 20 A, and 30 A circuits. Table 250-95.

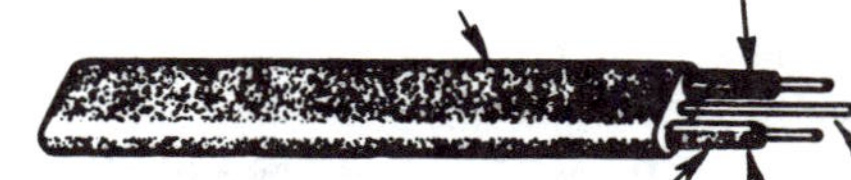

NONMETALLIC-SHEATHED CABLE (ROMEX) WITH GROUND WIRE

Type NM and NMC cables shall have ampacity of 60°C wire in Table 310-16, but conductors to have rating of 90°C. (336-26)

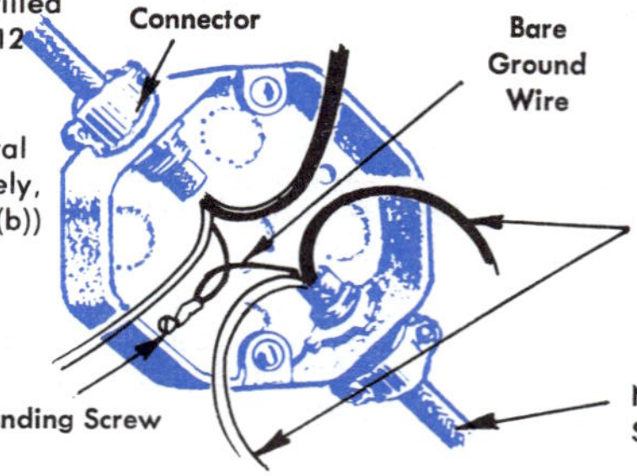

NEC RULES GOVERNING USE OF ARMORED-CABLE TYPE AC (BX) AND METAL-CLAD CABLE TYPE MC

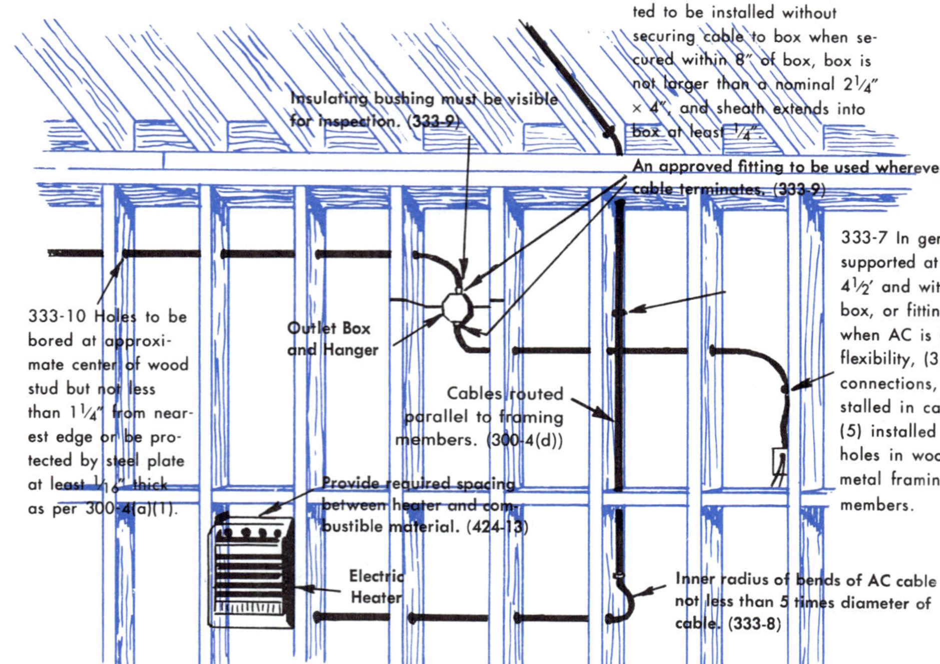

333-7 In general, AC cable shall be supported at intervals not exceeding 4½' and within 12" of every cabinet, box, or fitting. See Exceptions 1–4 when AC is (1) fished, (2) used for flexibility, (3) used for lighting fixture connections, (4) installed in cable tray, (5) installed through holes in wood or metal framing members.

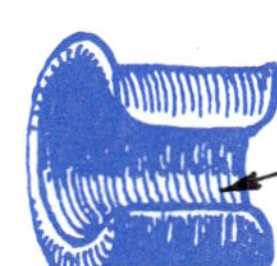

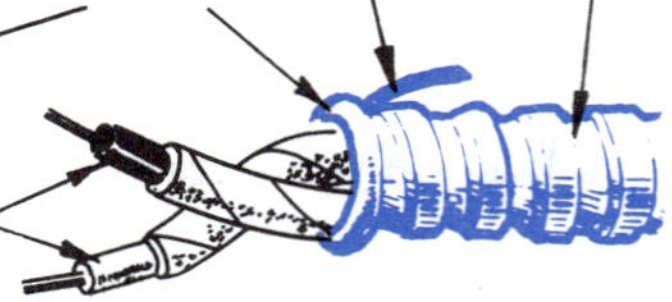

ARMORED CABLE—TYPE AC

AC cable installed in insulated walls, ceilings, attics shall have 60°C ampacity as per 310-15 but shall have 90°C conductors. (333-5)

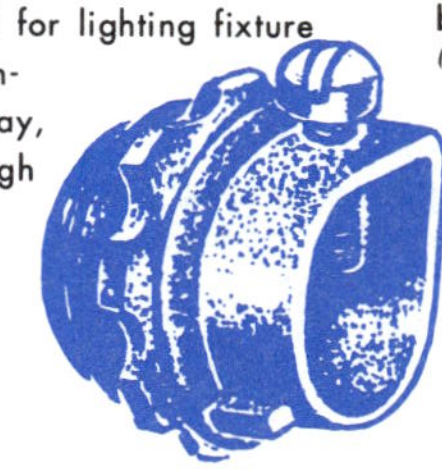

BX CONNECTOR

Type MC cable may be used for services, feeders, branch circuits, indoors or outdoors, and most other applications. (See 334-3 Uses Permitted)

Type MC cable not permitted for direct burial in earth, in concrete, or where exposed to deleterious chemicals, except where protected by suitable materials. (334-4)

334-11 Bends shall be such that cable will not be damaged. See a, b, c for radius of various cables.

Type MC cable is assembly of insulated wires enclosed in metal sheath of interlocking tape or a smooth or corrugated tube. (334-1)

METAL-CLAD CABLE—TYPE MC

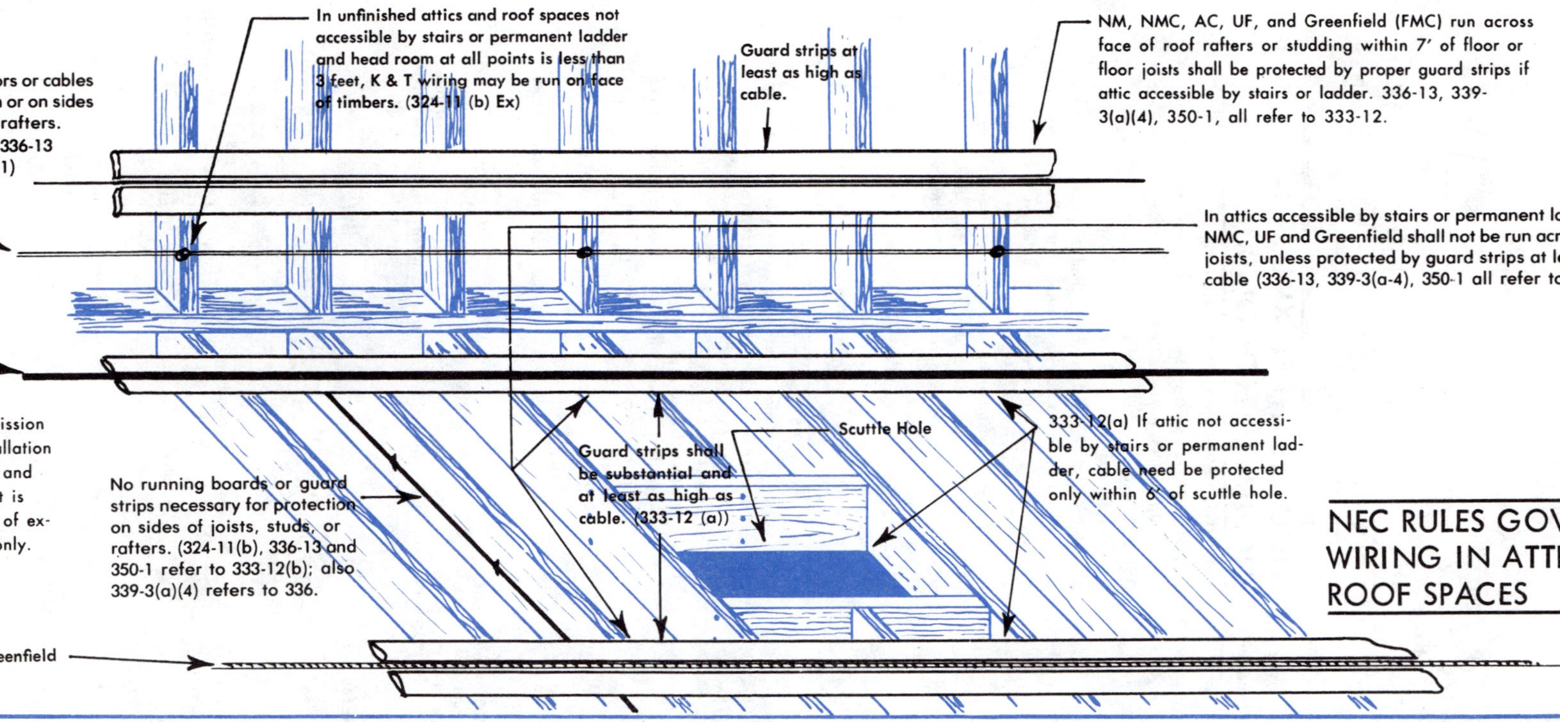

324-3 Special permission is required for installation of concealed knob- and tube-wiring unless it is used for extensions of existing installations only.

NEC RULES GOVERNING WIRING IN ATTIC OR ROOF SPACES

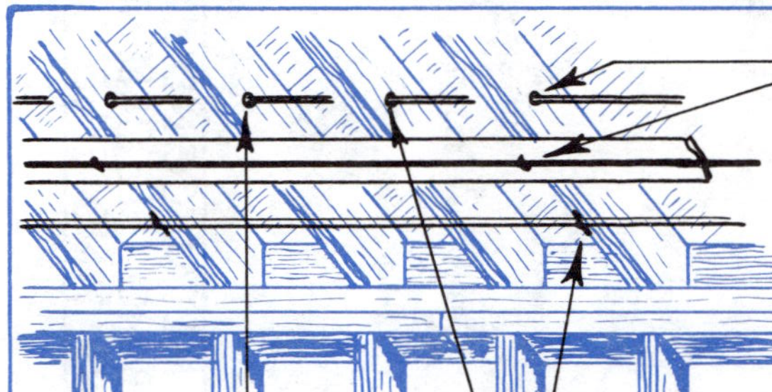

NEC RULES APPLYING TO NONMETALLIC SHEATHED CABLE (NM, NMC, AND NMS) AND UNDERGROUND FEEDER AND BRANCH-CIRCUIT CABLE (UF) RUN AT ANGLES TO JOISTS IN UNFINISHED BASEMENT

LEGEND

K & T—Knob-and-Tube Wiring (NEC 324)
BX—Armored Cable, Type AC (NEC 333)
NM, NMC (Romex) Nonmetallic Sheathed Cable, NM, NMC, and NMS (336), (Greenfield) Flexible Metal Conduit (350), (UF) Underground Feeder and Branch-Circuit Cable (339).

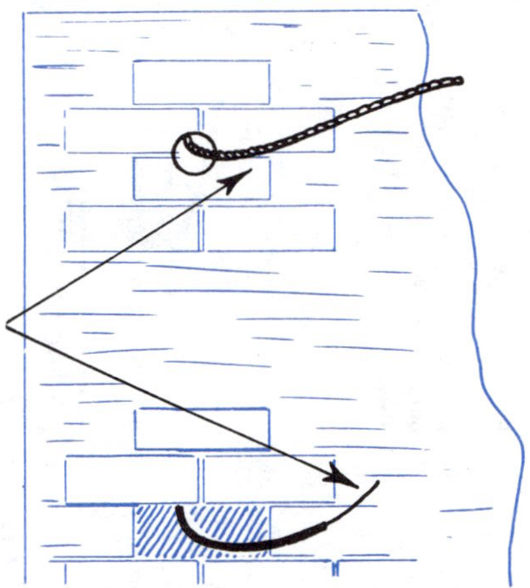

336-1 NMS is a factory assembled hybrid cable combining power and remote control signaling conductors under a common, moisture-resistant, flame-retardant, nonmetallic jacket.

333-3, 336-3(a) Types NM or AC may be fished in masonry walls if no moisture is present and above grade line.

333-3, 336-3(b), 339-3(a)(4) If moisture is present, Types NMC, UF, or AC shall be used.

336-3(c) NMS is permitted for both exposed and concealed installations in dry locations. NMS cable shall be installed per 780.

336-4(a)(8) NM, NMC, and NMS not permitted to be embedded in poured cement, concrete, or aggregate.

336-4(b) NM and NMS not permitted to be embedded in masonry, concrete, fill, or plaster, nor run in a shallow chase in masonry or concrete and covered with plaster or similar finish.

NEC RULES DEALING WITH CABLE FISHED INTO MASONRY WALLS

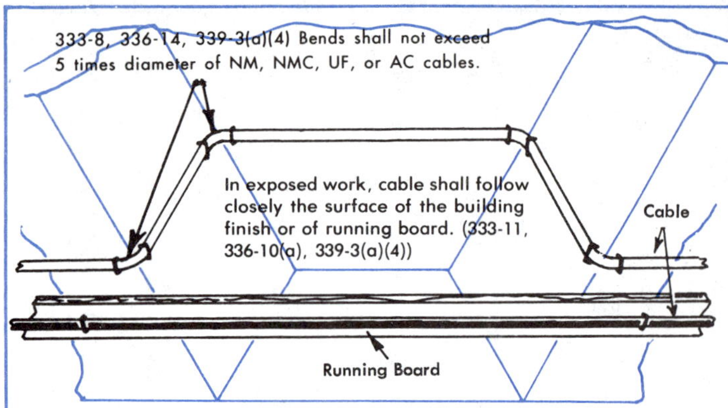

GENERAL NEC RULE FOR EXPOSED CABLE

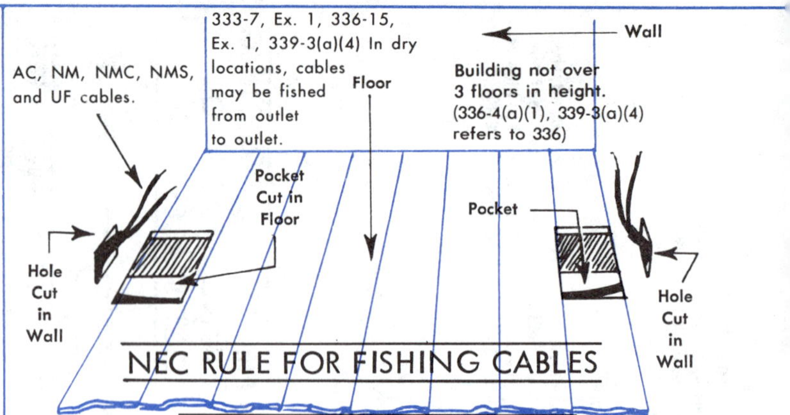

NEC RULE FOR FISHING CABLES

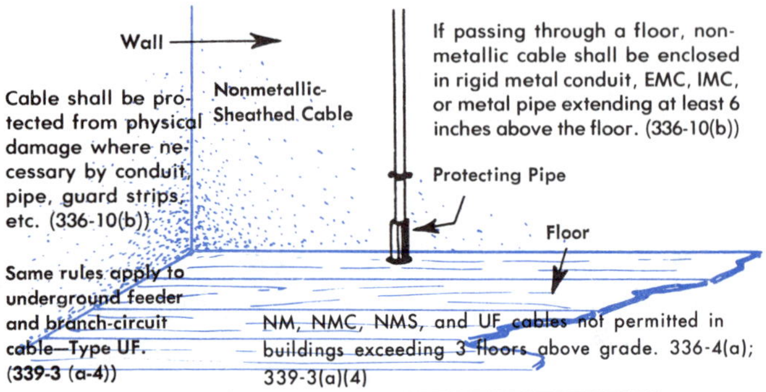

NEC RULE FOR NONMETALLIC-SHEATED CABLE (ROMEX)

LIQUIDTIGHT FLEXIBLE NONMETALLIC CONDUIT

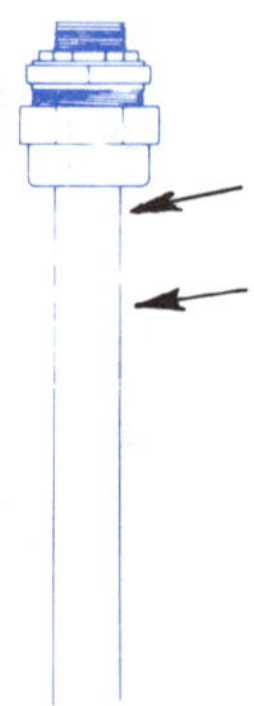

NEC 351-22 Definition and types of LTFNC.

NEC 351-26 Fittings shall be identified for such use.

Liquidtight flexible nonmetallic conduit—Article 351 Part B

351-24 LTFNC shall be installed in trade sizes from ½″ to 4″. See Exceptions 1 through 3 for ⅜″ trade size.

351-28 An EGC installed in LTFNC shall be permitted to be installed on inside or outside of conduit. If installed on outside, EGC shall not exceed 6′ and shall be routed with raceway.

351-23(a) Listed LTFNMC is permitted to be used (1) when flexibility is needed for installation, operation, or maintenance, (2) where physical protection of conductors is required, (3) outdoors, where listed and suitable for purpose, (4) for direct burial applications, where listed for purpose.

351-23(b) Not permitted to be used where: (1) subject to physical damage; (2) temperature rating of conduit is exceeded; (3) over 6′ except for special applications; (4) over 600 V except for 600-31(a).

NEC 345-6, 346-5, 348-5 No intermediate, rigid metal or nonmetallic conduit or EMT smaller than ½″. No intermediate metal conduit or EMT larger than 4″.

NEC 345-3 (b), 346-1, 348-1 No metallic conduit or fittings shall be installed in concrete in direct contact with earth or in corrosive locations unless suitably protected or judged suitable for condition. (300-6 (b))

NEC 345-11, 346-11, 348-10, 350-16 No more than equivalent of four quarter bends from outlet. This includes flexible metal conduit.

NEC 345-10, 346-10 Radius of inner bend for field bends in intermediate and rigid metal conduit not less than 4 inches for ½ inch, 5 inches for ¾ inch. (See Table 346-10.) Bends as per NEC 345-10 Ex permitted if one-shot bender used.

NEC 348-9 Field bends in EMT as per Table 346-10. Bends as per Table 346-10 Ex permitted if special bending machine used.

NEC 345-8, 346-7, 348-11 Ends shall be reamed to remove rough edges.

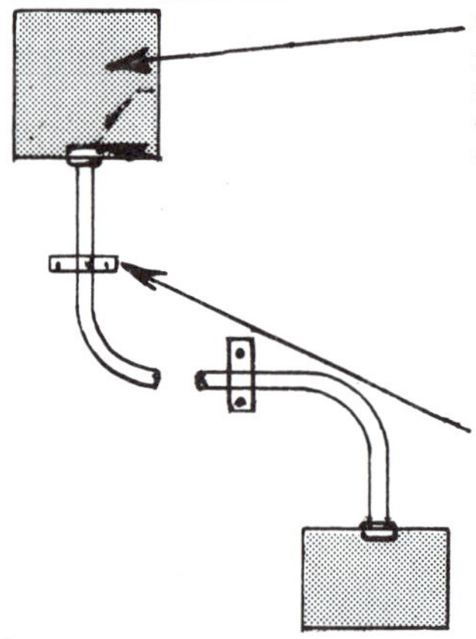

NEC 345-15, 346-8, Bushing required where conduit enters box or fitting unless box or fitting affords conductor protection. See 373-6 (c).

345-12, 346-12, 348-12 Intermediate and rigid metal conduit and EMT shall be supported within 3′ of outlet box, cabinet, and fitting and at least every 10′ of run. Fastening may increase to 5′ where structural members do not permit fastening within 3′.

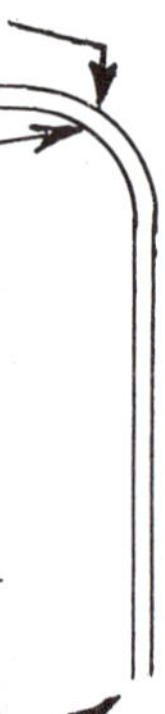

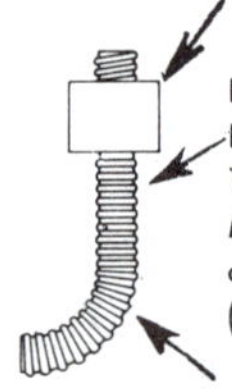

Electrical Nonmetallic Tubing Made of nonmetallic corrugated, moisture chemical and flame-retardant material. (331-1)

331-5 Size ½″ thru 2″

331-10 Limit: 360° of bends between boxes or conduit bodies.

331-15 Marked every 10′.

331-11 Supports every 3′. See Exceptions.

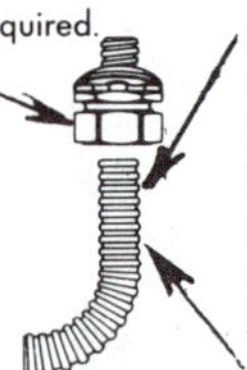

349-18 approved fittings required.

Size ½ inch through ¾ inch (See 349-10, Ex. 1, 2 for ⅜ inch) 349-4 Limited use and never over 6 ft. 349-16 Grounding 250-91(b), Ex. 2

Flexible Metallic Tubing 349-1 Similar to sealtight without nonmetallic jacket.

Bends per Table 349-20 (a) (b)

NEC 350-5 Flexible metal conduit shall not be used: (1) In wet location unless lead-covered or other special conductors used; (2) In hoistways except per 620-21; (3) In storage battery rooms; (4) In hazardous area except as per 501-4(b) and 504-20; (5) Rubber-covered wire used in areas where gas, oil, etc., will have deteriorating effect on rubber; (6) Underground or in concrete or aggregate; (7) Where subject to physical damage.

NEC 230-43 Six ft. of Greenfield permitted for service if bonded per 250-79.

350-10 Flexible metal conduit not permitted in sizes less than ½″ unless permitted for ⅜″ trade size (1) for enclosing motor leads per 430-145(b); (2) for listed assemblies not over 6′ per 410-67(c); (3) manufactured wiring systems per 604-6(a); (4) as permitted by 680-21; (5) as part of a listed assembly per 410-77(c).

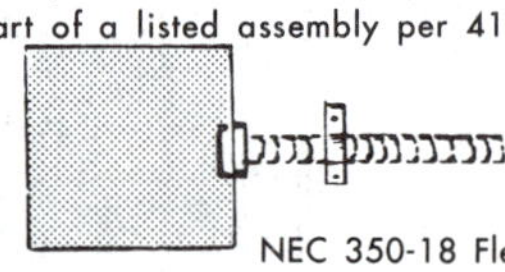

NEC 350-18 Flexible metallic conduit to be supported within 12″ of outlet box and every 4½′.
Exception 1: Where fished;
Exception 2: Not over 3 feet at terminals where flexibility is necessary;
Exception 3: Not over 6 feet from box to fixture as required by 410-67 (c).

See Table 350-12 for maximum conductors allowed in ⅜″ flexible conduit.

INTERMEDIATE METAL CONDUIT, RIGID METAL CONDUIT, ELECTRICAL METALLIC TUBING, FLEXIBLE METAL CONDUIT, FLEXIBLE METALLIC TUBING, AND ELECTRICAL NONMETALLIC TUBING

THREE EASY STEPS FOR FIELD BENDING

(Photos courtesy of Carlon, an Indian Head Co.)

Small Diameters (½" - 1½")   Large Diameters (2" - 6")

1. HEATING. Rotate the conduit to provide even heating. Time varies with conduit size. In large sizes insert bending plugs as shown before heating conduit.

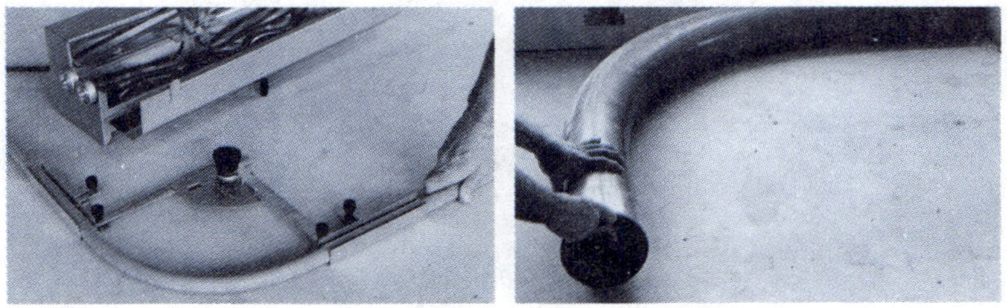

2. FORMING THE BEND. Use bend guide for small diameters, or a jig, or a line drawn on the work surface. Large sizes can be formed using a corner for an "extra set of hands." Plugs must remain tight during handling.

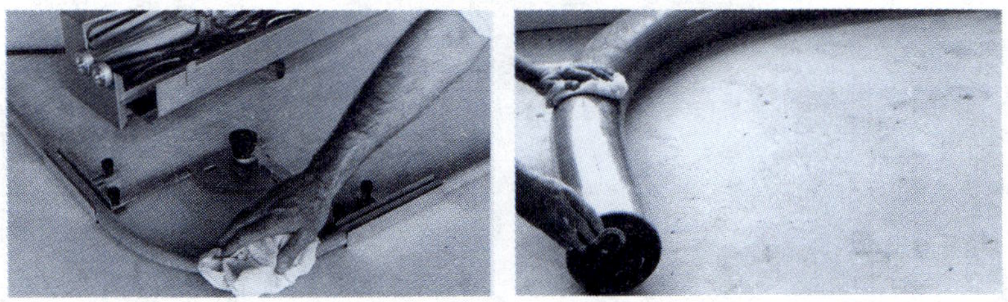

3. COOLING. The bend will "set" when cooled. Use a wet rag or sponge to speed up the cooling process. When "set", remove plugs. The bend is now ready to install.

NEC 347-2 USES PERMITTED:
(a) Concealed in ceilings, walls, and floors; (b) in corrosive areas as per 300-6 and where exposed to chemicals; (c) in cinder fill; (d) in wet locations where walls are washed, such as dairies or canneries; (e) in dry or damp areas where not prohibited by 347-3; (f) for exposed work if identified for such use and not exposed to physical damage; (g) underground if as per 300-5 and 710-4(b).

NEC 347-3 USES NOT PERMITTED:
(a) In hazardous locations except as permitted in 503-3(a), 504-20, 514-8, 515-5, and 501-4(b), Ex.; (b) To support fixtures and equipment. See Exceptions for nonmetallic conduit bodies; (c) Where exposed to physical damage unless identified for such use; (d) Where ambient temperatures exceed those of conduit; (e) Where conductor insulation temperature limitations exceed conduit insulation; (f) In theaters, except as permitted in 520.

sulation.

RIGID NONMETALLIC CONDUIT (NEC 347)

NEC 347-6 Joints shall be made by approved methods.

NEC 347-13 Field bends made only with identified bending equipment and radius of inner bend not less than in Table 346-10.

Table 300-5 Minimum depth of rigid nonmetallic conduit that is approved for direct burial without concrete encasement shall be at least 18 inches, except if 2-inch concrete pad over conduit, then need only be 12 inches deep. (Table 300-5)

NEC 347-5 Conduit cuts shall be trimmed inside and outside to remove rough edges.

CARLON P&C DUCT FITTINGS AND ACCESSORIES

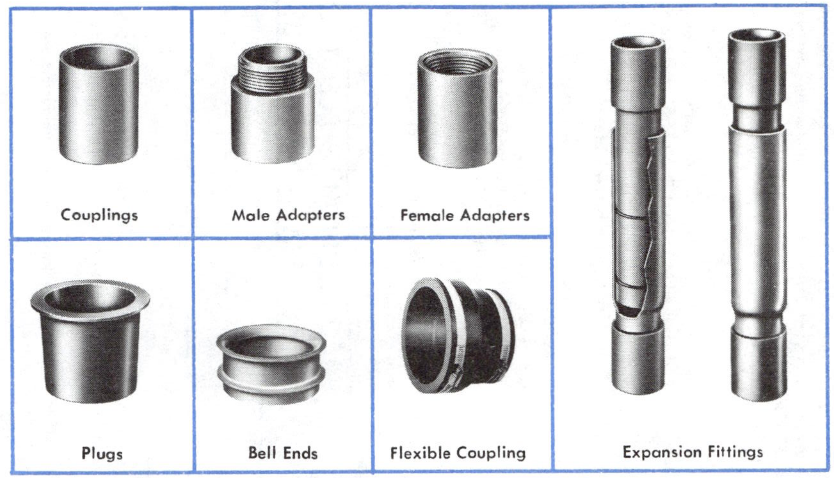

NEC 347-14 Not over four quarter bends between pulling points.

347-8 PVC shall be supported per Table 347-8 and within 3′ of each box or conduit termination. Installations of PVC shall allow for thermal expansion and contraction.

NEC 347-1 DESCRIPTION Shall be resistant to moisture and chemical atmospheres. For use above ground shall be flame retardant, resistant to impact and crushing, resistant to heat encountered, resistant to low temperature and sunlight. For use underground shall be resistant to moisture and corrosive agents, shall have strength to withstand abuse during and after installation.

NEC 347-17 CONSTRUCTION SPECIFICATIONS
(a) Each length clearly and durably marked at least every 10 ft as per 110-21 including type of material unless visually identifiable. Markings listed aboveground use to be permanent and for underground use only legible until installed.

NEC 347-9 Expansion joints shall be provided to compensate for thermal expansion and contraction.

Type EPC-40 (electrical plastic conduit) a semi-rigid conduit for virtually every application requiring RIGID conduit. Heavy Wall PV-Duit for underground encased and many above-ground installations. See NEC Article 347. UL listed for above ground and underground use.

Type A-EPT (electrical plastic tubing) Thin Wall PV-Duit. UL listed for underground concrete encasement only.

NEC 300-5 (d) PVC Schedule 80 or equivalent may be used for physical protection of conductors from ground up poles. Conduit shall extend 8 feet up pole.

Type EPC-80 (electrical plastic conduit) Extra Heavy Wall for special applications requiring additional mechanical protection such as pole risers and bridge crossings.

347-10(a)(b) Minimum allowable trade size for RNMC is ½". Maximum size permitted is 6".

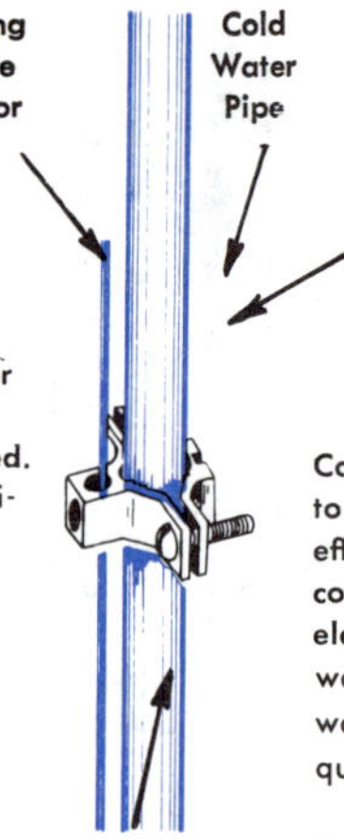

GROUNDING ELECTRODE SYSTEM (NEC 250-81) If available at building or structure, the following grounding electrodes SHALL be bonded together and used as the "Grounding Electrode System": (Sized as per 250-94, connected as per 250-115).

a. Metal water pipe with at least 10 feet in contact with earth (including well casing if bonded to pipe). Water pipe to be supplemented by additional electrode as per 250-81 or 250-83.

b. Effectively grounded metal frame of building.

c. Electrode consisting of re-bar or rod at least ½ inch diameter and 20 feet long or at least 20 feet of bare copper wire not smaller than No. 4 and encased in at least 2 inches of concrete at bottom of footing.

d. A ground ring encircling building, consisting of at least 20′ of bare copper wire not smaller than No. 2 and in direct contact with earth at least 2½ feet below surface.

NEC 250-115 Ground clamps suitable for conductor and electrode materials, and suitable for direct soil burial where buried. One conductor only unless listed for multiple conductors. See 250-115 (a, b, c, d) for connection methods.

250-81 Interior metal water piping more than 5′ from point of entrance to building shall not be used as GEC or as a conductor to interconnect electrodes that form GES.

Connection of grounding electrode conductor to grounding electrode shall be permanent, effective and accessible, except the connections to concrete encased or buried electrodes need not be accessible. Where water pipe is used, bonding around water meter, valves, unions, etc., is required. 250-112

Aluminum or copper-clad aluminum grounding electrode conductors not permitted in direct contact with masonry, earth or where subject to corrosive conditions. (250-92 (a))

Metallic enclosures for grounding conductors shall be continuous from cabinet to grounding electrode. Enclosures which are not continuous from cabinet to electrode can be bonded at each end to grounding conductor. (250-92 (a))

Aluminum or copper-clad aluminum conductors shall not be used within 18 inches of earth when used outside. (250-92 (a))

GROUNDING ELECTRODE SYSTEM -- NEC 250-81

Black

White

Red

3-Wire Circuit

Ungrounded conductors shall be clearly distinguishable from grounded and grounding conductors by colors or combination of color plus distinguishing marking, and in either case the color shall be other than white, natural gray, or green. (310-12 (c))

200-2 Grounded conductor, on premises, shall be identified as per 200-6. Grounded conductor, when insulated, to have suitable insulation as ungrounded conductor.

Conductors shall be insulated except where covered, or bare conductors are specifically permitted in Code. (310-2)

For covered service-drop conductors, see 230-22. For bare service entrance grounded conductors, see 230-41, Ex. a. For bare or covered overhead service grounded conductors, see 230-22. For bare or covered underground service grounded conductors, see 230-30. Conductors having a white or natural gray covering shall be used only as identified conductors. 200-7. See Exceptions.

200-7, Ex. 1 Where necessary to use these conductors otherwise than above, they may be reidentified to indicate use.

200-7, Ex. 2 Conductors that are part of a cable need not be reidentified if they are used as a supply, but not as a return, to single-pole, 3-way, or 4-way switch loops.

Branch circuits supplying lampholders or 125 V devices or receptacles shall not exceed 125 V between conductors or to ground. 210-6(a)

NEC 210-19(a) (Fine Print Note 4) Voltage drop should not exceed 3 percent to farthest outlet on branch circuit. Maximum voltage drop for feeders plus branch circuits should not exceed 5 percent. (Also see 215-2(b), FPN 2.)

Continuous loads shall not exceed 80 percent of branch-circuit rating, except: (1) Motor loads that have been computed as per Article 430; (2) Circuits supplied by assembly listed for continuous operation at 100 percent of rating. (384-16(c), Ex.)

CONDUCTORS -- GENERAL RULES

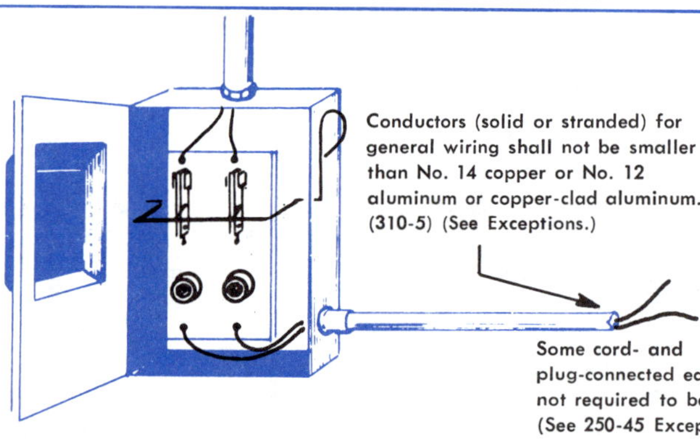

Conductors (solid or stranded) for general wiring shall not be smaller than No. 14 copper or No. 12 aluminum or copper-clad aluminum. (310-5) (See Exceptions.)

Some cord- and plug-connected equipment not required to be grounded. (See 250-45 Exceptions.)

Flexible cord approved for specific appliances considered protected by branch-circuit overcurrent device conforming to Article 210 as per following:

- 20-amp circuit – Tinsel cord No. 18 or larger
- 30-amp circuit – No. 16 or larger cord
- 40-amp circuit – 20 amp or more rated cord
- 50-amp circuit – 20 amp or more rated cord. (240-4)

See 240-4, Ex. 1.

Plug fuses and fuseholders shall not be used in circuits exceeding 125 volts between conductors except in systems having a grounded neutral and of which no conductor operates at more than 150 volts to ground. (240-50 (a))

Plug fuses of the Edison-base type shall be used only as a replacement item in existing installations where no evidence of overfusing or tampering. (240-51 (b))

NEC 310-4 Conductors 1/0 and larger may be run in parallel if of same length, same conductor material, same size, same insulation type, and terminated in the same manner. If in separate raceways or cables, these must have the same physical characteristics. (See Exceptions.)

NEC RULES GOVERNING CONDUCTORS AND CIRCUITS

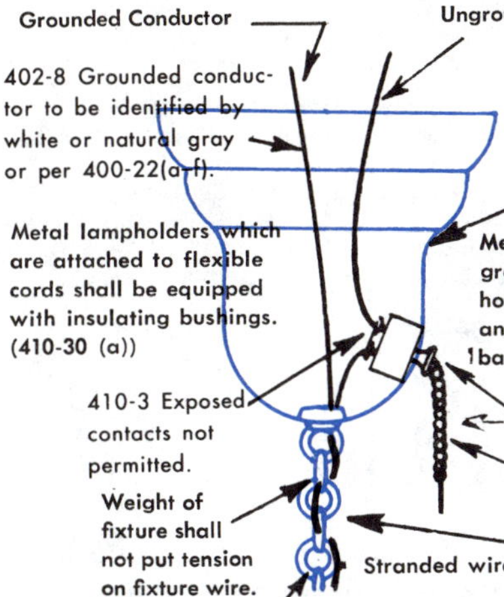

402-8 Grounded conductor to be identified by white or natural gray or per 400-22(a-f).

Metal lampholders which are attached to flexible cords shall be equipped with insulating bushings. (410-30 (a))

Metal fixture canopy or socket required to be grounded if within 8 feet vertically or 5 feet horizontally of ground or grounded objects and subject to contact by persons (such as in bathrooms, basements, etc.). (250-42) (a)

410-3 Exposed contacts not permitted.

Weight of fixture shall not put tension on fixture wire.

Stranded wire required on fixture chains. (410-28 (e))

Fixtures shall have no live parts normally exposed to contact—and lampholders, receptacles, and switches with exposed terminals shall not be installed in metal fixture canopies. (410-3)

410-4(d) Hanging fixtures and pendants not permitted within 8′ vertically or 3′ horizontally of a bathtub.

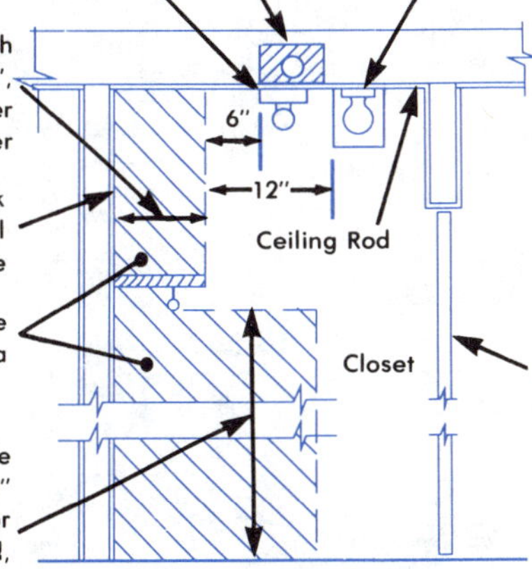

NEC 410-8 Storage is the space above the shelf and below the rod. Storage space above the rod is shelf width or 12 inches, whichever is greater. Storage space below the rod is 24 inches to the rod or 6 feet above the floor, whichever is greater.

FIXTURES IN CLOTHES CLOSETS

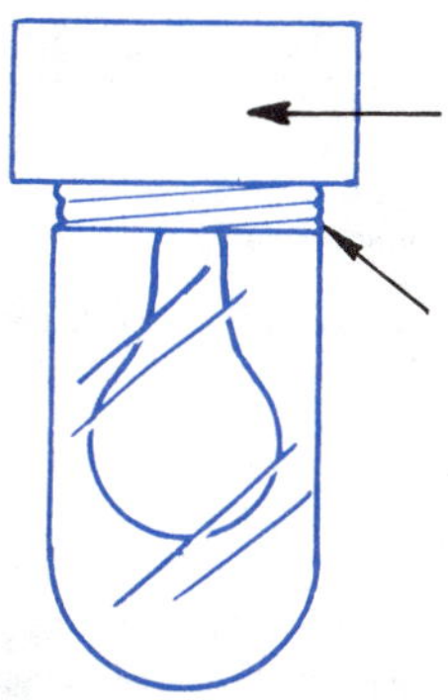

Fixtures installed in damp or wet locations shall be so constructed or installed so water cannot enter or accumulate. Fixtures in wet locations to be marked "Suitable for Wet Locations." Fixtures in damp locations shall be marked "Suitable for Wet Locations" or "Suitable for Damp Locations." (410-4 (a))

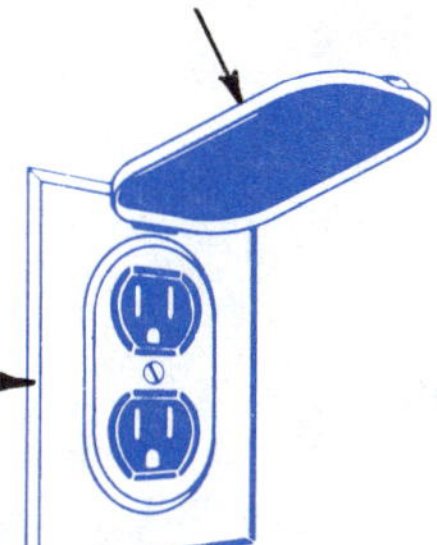

410-57(a)(b) Receptacles in damp locations shall be installed in an enclosure that is weatherproof only when the plug cap is not inserted. Receptacles in wet locations shall maintain weatherproof integrity when plug cap is inserted.

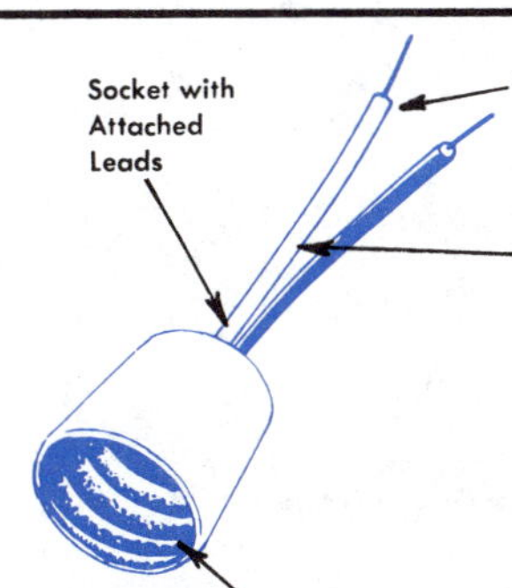

NEC 200-10 (d) For devices with attached leads, the conductor attached to the screw-shell shall have white or natural gray finish.

NEC 410-47 Grounded conductor to be connected to shell of screw-shell type lampholder. (See 410-23)

NEC 200-10 (c) In devices with screw-shells—except fuseholders—shall have the grounded conductor attached to the screw-shell.

NEC 240-50 (e) The screw-shell of plug type fuseholders shall be connected to the load side of the circuit.

Fixtures over 50 pounds to be supported independently of box. (410-16 (a))

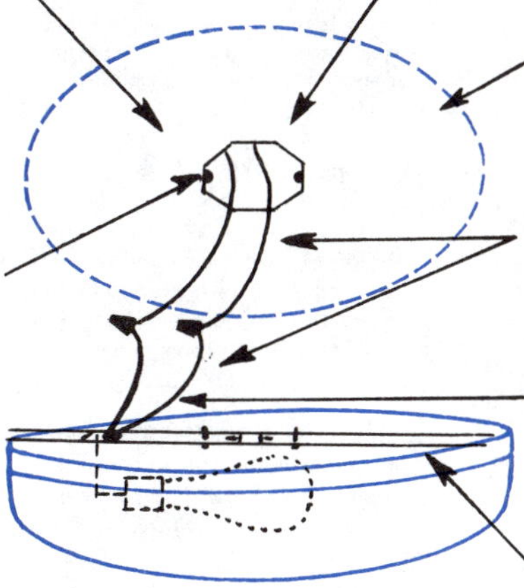

Identified conductor shall connect to wire from screw-shell of socket. (200-10 (c))

Fixture conductors shall not be smaller than No. 18. (410-24 (b))

Combustible wall or ceiling finish exposed between the edge of a fixture pan or canopy and an outlet box shall be covered with noncombustible material. (410-13)

Weatherproof lampholders required in wet or damp locations. (410-49)

Socket support for messenger wire in festoon lighting. (225-13)

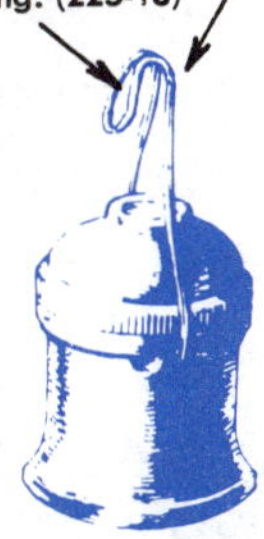

WEATHERPROOF PIN-TYPE SOCKET

WEATHERPROOF PIGTAIL SOCKET

Metal lampholders which are attached to flexible cords shall be equipped with insulating bushings. [410-30 (a)]

LIGHTING FIXTURES AND LAMPHOLDERS

This fixture UL listed and box marked to indicate No. of wires.

Branch-circuit wires suitable for temperature encountered shall be used. (410-67 (b))

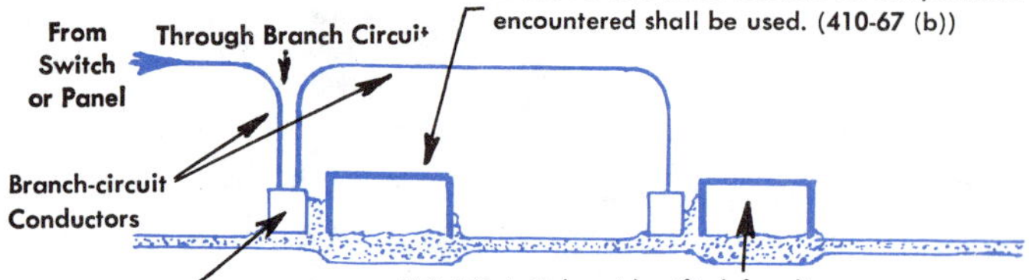

Wires not permitted through outlet box unless identified for such use. (410-11)

410-66(a) Unless identified for direct contact, recessed parts of fixture enclosures shall maintain at least ½" spacing from combustible materials.

RECESSED INCANDESCENT FIXTURES

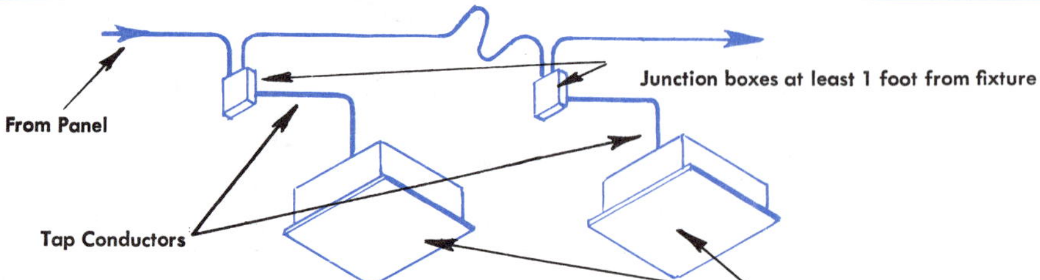

Not less than 4 feet or more than 6 feet of suitable raceway with conductors having an insulation suitable for temperature of fixture. (410-67(c))

Metal face plates shall be grounded (410-56(d))

Shall be installed to completely cover wall opening and seat against wall surface.

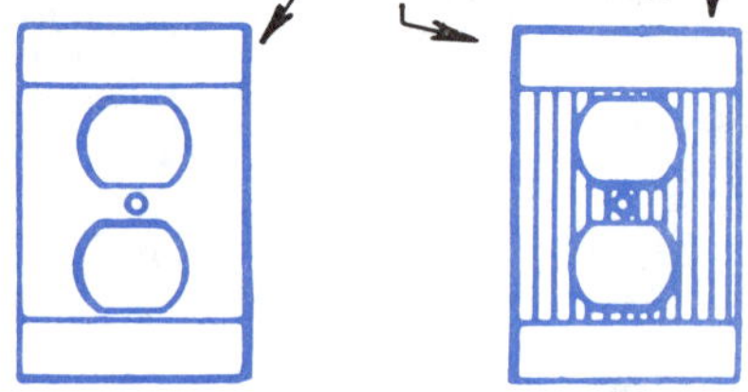

Face plates of nonferrous metal shall be not less than .04 inch thick, ferrous .03 inch thick, and composition .10 inch thick, unless formed or reinforced to provide adequate mechanical strength (410-56(d))

Face plates to completely cover opening and seat against mounting surface (410-56(e))

FACE PLATES
NEC 410-56(d & e)

Fixtures in suspended ceilings subject to requirements of Article 410-65 through 410-72. (410-64)

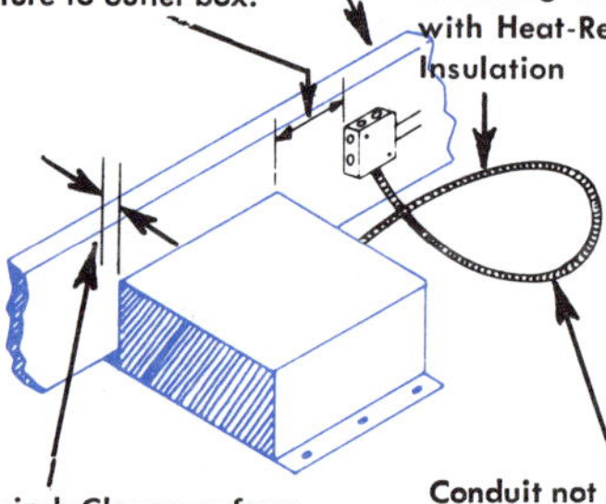

Typical Recessed Fixture

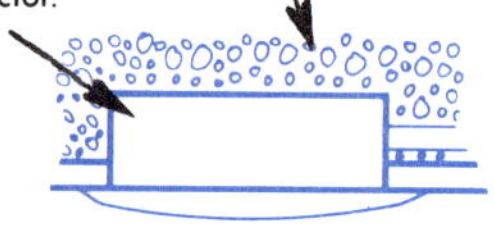

Type IC (Insulated Ceiling) Fixture

GENERAL RULE: Conductors to flush and recessed fixtures shall have insulation suitable for temperature encountered. 410-67(a).

RECESSED INCANDESCENT FIXTURES TO HAVE THERMAL PROTECTION AND BE SO IDENTIFIED - Ex. 1 if installed in poured concrete and identified for such use. (410-65(c))

Recessed portions of enclosures, other than at points of support, shall be spaced at least ½ inch from combustible material. [410-66 (a)]

Thermal insulation not permitted within 3 inches of fixture and not above fixture so as to entrap heat and prevent free air circulation except when identified for the purpose. [410-66 (b)]

Branch-circuit wires permitted to fixture if insulation suitable for fixture temperature. (410-67 (b))

Tap wires with insulation suitable for fixture temperature permitted from fixture terminal to outlet box if in suitable raceway. Outlet box at least 1 foot from fixture and raceway shall be at least 4 feet long but not longer than 6 feet. (410-67(c))

RECESSED FIXTURES

Circuit breakers used as disconnect means shall be indicating type (422-26)

Circuit breaker may serve as disconnect means for a permanently connected motor-driven appliance. (422-27)

Branch-circuit switch or circuit breaker, where readily accessible, may serve as disconnecting means for permanently connected appliances not over 300 VA. See 422-21(b).

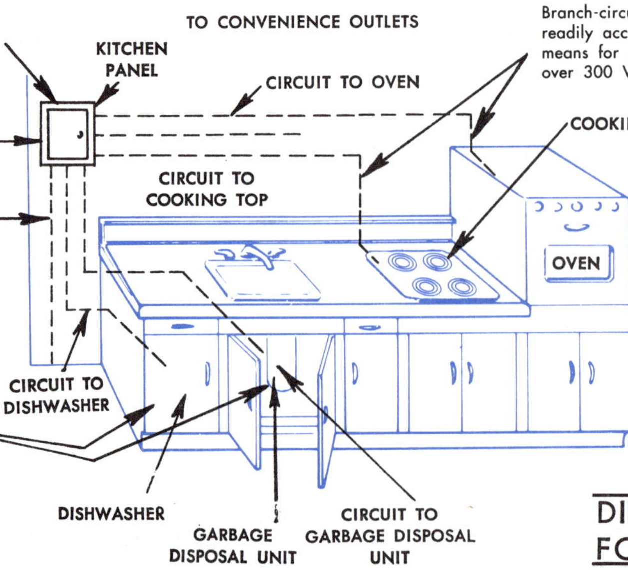

Wall-mounted ovens and counter-mounted cooking units permitted to be permanently wired or for ease in servicing or installation a cord- and plug-connection permitted 422-17

422-17(b) A separable connector or a plug and receptacle combination permitted to wall-mounted oven or cooking unit but (1) not considered as disconnect means as per 422-20, and (2) be approved for temperature it is exposed to.

If branch-circuit switch or circuit breaker is not in sight of motor driven appliance of over 1/8 hp, then a disconnect means must be installed at unit as per Article 430. (422-27)

Switch or circuit breaker used as disconnect may be out of sight if appliance has unit switch with OFF position, and unit switch disconnects all ungrounded wires. (422-27, Ex.)

DISCONNECTING MEANS FOR KITCHEN APPLIANCES

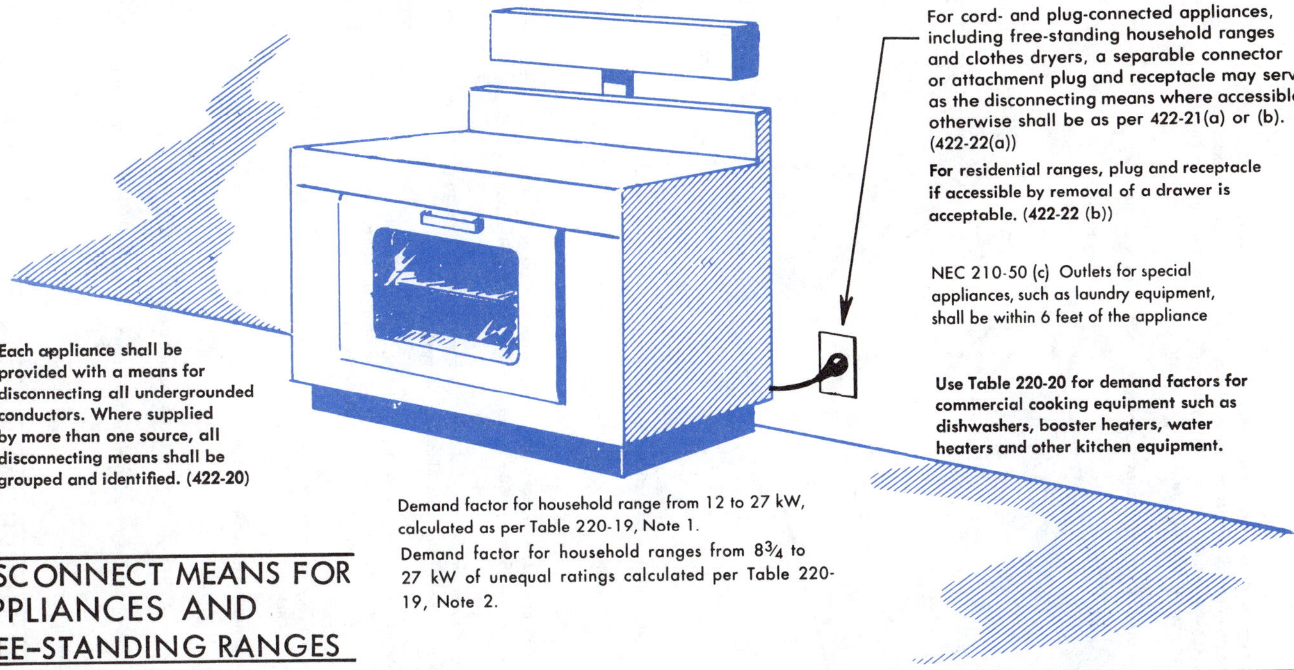

For cord- and plug-connected appliances, including free-standing household ranges and clothes dryers, a separable connector or attachment plug and receptacle may serve as the disconnecting means where accessible, otherwise shall be as per 422-21(a) or (b). (422-22(a))

For residential ranges, plug and receptacle if accessible by removal of a drawer is acceptable. (422-22 (b))

NEC 210-50 (c) Outlets for special appliances, such as laundry equipment, shall be within 6 feet of the appliance

Use Table 220-20 for demand factors for commercial cooking equipment such as dishwashers, booster heaters, water heaters and other kitchen equipment.

Each appliance shall be provided with a means for disconnecting all undergrounded conductors. Where supplied by more than one source, all disconnecting means shall be grouped and identified. (422-20)

Demand factor for household range from 12 to 27 kW, calculated as per Table 220-19, Note 1.

Demand factor for household ranges from 8¾ to 27 kW of unequal ratings calculated per Table 220-19, Note 2.

DISCONNECT MEANS FOR APPLIANCES AND FREE-STANDING RANGES

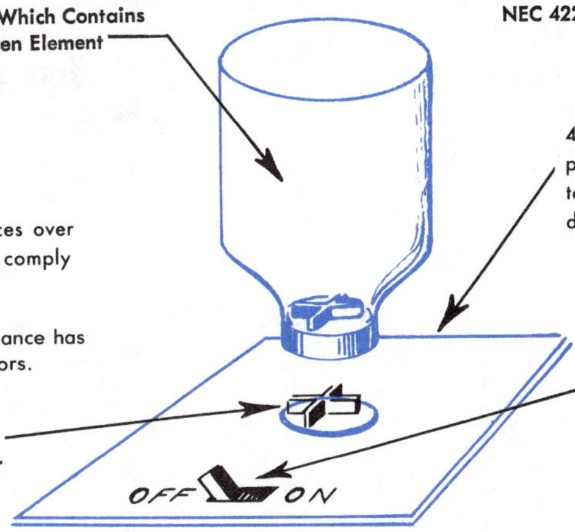

NEC 422-20 Each appliance shall have a disconnect.

For permanently connected appliances over 300 VA or ⅛ hp, the branch-circuit switch or circuit breaker may serve as the disconnecting means, if readily accessible to user of the appliance if it can be locked open per 422-21(b).

422-27 Disconnect means for motor-driven appliances over ⅛ hp shall be within sight of motor controller and comply with 430, Part I.

Disconnect may be out of sight if unit switch on appliance has OFF position and disconnects all ungrounded conductors. (422-27, Ex.)

422-22 For appliances that are cord-and-plug connected, an accessible plug and receptacle shall be permitted to be utilized as disconnecting means.

General Rule: Switches which are a part of the appliance shall not be considered as taking the place of the single disconnecting means when such is required by this section. (422-25)

Exception to the General Rule: In a single-family dwelling the service switch is acceptable as the single disconnecting means if the unit switch on the appliance has an OFF position and disconnects all ungrounded conductors. (422-25(c)) (Also see 422-25(a, b, d) for other occupancies.)

DISCONNECTING MEANS FOR APPLIANCES

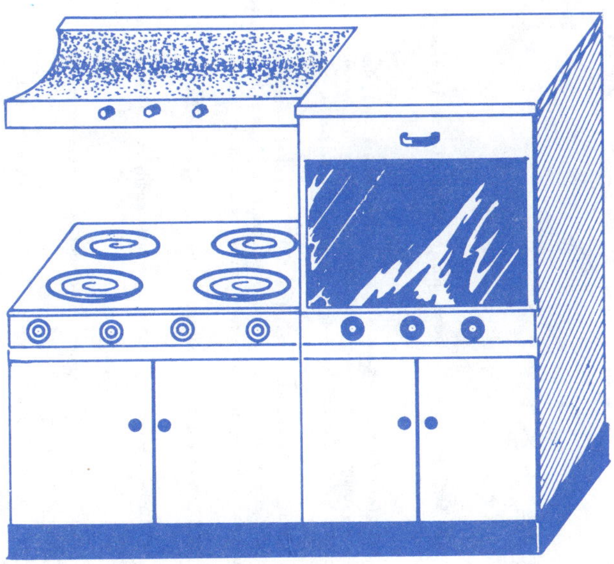

No fixtures permitted in nonresidential hoods unless identified for the purpose. [410-4 (c)]

Where tap conductors supply wall-mounted oven or countertop from 50-amp circuit, they shall have not less than 20-amp rating and shall not be longer than necessary for servicing. (210-19(b), Ex. 1)

Table 220-19 Note 3 in lieu of values given in Column 1, the load for ranges over 1¾ kW but not over 8¾ kW may be taken as the sum of nameplate ratings multiplied by demand factors of Columns B and C.

Table 220-19 Note 4 The branch-circuit load for one wall-mounted oven or countertop unit shall be taken as the nameplate rating. Where countertop and not more than two wall-mounted ovens are in the same room, the nameplate ratings shall be added and the load treated as that of a single range.

Separable connector or plug and receptacle used for servicing, not acceptable as disconnecting means. (422-17 (b-1))

WALL-MOUNTED OVEN AND COUNTERTOP RANGES

MANUFACTURED BUILDINGS

NEC 545-3 Definition of a Manufactured Building: A Manufactured Building is of closed construction, normally factory-made or assembled for installation, or assembly and installation on building site. (Does not include mobile home or recreational vehicles.)

Manufactured buildings not restricted to dwellings; could be Industrial or Commercial.

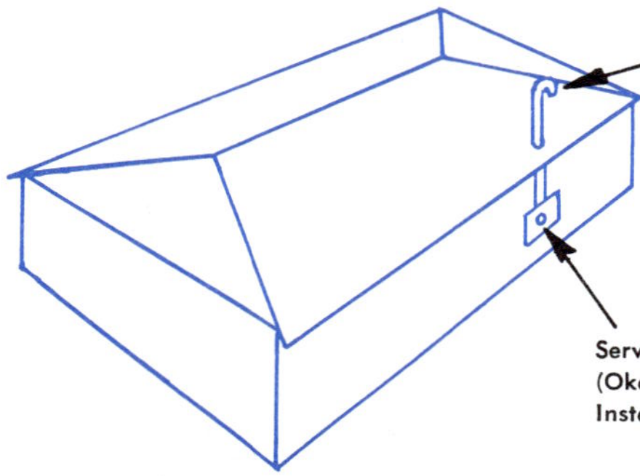

NEC 545-6 Service-entrance conductors to be installed at building site, except where point of attachment is known before manufactured.

NEC 545-4 (a) All wiring methods in code permitted for use, and other wiring systems and fittings specifically approved for use with manufactured buildings also permitted. Wiring devices with integral enclosures permitted if sufficient length of conductor provided for replacement of devices.

NEC 545-4 (b) Concealed cables of No. 10 and smaller permitted to be installed without supports, such as staples or straps, as required in normal construction, if cables are secured at cabinets, boxes, and fittings and protected from physical damage as per 300-4.

NEC 545-5 Service-entrance conductors to be installed as per 230.

NEC 545-7 Service equipment located at nearest, readily accessible point, either inside or outside of building.

NEC 545-8 Protection of equipment and conductors that are exposed during manufatcuring, transit, and erection at building site is required.

NEC 545-9(a) Boxes other than those in Table 370-6(a) permitted if listed, identified, and tested. (b) Boxes under 100 cu. in. for mounting in closed construction shall have anchors or clamps for rigid and secure installation.

NEC 545-10 Switch or receptacle with integral enclosure permitted when tested, identified, and listed.

NEC 545-11 Bonding and grounding of prewired panels and building components to conform to 250, Parts E, F, and G.

NEC 545-12 Means to be provided to route grounding electrode conductor from service equipment to grounding electrode and to conform to 250, Part J.

NEC 545-13 Where on-site interconnection of modules or other building components is necessary, fittings and connectors when tested, identified, and listed are permitted to be used. These fittings to be equal to wiring method and may be concealed at time of on-site assembly.

AGRICULTURAL BUILDINGS

547-1(a)(b) Agricultural buildings that contain excessive dust, dust with water, or corrosive atmospheres are covered by 547.

NEC 547-1 (b) This Article also applies to agricultural buildings where corrosive fumes exist, such as totally enclosed and environmentally controlled areas where: (1) animal and poultry excrement causes corrosive vapors, (2) corrosive particles mix with water, (3) periodic washing and sanitizing causes area to be damp or wet, (4) other similar conditions exist.

NEC 547-2 For buildings other than 547-1 above applicable code rules shall be used.

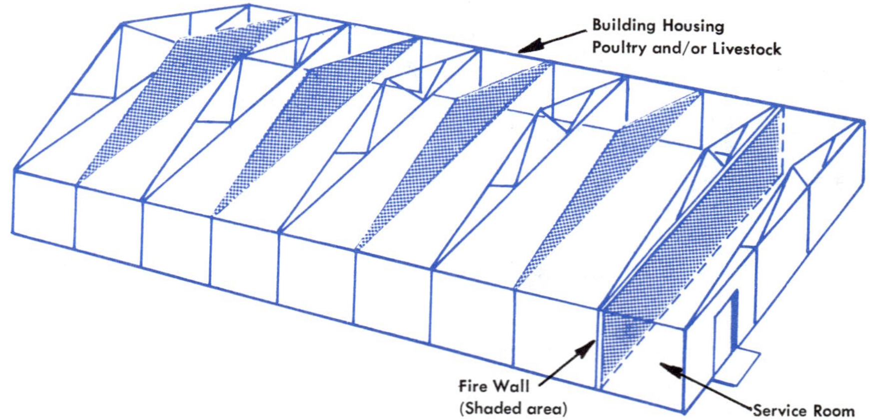

NEC 547-3 Electrical equipment and devices to be installed to function at full rating without overheating.

NEC 547-4 Wiring methods to be UF, NMC, SE (Cu) or other cables or raceways suitable for location with approved fittings.

NEC 547-4(a) All boxes containing devices, taps, joints, or terminal connections shall be dust- and watertight and made of corrosion-resistant material (FPN See 300-7 and 347-9 for seals and expansion fittings).

NEC 547-4(b) Flexible connections to be made with dust-tight flexible connectors, liquid-tight flexible metal conduit, or hard usage flexible cord, with listed and identified fittings.

NEC 547-5 Devices, such as switches, fuses, circuit breakers, controller, pushbuttons, relays, etc., shall be installed in dustproof, weatherproof, watertight, or corrosion-resistant enclosures designed to keep out dust, water, and corrosive elements, and shall have telescoping or close-fitting covers.

NEC 547-6 Rotating electrical machinery and motors to be designed to limit entrance of dust, moisture, or corrosive particles, or to be totally enclosed.

NEC 547-7(a) Lighting fixtures installed to minimize entrance of dust, foreign matter, moisture, and corrosive material.

NEC 547-7(b) Light fixture exposed to physical damage to have suitable guard.

NEC 547-7(c) Light fixture to be water-tight if exposed to water from condensation and/or cleansing water or solution.

NEC 547-8 Grounding and bonding per 250.

Trade Competency Test 1

Name ______________________ Date ______________

Section	Answer	
______	T F	**1.** The term "service-entrance conductors" includes the term "service drops."
______	______	**2.** The general rule with respect to the service disconnecting means states that a grounding connection shall not be made to any grounded circuit on the ______. A. supply side C. either A or B B. load side D. both A and B
______	______	**3.** The smallest stranded copper conductor permitted for general wiring is No. ______.
______	T F	**4.** The smallest permissible size of hard-drawn copper service drop is No. 10.
______	T F	**5.** Service drops shall clear the ground by not less than 10′.
______	______	**6.** The GEC for an AC ungrounded system need not be larger than No. ______ copper when it is the sole connection to a made electrode.
______	T F	**7.** The disconnecting means for a service shall disconnect all ungrounded conductors simultaneously.
______	______	**8.** Nonmetallic boxes may have nonmetallic or ______ covers.
______	______	**9.** An equipment bonding jumper is permitted outside a raceway when not over ______′ long. A. 6 C. 10 B. 8 D. none of the above
______	______	**10.** An outdoor fence at least ______′ in height can be used to enclose electrical equipment over 600 V.
______	______	**11.** The free length of conductor at a switch outlet shall be not less than ______″.
______	______	**12.** The service disconnecting means need not be greater than 30 A where there are not more than ______ 2-wire branch circuits.
______	______	**13.** The clearance of service drops above a flat roof shall be not less than ______′.
______	T F	**14.** The service disconnecting means may consist of not more than eight switches or circuit breakers.
______	______	**15.** The clearance of service drops from the side or bottom of windows that are designed to open, shall not be less than ______′.
______	______	**16.** The minimum clearance of service drops above a public driveway on other than residential property is ______′. A. 10 C. 16 B. 12 D. 18
______	______	**17.** Not more than six switches at one service location shall be used as the service disconnecting means if they are each ______ marked.

_________ _________ **18.** In residential occupancies, the Code requires one 15 A lighting circuit for every _________ sq ft.

_________ _________ **19.** The service disconnect rating for a one-family dwelling shall be not less than 100 A when the net computed load is _________ kVA.

A. 7.5 C. 12.5
B. 10 D. 15

_________ _________ **20.** One service lateral can consist of six underground sets of multiple conductors sized at No. _________ or larger.

_________ _________ **21.** In determining VA/sq ft area, the _________ dimensions of the building shall be used.

_________ _________ **22.** A bare copper neutral service-entrance conductor is permitted only if the voltage is not over _________ V to ground.

A. 150 C. 600
B. 300 D. all of the above

_________ _________ **23.** Under the optional calculation for a one-family dwelling, the air conditioning load is assessed at _________%.

_________ _________ **24.** Conduit shall not be connected to the _________ of round boxes.

_________ _________ **25.** Unless otherwise marked, a CB with a 60°C terminal shall be loaded per the _________°C Column of Table 310-16.

_________ _________ **26.** A metal fixture shall be grounded if located within _________′ vertically or 5′ horizontally of a kitchen sink.

_________ _________ **27.** Unused openings in boxes and fittings shall be effectively _________.

_________ _________ **28.** Disconnecting means for motor-driven appliances over _________ hp shall be located within sight of the motor controller.

_________ _________ **29.** Metal enclosures for grounding electrode conductors shall be _________.

A. rigid conduit C. electrically continuous
B. isolated D. none of the above

_________ _________ **30.** A new receptacle outlet installed on a 15 A or 20 A branch circuit shall be of the _________ type.

_________ _________ **31.** The front edge of an outlet box in a noncombustible ceiling may be _________″ back from the surface.

_________ _________ **32.** For the purpose of determining the maximum number of conductors in boxes, two conductors shall be counted for each yoke or _________.

_________ _________ **33.** A receptacle outlet, in a dwelling, within 6′ of a wet bar sink shall be _________ protected.

_________ _________ **34.** Receptacles in living areas are spaced so that no point along the available floor line perimeter shall be more than _________′ from a receptacle outlet.

_________ _________ **35.** In general, the minimum ampacity of the ungrounded service-entrance conductors of a service is _________ A.

A. 50 C. 80
B. 60 D. 100

_________ _________ **36.** Where necessary, a shallow outlet box not less than _________″ deep may be used.

_________ _________ **37.** An additional service drop can be added to building where the total service capacity exceeds _________ A at 600 V or less.

Trade Competency Test 2

Name ______________________ Date ______________

Section	Answer	
______	______	**1.** The unit lighting load for a dwelling unit, expressed in VA/sq ft, shall be ______VA. A. 1½ C. 3 B. 2 D. 5
______	______	**2.** The cubic inch rating of a No. 18 fixture wire is ______ cu in.
______	T F	**3.** An insulated copper grounding electrode conductor used with a single service disconnecting means may be spliced.
______	______	**4.** Rooftop receptacle outlets used to service HACR equipment in other than dwelling units, shall be ______ protected.
______	T F	**5.** The sheet-metal-straps type of ground clamp constructed solely of strong sheet metal is not permitted for connecting the grounding electrode conductor to the grounding electrode.
______	______	**6.** A receptacle outlet is required for a hallway that has a length of ______′ or more. A. 4 C. 8 B. 6 D. 10
______	______	**7.** Water heaters rated over ______ VA shall have a disconnecting means located within 50′ and within sight.
______	T F	**8.** Additional plug receptacles in bedrooms, above the number required by the Code, shall be assessed at ½ A each.
______	______	**9.** The maximum number of conductors permitted, at any cross section, in a metal auxiliary gutter is ______.
______	______	**10.** The minimum clearance between the lug and wall of an enclosure for a No. 4/0 conductor being terminated as an S or Z bend is ______″.
______	______	**11.** The service or feeder capacity necessary to provide for six 5 kVA household electric clothes dryers is ______kVA. A. 19.5 C. 24 B. 21 D. 30
______	______	**12.** The current carried continuously in bare aluminum bars installed in metal auxiliary gutters shall not exceed ______ A per sq in. of the conductor.
______	______	**13.** When using the optional calculation for a dwelling service, all other loads above the initial 10 kVA are to be assessed at ______%. A. 40 C. 60 B. 50 D. 75
______	______	**14.** The branch-circuit switch or circuit breaker may serve as the disconnecting means for a permanently connected appliance of over 300 VA or ⅛ hp, where ______ and readily accessible to the user.

__________ T F **15.** A No. 6 grounding electrode conductor that is secured to the building surface and is exposed to moderate physical damage requires metal covering or protection.

__________ __________ **16.** The rating of a cord-and-plug appliance used on a 20 A branch circuit having two or more outlets shall not exceed _________ A.

A. 12
B. 15
C. 16
D. 18

__________ __________ **17.** In general, the voltage to ground on a branch circuit supplying 15 A receptacles shall be not greater than _________ V.

__________ T F **18.** In a bedroom, if a point on any available wall is 7′ from the nearest plug receptacle, an additional receptacle is needed.

__________ __________ **19.** For outside branch circuits and feeders, the 8′ vertical clearance above the roof level shall be maintained for _________′ in all directions from the edge of the roof.

__________ __________ **20.** A service drop conductor shall not be permitted to exceed _________′ of length with 18″ of clearance above the overhang portion of the roof.

__________ __________ **21.** The maximum rating of the overcurrent protection device on a 17 A noncontinuous, single nonmotor appliance circuit is _________ A.

__________ T F **22.** Track lighting installed in dwelling units shall be calculated at 180 VA for each 2′.

__________ __________ **23.** The minimum clearance between the lug and wall of an enclosure for a No. 4/0 conductor being terminated as a L bend is _________″.

__________ __________ **24.** For the kitchen small appliance load in dwellings, the Code requires not less than _________ A circuit(s).

A. one 15
B. one 20
C. two 15
D. two 20

__________ __________ **25.** A single gang FS box can house six No._________ conductors.

__________ __________ **26.** Nonmetallic sheathed cable shall be installed so no bend will have a radius less than _________ times the diameter of the cable.

__________ __________ **27.** The number of No. 12 conductors allowed in a 4″ × 1¼″ octagon box is _________.

__________ T F **28.** A plug receptacle in the hallway outside of a kitchen shall not be connected to the small appliance circuit.

__________ __________ **29.** The cubic inch rating of a No. 10 conductor is _________ cu in.

__________ __________ **30.** A total of _________ No. 14 conductors is permitted in a 3″ × 2″ × 2″ device box.

__________ T F **31.** A receptacle in the middle of the dining room floor is not counted toward fulfilling minimum Code requirements.

__________ __________ **32.** In dwelling occupancies, the voltage between conductors supplying screw shell lampholders shall not exceed _________ V.

__________ __________ **33.** The current carried continuously in bare copper bars installed in metal auxiliary gutters shall not exceed _________ A per sq in. of the conductor.

__________ __________ **34.** Metal auxiliary gutters shall be supported at intervals not exceeding _________′.

Trade Competency Test 3

Name ______________________ Date ______________

Section	Answer	
______	______	**1.** Outlet boxes for recessed lighting fixtures shall not be located closer than ______′ from the fixture and raceway.
______	T F	**2.** A general purpose branch circuit supplies outlets for both lighting and appliances.
______	______	**3.** If a $1\frac{1}{2}'' \times 4''$ octagonal box contains a fixture stud and a cable clamp, the maximum number of No. 14 conductors shall be reduced to ______. A. 4 B. 5 C. 7 D. 8
______	______	**4.** A ground ring conductor shall be installed at least ______′ below the earth's surface.
______	T F	**5.** Junction boxes shall be readily accessible.
______	______	**6.** An approved, listed stainless steel rod shall be not smaller than ______″ in diameter.
______	______	**7.** Motors rated less than ______ hp can have their disconnecting means located in the service panel.
______	______	**8.** Hanging lighting fixtures are not permitted within ______′ vertically from the rim of a bathtub.
______	T F	**9.** The screw shell of a plug-type fuseholder shall be connected to the load side of the circuit.
______	______	**10.** The demand load for a 16 kVA electric range is assessed at ______VA. A. 8200 B. 9600 C. 12,600 D. 14,400
______	______	**11.** To be considered as a grounding electrode conductor, a metal water pipe shall be in contact with the earth for at least ______′.
______	______	**12.** Nonmetallic-sheathed cable (Romex) shall be protected at a height of to least ______″ above the floor.
______	T F	**13.** Service equipment shall be identified as suitable for use.
______	______	**14.** In combustible walls or ceilings, the front edge of an outlet box or fitting may set ______″ back of the finished surface. A. ⅛ B. ¼ C. ½ D. none of the above
______	______	**15.** EMT shall be supported within ______′ of outlet boxes.
______	______	**16.** Liquidtight flexible nonmetallic conduit shall not be used for systems having conductors operating at over ______ V.
______	T F	**17.** Wooden supports for boxes in concealed work shall not be less than nominal $1'' \times 2''$.

__________ T F **18.** When the neutrals of two branch circuits that are derived from two different systems enter the same enclosure, one neutral may be white and the other natural gray.

__________ __________ **19.** Plug fuses are never permitted in circuits that exceed ________.

A. 125 V between conductors
B. 125 V to ground
C. 150 V between conductors
D. 150 V to ground

__________ __________ **20.** Receptacle outlets for special appliances such as laundry equipment shall be located within ________′ of the appliances.

__________ __________ **21.** Recessed lighting fixtures shall have at least ________″ clearance from combustible material.

__________ __________ **22.** The grounding electrode conductor can be spliced by means of the ________ welding process.

__________ __________ **23.** Securely installed boxes and fittings that are threaded into two or more conduits from two or more sides and that do not support fixtures or contain devices shall not exceed ________ cu in.

A. 100
B. 125
C. 150
D. 200

__________ __________ **24.** Surface-mounted fluorescent lighting fixtures shall be located at least ________″ from the storage space in a clothes closet.

__________ __________ **25.** Nonmetallic-sheathed cable (Romex) shall not be used to wire buildings exceeding ________ floors above grade.

__________ __________ **26.** The minimum size conductor permitted to be used as an encased electrode is No. ________ copper.

__________ T F **27.** When two switches are mounted on the same strap, the number of conductors allowed in the box shall be reduced by one.

__________ __________ **28.** The maximum number of No. 14 conductors permitted in a $1\frac{1}{2}'' \times 4''$ octagonal outlet box is ________.

A. 5
B. 7
C. 8
D. 9

__________ __________ **29.** The internal depth of outlet boxes intended to enclose flush devices shall be at least ________″.

A. $\frac{1}{2}$
B. $\frac{7}{8}$
C. $\frac{15}{16}$
D. $1\frac{1}{2}$

__________ __________ **30.** Aluminum conductors shall not be installed within ________″ of the earth when used outside as a GEC.

__________ __________ **31.** Hanging lighting fixtures are not permitted within ________′ vertically from the rim of a bathtub.

__________ __________ **32.** A metal underground gas pipe shall not be used as a grounding ________.

__________ __________ **33.** Driven rods made of ________ shall not be used as electrodes.

__________ __________ **34.** The minimum size rebar permitted to be installed in concrete at the bottom of a footing is ________″ in diameter and 20′ long.

__________ __________ **35.** The smallest service drop permitted by the Code is ________ copper or No. 6 aluminum.

__________ __________ **36.** The identified conductor shall be connected to a wire from the ________ of a lighting socket.

Trade Competency Test 4

Name ______________________________ Date ______________

Section	Answer	
________	T F	**1.** All 125 V, 1-phase, 20 A receptacles installed in bathrooms of dwelling units shall have ground-fault circuit-interrupter protection for personnel.
________	________	**2.** Metal auxiliary gutters shall be supported throughout their entire length not exceeding ________′ intervals. A. 3 C. 8 B. 5 D. 10
________	________	**3.** Conductors installed inside AC cable (BX) shall have ampacity values rated at ________°C per Table 310-16.
________	T F	**4.** Receptacle outlets located above 5½″ from the finished floor can be counted as a required outlet.
________	________	**5.** Where several cables are routed through the same hole at a length of more than ________″ without spacing, derating factors shall be applied.
________	T F	**6.** Parallel conductors in a conduit run shall have an equipment ground pulled through each conduit.
________	________	**7.** Corner-grounded systems have one phase connected to ________.
________	________	**8.** At least ________″ of free conductor shall be left at each outlet.
________	T F	**9.** Where one or more fittings are within a box, the count toward the fill is based on the largest conductor entering the box.
________	________	**10.** The high leg in a delta system is connected to the ________ on the customer's side of the service equipment.
________	________	**11.** Conductors installed inside nonmetallic-sheathed cable shall have a rating of at least ________°C.
________	________	**12.** When 100′ of No. 16 fixture wire is connected to a standard 120 V circuit, the branch-circuit overcurrent protection device shall not exceed ________ A. A. 15 C. 30 B. 20 D. 40
________	________	**13.** Fittings such as capped elbows and service entrance elbows can be classified as ________. A. conduit bodies C. junction boxes B. switch boxes D. gutter boxes
________	T F	**14.** Circuit breakers that are not marked with an IC rating are rated at 5000 A.
________	________	**15.** Conductors or cables are not required to be connected in a nonmetallic box where secured within ________″ of the box.
________	________	**16.** A 4″ × 1¼″ round box normally has ________ cu in. of fill space area.

_________ _________ **17.** Flexible metal conduit routed over 6′ in length shall be provided with a(n) _________.

_________ _________ **18.** Under ordinary circumstances, the Code requires that the distance between supports for open wiring on insulators shall not exceed _________′.

A. 3 C. 4½
B. 4 D. 7½

_________ _________ **19.** The metal raceway enclosing tap conductors for recessed fixtures shall not be less than _________′ or more than _________′ long.

A. 2; 6 C. 3; 6
B. 3; 5 D. 4; 6

_________ _________ **20.** A fuse which does not have an IC rating marked on its surface shall be considered having an IC rating of _________ A.

_________ _________ **21.** Conductors connected in parallel and general wiring systems shall be at least No. _________ or larger.

A. 1/0 C. 2
B. 1 D. 3

_________ _________ **22.** Two No. 250 kcmil conductors can be terminated to one lug if there is a clearance of at least _________″ from the lug to the opposite wall.

A. 5 C. 8
B. 6 D. 10

_________ T F **23.** Where not permanently accessible, nonmetallic-sheathed cable within 5′ of a scuttle hole does not require protection.

_________ _________ **24.** When it is impractical to install the service drop below the weatherhead, the point of attachment can be located _________″ from the weatherhead.

_________ _________ **25.** Only service drop conductors should be attached to a service _________.

_________ T F **26.** Lighting switches mounted in or on the wall can be located above 6′- 7″ from the finished floor.

_________ _________ **27.** Service drop conductors shall have a clearance of not less than _________′ from a window that opens and closes.

_________ _________ **28.** The Code requires that recessed portions of fixture enclosures shall be spaced at least _________″ from combustible material.

A. ⅜ C. ⅝
B. ½ D. ¾

_________ _________ **29.** AC cable (BX) does not have to be supported where used in _________′ lengths for flexibility.

_________ T F **30.** Underground covers over metal boxes, that weigh over 100 lb, containing feeders rated over 600 V shall be marked "DANGER—HIGH VOLTAGE—KEEP OUT."

_________ _________ **31.** All receptacle outlets installed within _________′ of a wet bar sink in a dwelling unit shall be GFCI-protected.

_________ _________ **32.** A Class I circuit operates at _________ VA and 30 V or less.

_________ T F **33.** The metal boxes listed in Table 370-16(a) are permitted to have conductors contained within rated at sizes No. 4 and larger.

_________ _________ **34.** The general rule states that cablebus shall be securely supported at intervals not exceeding _________′.

Multifamily Dwellings 2

WIRING PLANS

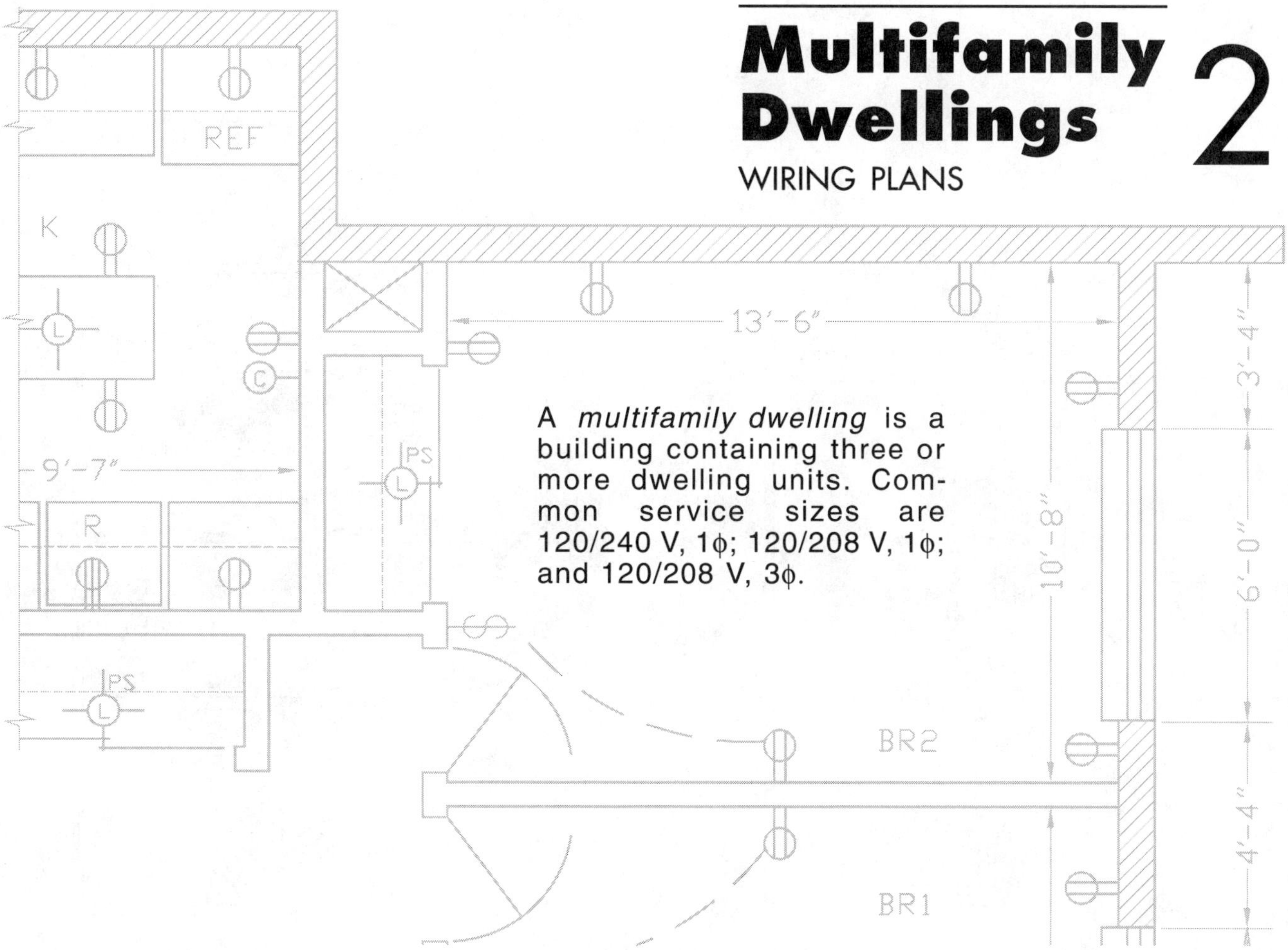

A *multifamily dwelling* is a building containing three or more dwelling units. Common service sizes are 120/240 V, 1ϕ; 120/208 V, 1ϕ; and 120/208 V, 3ϕ.

CONDUCTORS

Table 310-16 lists current-carrying values for copper and aluminum wires when not more than three conductors are grouped in raceway, cable, or buried in the ground. Table 310-17 lists values for copper and aluminum conductors in free air. Tables 310-16 through 310-19 list ampacities of conductor insulation for specialized application.

Table 310-16

The upper portion of this table deals with ampacities of conductors that are used at a nominal room temperature of 30° Celsius (C), or 86° Fahrenheit (F). The lower part of Tables 310-16 through 310-19 supplies correction factors for limiting ampacities of conductors at room temperatures greater than 30°C or 86° F.

Column 1 in the main part of the table lists AWG sizes of conductors from No. 18 to 2000 kcmil inclusive. Succeeding columns show ampacities for each size of conductor according to the type of insulation it bears. The columns relating to 60°C, 75°C, and 90°C insulation in Tables 310-16 and 310-17 are most commonly used for inside wiring.

With size No. 8, for example, Column 2 shows that with 60° insulation, the conductor will carry 40 A without subjecting insulation to damaging heat. If the No. 8 conductor has 75° insulation (Column 3) it may be loaded to 50 A without harmful result. This size conductor may carry 55 A with insulation in Column 4.

Suppose that the room temperature for this particular application is 50°C (122°F) instead of 30°C (86 °F). Referring to the lower part of Table 310-16, Ampacity Correction Factors, a multiplying factor of .58 shall be used to determine the ampacity of a conductor having any of the Column 2 insulations. Use multipliers of .75 for Column 3 insulations and .82 for Column 4 insulations. The absence of any

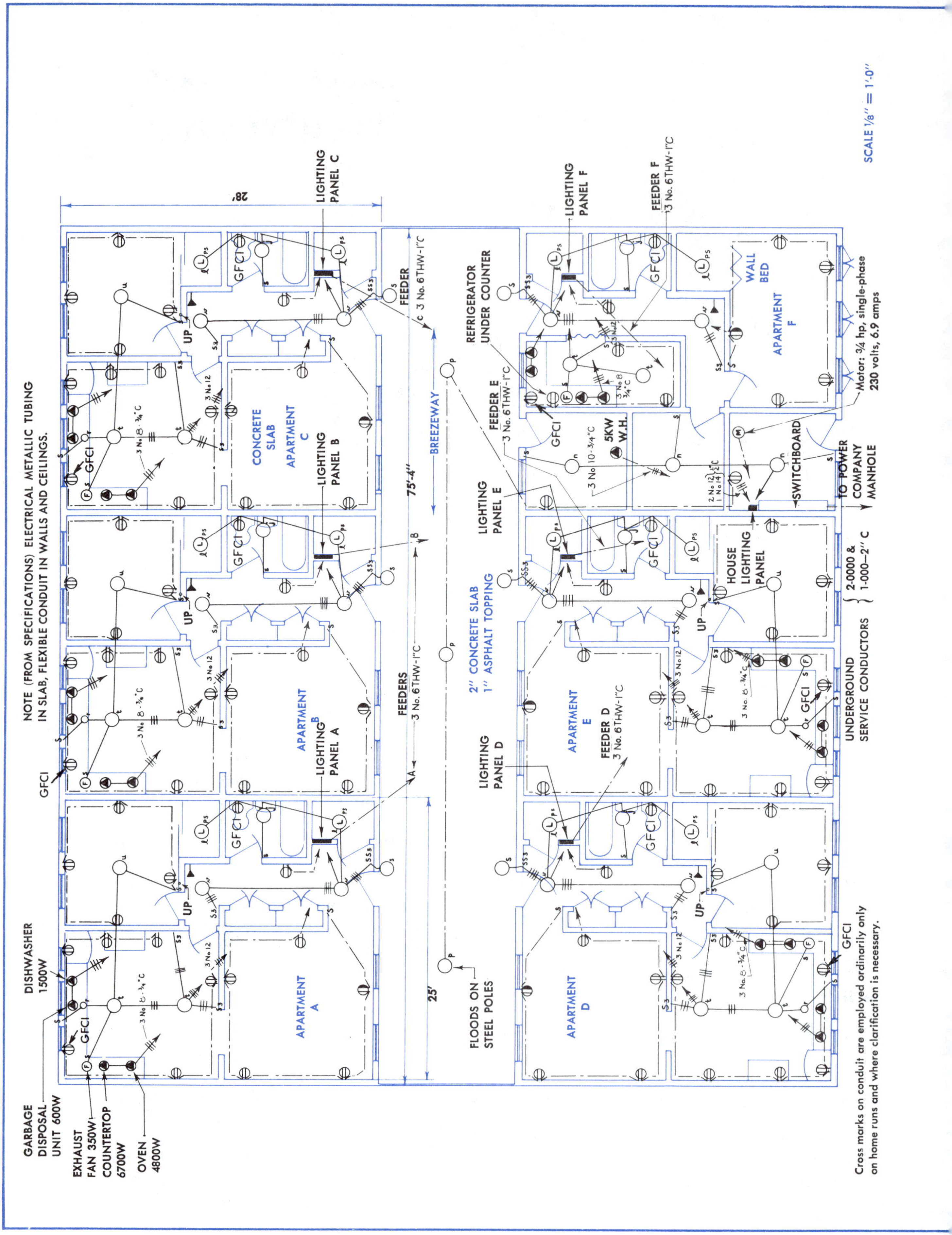
NOTE (FROM SPECIFICATIONS) ELECTRICAL METALLIC TUBING IN SLAB, FLEXIBLE CONDUIT IN WALLS AND CEILINGS.
SCALE 1/8" = 1'-0"
LIGHTING PANEL C
LIGHTING PANEL B
LIGHTING PANEL A
LIGHTING PANEL D
LIGHTING PANEL E
LIGHTING PANEL F
FEEDER C 3 No. 6THW-1"C
FEEDERS 3 No. 6THW-1"C
FEEDER D 3 No. 6THW-1"C
FEEDER E 3 No. 6THW-1"C
FEEDER F 3 No. 6THW-1"C
BREEZEWAY
2" CONCRETE SLAB 1" ASPHALT TOPPING
CONCRETE SLAB
APARTMENT A
APARTMENT B
APARTMENT C
APARTMENT D
APARTMENT E
APARTMENT F
WALL BED
REFRIGERATOR UNDER COUNTER
5KW W.H.
HOUSE LIGHTING PANEL
SWITCHBOARD
TO POWER COMPANY MANHOLE
UNDERGROUND SERVICE CONDUCTORS
2-0000 & 1-000—2" C
Motor: 3/4 hp, single-phase 230 volts, 6.9 amps
FLOODS ON STEEL POLES
DISHWASHER 1500W
GARBAGE DISPOSAL UNIT 600W
EXHAUST FAN 350W
COUNTERTOP 6700W
OVEN 4800W
GFCI
UP
28'
75'-4"
25'
Cross marks on conduit are employed ordinarily only on home runs and where clarification is necessary.

multiplying factor in some columns indicates that the insulations shall not be used for the listed ambient temperatures.

Using the indicated factors, the ampacity of No. 8 Type TW conductor used at a room temperature of 50°C (122°F) equals .58 × 40 A, or 23 A. No. 8 Type THWN conductor is equal to .75 × 50 A, or 37.5 A.

The table shows that Column 2 insulations are not to be used at room temperatures greater than 140°F or Column 4 insulations at temperatures greater than 194°F. The notes below the table indicate the rating of No. 14, 12, and 10 copper is 15, 20, and 30 A and No. 12 and 10 aluminum is 15 and 25 A.

Table 310-17

This table is similar to Table 310-16 except that it deals with the ampacities of single insulated conductors rated 0 V through 2000 V in free air. Table 310-17 is based on an ambient air temperature of 30°C (86°F.) Ampacities of conductors in free air are greater than those of conductors in raceways, cables, or earth.

Tables 310-18 and 310-19

These tables are similar in construction and they deal with allowable ampacities of conductors rated 150°C to 250°C. The correction factor tables are also similar, but they also relate to higher temperatures.

Notes to the Tables

There are eleven notes connected with Tables 310-16 through 310-19, which state specific applications. These notes, Notes to Ampacity Tables of 0 to 2000 Volts, should be carefully evaluated. Although the facts are clearly stated, notice No. 5, which states that bare conductors have the same ampacity as insulated conductors with which they are used, and No. 8, which lists derating factors where more than three current-carrying conductors are installed in a raceway or cable.

Table 310-13—Conductor Application and Insulations

This table lists conductor applications and the physical characteristics of the various types of insulated conductors. It should be noted that 310-13, Conductor Constructions and Applications, states that the conductors listed in this table may be installed for any of the wiring methods recognized by Chapter 3 as specified in the tables. It should also be noted that 310-13 states that insulated conductors used for 600 V or less shall comply with Table 310-13.

There are several types of conductors that are given two maximum operating temperatures depending upon how they are used, such as Type THW. Type THW has a 75°C rating when used as a general purpose conductor and installed in a wet or dry location, but when it is installed within 3″ of a ballast it is given a 90°C rating as required in 410-31.

Table 310-13 also lists several types of conductors for specialized applications as noted, such as: MI cable, which has an outer metal sheath; Types TA, TBS, and SIS, which are limited to switchboard wiring only; and Type TFE, which is limited to leads within apparatus in a dry location. Type THNN is a 90°C conductor that can be used in dry and damp locations. The ampacity is used with 60°C or 75°C based on equipment terminals.

Table 402-3—Fixture Wire

The operating temperature of fixture wires shall not exceed the temperatures listed in Table 402-3.

Table 402-5—Ampacity of Fixture Wires

Table 402-5 lists the allowable ampacities of fixture wires in sizes 18 through 10 AWG. The ampacities of the fixture wires listed in this table are based on an ambient temperature of 30°C (86°F).

Flexible Cords

Table 400-5(A) shows ampacities of flexible cords where not more than three conductors are present. If there are more than three conductors in the group, ampacities given here are to be reduced to 80% of listed values. It may be noted that ordinary portable cord size No. 18 will carry 7 A and size No. 16 will carry 10 A. Notes 1 through 12 of Table 400-4 should be read carefully.

Table 400-4—Flexible Cords and Cables

This table sets forth construction features of the various cords and their conditions of use and ap-

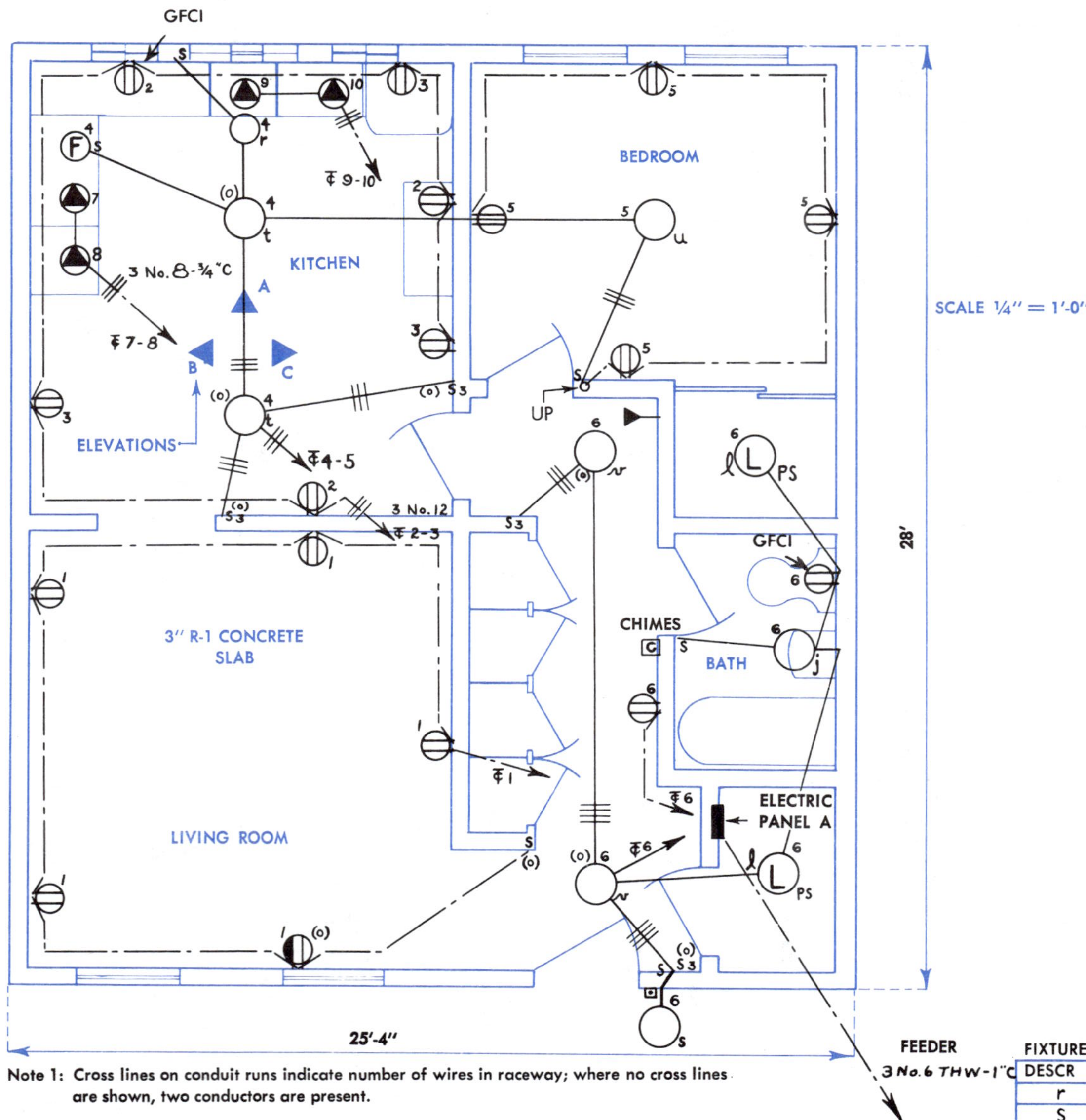

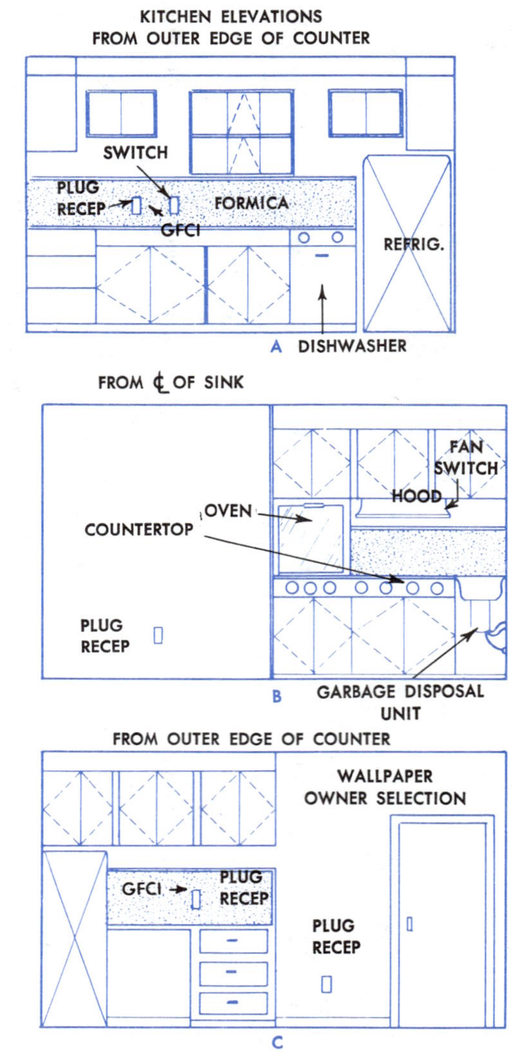

Note 1: Cross lines on conduit runs indicate number of wires in raceway; where no cross lines are shown, two conductors are present.

Note 2: "Up" notation in bedroom partition shows conduit riser from plug receptacle outlet box to switch box.

Note 3: "(o)" notation on switches and outlets of same circuit indicates which switches control the outlets.

Note 4: Notation "₵" sometimes used as symbol for "circuit."

FIXTURE LIST

DESCR	QUANT	RATING	MFGR	CAT.	TYPE
r	5	100W	Morbrite	612	Recessed
S	6	50	Allray	1117	Bracket
t	12	150	Allray	1129	Drum
u	5	100	Morbrite	700	Drum
~	12	75	Allray	1119	Ceiling
J	6	75	Morbrite	804	Bracket
ℓ	12	50	Titan	322	Porc. P.C. Socket
n	3	50	Titan	323	Porc. Socket
p	3	200	Franklin	1791	W.P. Flood

plication. Flexible cords and cables shall conform to the description provided in Table 400-4. If flexible cords and cables not included in Table 400-4 are to be used, they must undergo a special investigation as to their suitability for the intended use.

General Print, Floor Plan

This print, drawn originally to a 1/8″ = 1′-0″ scale, shows a six-unit (Apartments *A–F*), one-story multifamily dwelling. Five of the apartments are identical, having living room, kitchen-dinette, bedroom, and bath. The lighting panel, a flush type, is placed in the entrance storeroom. The kitchen has a wall-mounted oven, a counter-mounted cooktop, dishwasher, garbage disposal, and exhaust fan.

The caretaker's apartment, Apartment *F*, has a living room, kitchen-dinette, and bath. It has no separate bedroom, a wall bed being used in the living room. Kitchen equipment is similar to that in the larger apartments.

There is a washroom, a storeroom, and a combination heating equipment and switchboard room. There is a house receptacle circuit, an electric water heater circuit, and a pump motor. A house lighting circuit supplies the three work rooms and three night-light standards. All these circuits originate in a special house panel which is adjacent to service equipment in the switchboard room.

Arrows pointing in the direction of the lighting panel indicate circuit runs. The number of cross marks on the arrow shows the number of wires in the particular home run. Where no cross marks are found, it is understood that the circuit consists of two wires. Conductors are assumed to be No. 14 unless referred to in the specifications or marked adjacent to the arrow.

Dot-and-dash lines denote branch-circuit raceways concealed in the floor. Solid black lines mark branch-circuit conduit concealed in the wall or ceiling. Designating signs for various types of outlets including lights, plug receptacles, recessed units, special outlets, telephones, and fans are those recommended by the list of standard symbols. Feeders are shown by arrows originating at the various panels and pointing in the direction of the switchboard.

It should be observed that a note taken from the electrical specifications states that electrical metallic tubing is to be used in concrete floor slabs. Flexible conduit is to be used in walls and ceilings.

LOAD CALCULATIONS

The unit load specified for dwelling occupancies in Table 220-3(b) is 3 VA per square foot. The allowance for two small appliance or utility circuits per 220-16(a) is 3000 VA. The load required for two cooking units, as per Column C of Table 220-19, is 65% of the sum of the nameplate ratings. On the basis of above Code sections, the load for each of the five apartments, *A* to *E* inclusive, is determined as follows:

LOAD CALCULATIONS FOR APARTMENTS *A–E*

Lighting and Appliance		
710 sq ft × 3 VA	2130 VA	
Two appliance circuits	3000 VA	
Total	5130 VA	
Feeder Demand Factors		
1st 3000 VA at 100%	3000 VA	
Remainder, 2130 VA at 35%	745 VA	
Total	3745 VA	**3745 VA**
Oven and Cooktop		
Oven rated	4800 VA	
Cooktop rated	6700 VA	
Total	11,500 VA	
.65 × 11,500 VA =	7475 VA	**7475 VA**
Dishwasher		**1500 VA**
Garbage Disposal		**600 VA**
Exhaust Fan		**350 VA**
	Total	**13,670 VA**

FEEDER SIZE (120/240 V service)
13,670 VA ÷ 240 V = **56.9 A**

Table 310-16 shows that No. 6 Type THW copper conductor is needed. Table 3A, Chapter 9, shows that a 1″ conduit is required for three No. 6 Type THW conductors.

LOAD CALCULATIONS FOR APARTMENT *F*

Lighting and Appliance	
448 sq ft × 3 VA	1344 VA
Two appliance circuits	3000 VA
Total	4344 VA

Feeder Demand Factor		
1st 3000 VA at 100%	3000 VA	
Remainder, 1344 VA at 35%	470 VA	
Total	3470 VA	**3470 VA**
Oven and Ccooktop		**7475 VA**
Dishwasher		**1500 VA**
Garbage Disposal		**600 VA**
Exhaust Fan		**350 VA**
	Total	**13,395 VA**

FEEDER SIZE (120/240 V service)
13,395 VA ÷ 240 V = **55.8 A**

Table 310-16 shows that No. 6 Type THW copper conductor is needed. Table 3A, Chapter 9, shows that a 1″ conduit is required for three No. 6 Type THW conductors.

In some cases, the neutral feeder may be smaller than the outer conductors. NEC 220-22 allows a 70% demand factor on the portion of the neutral feeder load that supplies an electric range or equivalent cooking appliance. Although the specifications and print state that three No. 6 Type THW copper conductors shall be installed to each panel, it is useful to perform calculations to see if a smaller neutral conductor might have been otherwise admissible. Calculations for Apartment *A* are determined as follows:

ADDITIONAL LOAD CALCULATIONS FOR APARTMENT *A*

Oven and Ccooktop = 7475 VA.		
.7 × 7475 VA = 5235 VA		**5235 VA**
Assumed Feeder Load		
Lighting and Appliance		**3470 VA**
Dishwasher		**1500 VA**
Garbage Disposal		**600 VA**
Exhaust Fan		**350 VA**
	Total	**11,155 VA**

NEUTRAL FEEDER SIZE
11,155 VA ÷ 240 V = **46.5 A**
Per Table 310-16, No. 8 THW copper conductor is acceptable for the neutral. Maximum unbalance of 50 A per 220-22.

Another factor that is sometimes important when the load appears close to the current-carrying capacity of a particular size conductor is that of balancing circuits. In Apartment *A*, for example, a 3-wire lighting circuit, a 3-wire utility circuit, and a 3-wire cooking circuit are balanced across the ungrounded conductors and the neutral wire. But the 1500 VA, 120 V dishwasher, which is connected between one ungrounded conductor and the neutral, does not balance the 950 VA garbage disposal and exhaust fan load, which is connected between the other ungrounded conductor and the neutral.

In such a case, an imaginary (artificial) load may be assumed in parallel with the smaller load in order to gain balance, its value here being equal to 1500 VA minus 950 VA, or 550 VA. The listing is then:

LOAD CALCULATION ON APARTMENT *A* WITH ARTIFICIAL LOAD

Lighting and Appliance		**3745 VA**
Oven and Cooktop		**7475 VA**
Dishwasher		**1500 VA**
Garbage Disposal		**600 VA**
Exhaust Fan		**350 VA**
Artificial Load		**550 VA**
	Total	**14,220 VA**

ADJUSTED LOAD
14,220 VA ÷ 240 V = **59 A**
Per Table 310-16, No.6 THW copper conductors may be used.

Although the calculation has not altered the result here, the method is worth noting. Some inspection authorities insist upon balancing loads; others do not.

SERVICE CALCULATIONS

NEC 220-3(c)(7) provides that a load of 180 VA per outlet shall be allowed for outlets other than for general illumination. The two plug receptacles in the work area come within such classification. NEC 220-13 permits this type of load to be included with the general lighting load in applying demand factors. These plug receptacles and the three 200 VA night-light standards come within scope of the section.

NEC 220-17 states that where four or more fixed appliances other than ranges, clothes dryers, space heating units, or A/C equipment are connected to the same feeder or service in a multifamily dwelling, a demand factor of 75% may be applied to the load. The six dishwashers, six garbage disposals, six exhaust fans, and electric water heater qualify here. The service conductor sizes are governed by 230-42, which refers to 220, so that service conductor sizes are calculated as follows:

SERVICE CALCULATIONS

Lighting and Appliance		
4220 sq ft × 3 VA =	12,660 VA	
Six Appliance Circuits		
= 6 × 3000 VA =	18,000 VA	
Laundry Circuit	1500 VA	
Three Night-light Standards		
= 3 × 200 VA =	600 VA	
Two Receptacle Outlets in Work Area		
= 2 × 180 VA =	360 VA	
Total	33,120 VA	
1st 3000 VA at 100%	3000 VA	
Remainder 30,120 VA at 35%	10,542 VA	
Total	13,542 VA	**13,542 VA**
Ovens and Cooktops		
6 × 4800 VA =	28,800 VA	
6 × 6700 VA =	40,200 VA	
Total	69,000 VA	
Service Demand Factor		
32% for 12 ranges = .32 × 69,000 VA =		**22,080 VA**
Other Loads		
Dishwashers 6 × 1500 =	9000 VA	
Disposals 6 × 600 =	3600 VA	
Exhaust fans 6 × 350 =	2100 VA	
Water heater 1 × 5000 =	5000 VA	
Total	19,700 VA	
.75 × 19,700 VA =		**14,775 VA**
¾ hp, 1ϕ, 230 V × 6.9 A = 1600 VA		
Motor Load		
1600 VA × 125% =		**2000 VA**
	Total	**52,397 VA**

SERVICE ENTRANCE CONDUCTORS

52,397 VA ÷ 240 V = **218 A**

Table 310-16 shows that No. 4/0 Type THW copper conductors are acceptable.

NEUTRAL CONDUCTOR

Lighting and Appliance		**13,542 VA**
Cooking Units		
.7 × 22,080 VA =		**15,455 VA**
Fixed appliances		**14,775 VA**
	Total	**43,772 VA**

43,772 VA ÷ 240 V = **182 A**

Table 310-16 shows that No. 3/0 Type THW copper conductor is acceptable as the neutral.

SERVICE SIZE

Two No. 4/0 and one No. 3/0 Type THW conductors, all in 2" conduit are acceptable. *Note:* The optional calculation per Table 220-32 may be applied.

Note: The method of determining the conduit size for different sizes of conductor in the same raceway will be considered in a later assignment.

DETAILS AND ELEVATIONS

The larger print, drawn to a scale of ¼" = 1'-0", shows one of the five similar apartments. The plan view brings out details to better advantage than is possible on the general print, so that the position of equipment may be found with greater accuracy.

The conduit in the floor slab is indicated by a dot-and-dash line. The conduit in a wall or ceiling is indicated by solid black lines. All wires are No. 14 except the 3-wire No. 8 circuit to the cooking units and the 3-wire No. 12 circuit to the kitchen appliance receptacles.

Circuits may be traced out to learn the reason for three or four wires, as the case may be, in various conduits. Certain runs that might otherwise have been combined are kept separate to avoid the penalty of derating where more than three wires are installed in one raceway.

One point that should not be overlooked is the arrangement for connecting bedroom plug receptacles. They are supplied by conduit laid in the floor slab. This run continues up to the switch outlet box in the wall at the side of the door and then to the ceiling outlet. A small circle and the word *UP* indicate this procedure. It should also be noted that one half of the symbol designating the duplex receptacle outlet in the living room is a solid color. This indicates that one half of this receptacle outlet is controlled by a wall switch.

Triangles *A*, *B*, and *C* marked in a semicircle on the kitchen floor represent one scheme for directing attention to elevations seen at the right of the drawing. From these elevations, it is possible to determine exact locations of outlets such as those for plug receptacles and for cooking units. The location of the garbage disposal and its method of connection are shown in elevation *B*.

SERVICE--MULTIPLE OCCUPANCY BUILDING

One service drop or lateral to a single-occupant building may feed 2 to 6 services that supply separate loads. (230-40 Ex. 2)

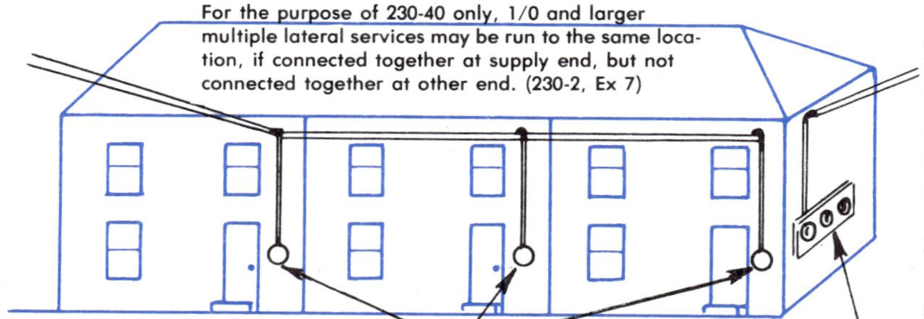

NEC 230-40 Each lateral or service drop to supply one set of service-entrance conductors except; (1) one set of service-entrance conductors permitted for each occupancy or group of occupancies in multiple occupancy bldgs.; (2) where 2 to 6 disconnects are grouped and in separate enclosures in one location and supply separate loads; (3) one-family dwelling unit and separate structure.

NEC 230-2(a)(b) A building shall be served by one service. Where exceptions permit more than one service, a permanent plaque or directory required at each, indicating location of all others and area served by each.

225-8 In general, when more than one building is on the same property, each building or structure shall have a separate disconnecting means. Disconnecting means shall be suitable for use as service equipment except for 3-way and 4-way snap switches for garages and out buildings on residential property. See Exceptions.

NEC RULES CONCERNING SERVICES AND FEEDERS

NEC 230-90 (b) No overcurrent device shall be inserted in a grounded service conductor except a circuit breaker which simultaneously opens all conductors.

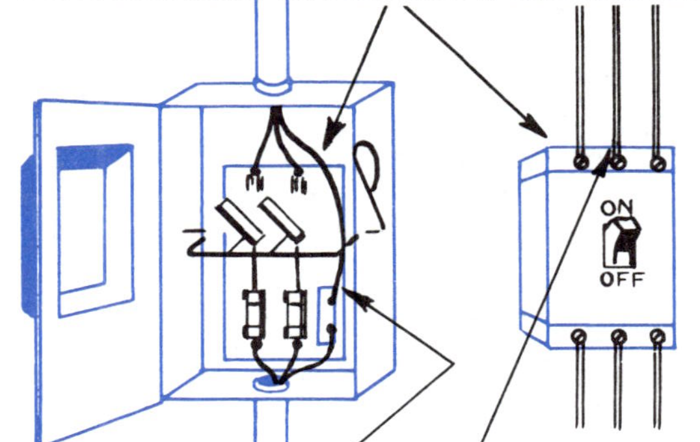

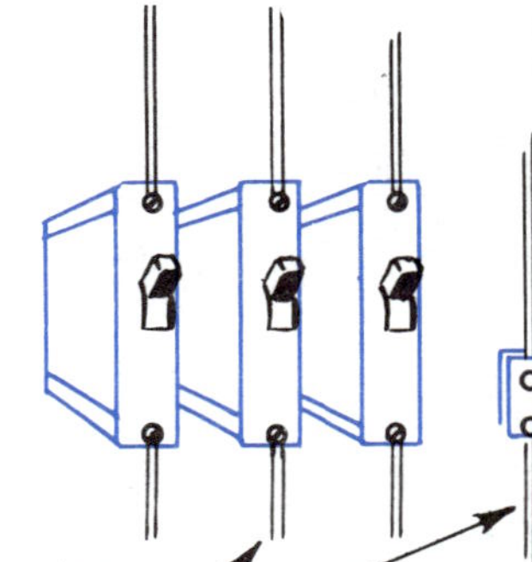

NEC 220-22 The neutral conductor shall be large enough to carry the maximum unbalanced load.

NEC 230-42 (c) The grounded (neutral) conductors shall not be smaller than required by 250-23 (b)

NEC 250-23 (b) Service grounded conductor not smaller than grounding electrode conductor as per Table 250-94 and if phase conductors are larger than 1100 kcmil copper or 1750 kcmil aluminum, the grounded conductor shall not be smaller than 12½ percent of area of largest phase conductor **except need never be larger than largest service ungrounded conductor.**

240-20 CBs are permitted as overcurrent devices and shall open all ungrounded conductors. Individual single-pole CBs may be used with handle ties in grounded systems in (1) 1ϕ circuits, (2) 3-wire DC circuits, (3) lighting and appliance 3ϕ, 4-wire or 2ϕ, 5-wire branch circuits supplied from a system with a grounded neutral and not operating at over 150 V to ground.

NEC 110-10 Impedance and other circuit characteristics must be considered so that selected overcurrent device will clear the fault before serious damage results.

See Appendix for typical Fault Current Form.

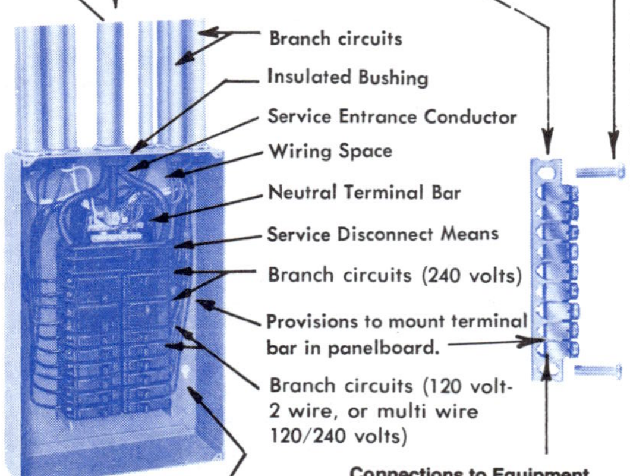

230-43 Wiring methods for service conductors are limited to (1) open wiring on insulators; (2) IGS cable; (3) Rigid Metal Conduit; (4) IMC; (5) EMT; (6) ENT; (7) SE Cable; (8) Wireways; (9) Busways; (10) Auxiliary Gutters; (11) Rigid Nonmetallic Conduit; (12) Cablebus; (13) MC Cable; (14) MI Cable; (15) Flexible Metal or Liquidtight Flexible Metal Conduit, 6′ maximum length, with equipment bonding jumper; (16) Liquidtight Flexible Nonmetallic Conduit.

NEC 373-5(c) Where cable is used, each cable required to be connected to cabinet. Two cables from one connector not permitted unless connector approved for two cables.

110-14(a) One conductor per terminal in equipment grounding terminal bar, unless terminal identified for two conductors.

SERVICE-LIGHTING AND APPLIANCE PANELBOARD

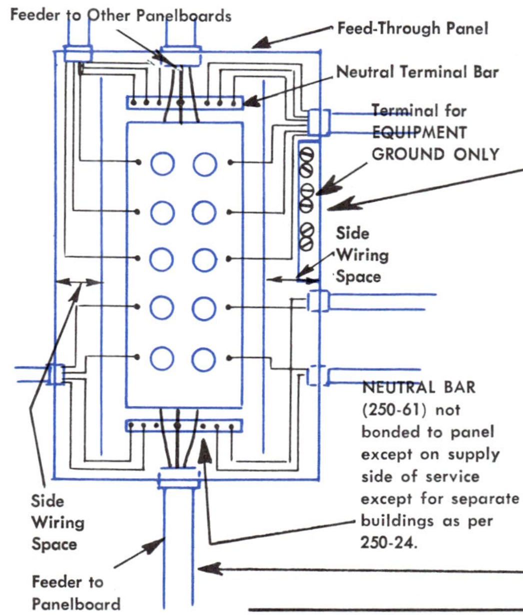

384-20 Identified terminal bar shall be provided inside cabinet and secured to cabinet for connection of all feeder and branch-circuit EGCs.

Terminal bar for equipment grounding conductors shall be secured to cabinet. (384-20)

Connection to neutral bar required in separate building as per 250-24(a). (384-20)

NEC 373-6 (b) **Conductors at terminals on entering or leaving cabinet shall have bending space as per Tables 373-6 (a) and 373-6 (b).**

Ungrounded conductors #4 and larger must be protected by a substantial insulating material, securely fastened in place. 373-6(c); 300-4(f)

Insulating bushing shall have temperature rating not less than temperature rating of conductors. (373-6 (c))

NEC 373-11 (c) Cabinets and cutout boxes which contain devices or apparatus connected to more than 8 conductors, not including the supply circuit or a continuation thereof shall have back or side wiring spaces.

Panelboard protected on supply side by overcurrent devices not exceeding rating of panelboard. (384-16 (a), Ex 1)

DISTRIBUTION PANELBOARDS

NEC 250-81 (c) At least 20 feet of bare copper not smaller than No. 4 or steel reinforcing rod or bar encased by at least 2 inches of concrete within bottom of concrete foundation footing in contact with the earth. NEC 250-83 Where none of the electrodes listed in 250-81 are available at a building, one or more of the "Made and Other Electrodes" listed in 250-83 shall be used.

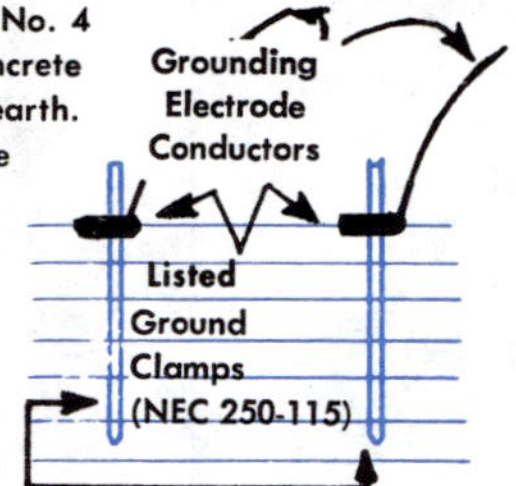

250-83(d) Plate electrodes shall be a minimum of .25" thick if iron or steel, and .06" thick if constructed of nonferrous metal.

Buried metallic plate, surface area 2 square feet.

Aluminum wire permitted if used with proper UL listed ground clamp. Not permitted outside within 18 inches of earth. (250-92 (a) and 250-115))

Only one wire per clamp unless listed for purpose.

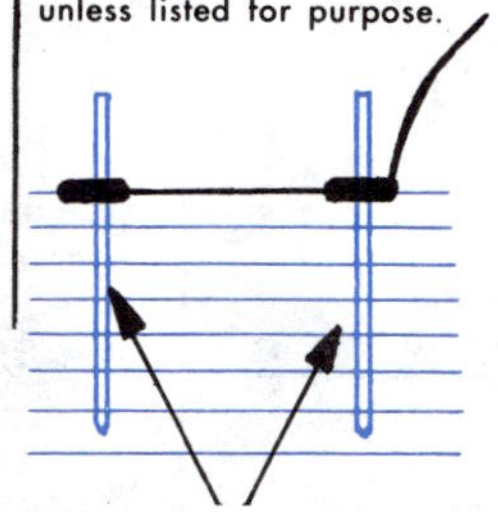

NEC 250-83 (c) (3) Driven rod at least 8 ft except where rock encountered, then rod driven at not more than 45° angle or buried in a trench at least 2½ ft deep. Upper end of rod to be flush or below ground [with suitable clamp (250-115)] level unless rod and clamp are protected from damage as per 250-117.

NEC 250-83 (c-2) Rod electrode ⅝ inch iron, or ½ inch nonferrous.

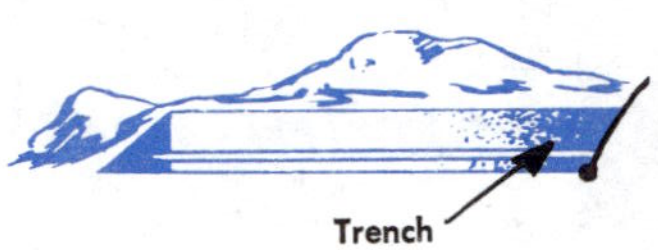

NEC 250-84 If resistance of made electrode exceeds 25 ohms, one additional electrode required and it shall be at least 6 ft away.

NEC 250-83 Rod or pipe electrodes of different systems shall be separated at least 6'. Lightning rod electrode may be bonded to other rods as permitted by 250-86. Also see 250-46, 800-41, and 820-40(d)

NEC 250-83 (c) (1) Galvanized iron pipe, ¾ inch trade size.
NEC 250-83 Where more than one electrode system is used, they shall be separated at least 6 ft. (Two or more bonded electrodes treated as one system. (FPN))

NEC 250–83 "MADE" ELECTRODES

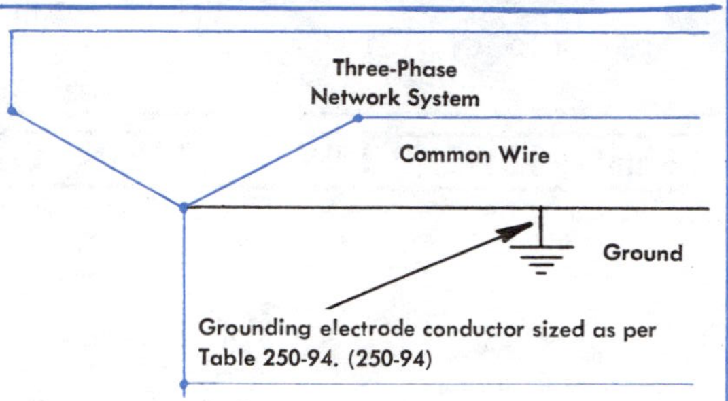

NEC 250-25 (3) For multiphase systems having a conductor common to all phases, the common conductor shall be grounded.

Same conductor to be grounded for separately derived AC systems as per NEC 250-26.

NEC RULE PERTAINING TO GROUNDING IN MULTIPHASE SYSTEM

NEC 250-83(a) Gas pipe shall not be used as a grounding electrode.

NEC 250-112 Attachment made in a manner to assure a permanent and effective ground.

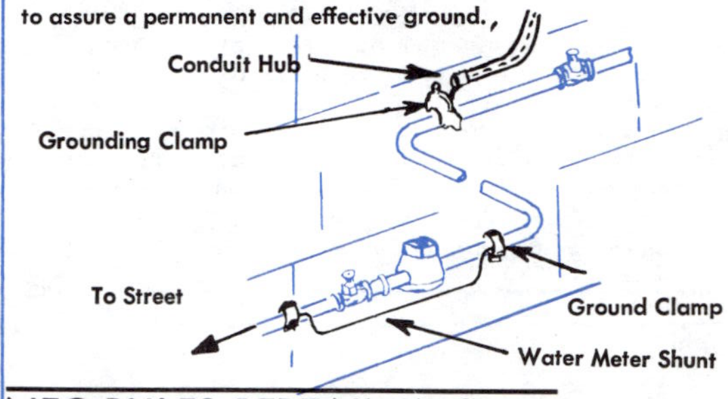

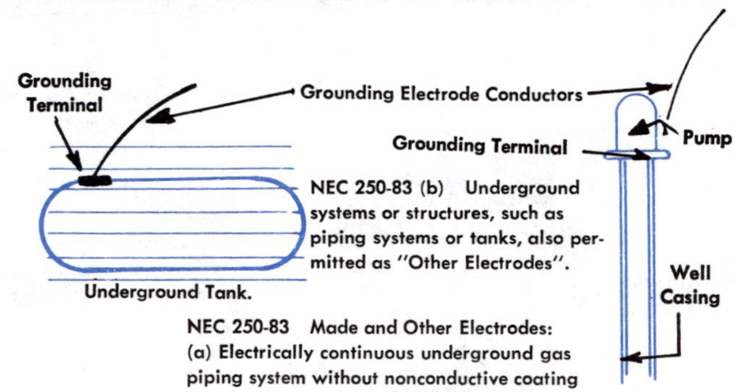

NEC 250-83 (b) Underground systems or structures, such as piping systems or tanks, also permitted as "Other Electrodes".

NEC 250-83 Made and Other Electrodes:
(a) Electrically continuous underground gas piping system without nonconductive coating permitted only when expressly permitted by gas supplier and authority enforcing the Code.
(b) Underground systems or tanks.
(c) Rods and pipes at least 8 ft long: (1) Galvanized pipe at ¾ inch size; (2) Rod electrode at least ⅝ inch iron or ½ inch nonferrous. (3) Pipes and rods driven to depth of at least 8 ft unless rock, then buried in trench at least 2½ ft deep or driven at oblique angle not over 45 degrees.
(d) Buried plate of iron or steel at least ¼ inch thick or 0.06 inch thick if of nonferrous metal. Plate electrode required to expose 2 square foot surface area to soil.

250-112 Grounding electrode conductor shall be accessible and installed in a manner that makes connection both permanent and effective. Bonding around insulated joints, like water meters, helps to ensure an effective ground path.

NEC RULES PERTAINING TO GROUNDING ELECTRODES

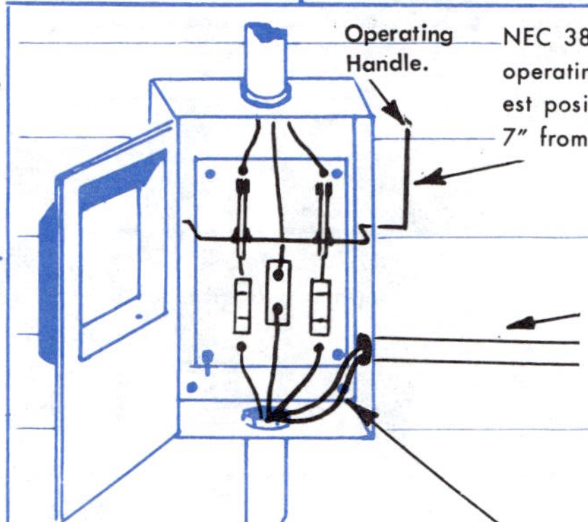

NEC 380-8(a) Center of grip of operating handle when in highest position not more than 6'-7" from floor or platform.

NEC 373-8 These conductors permitted if not more than 40% of cross-sectional area is used. For splices and taps, not over 75% of area may be used.

NEC 373-8 Switch enclosures are not to be used as auxiliary gutters, raceways, or junction boxes unless adequate space is provided. See 373-8.

NEC 110-12 Equipment shall be installed in a neat manner.

NEC 110-13(a) Electrical equipment to be securely fastened to mounting surface. Wooden plugs driven into masonry or concrete are not acceptable for support.

NEC 110-3(b) Evidence of suitability of equipment may be by listing or labeling.

SWITCH ENCLOSURES

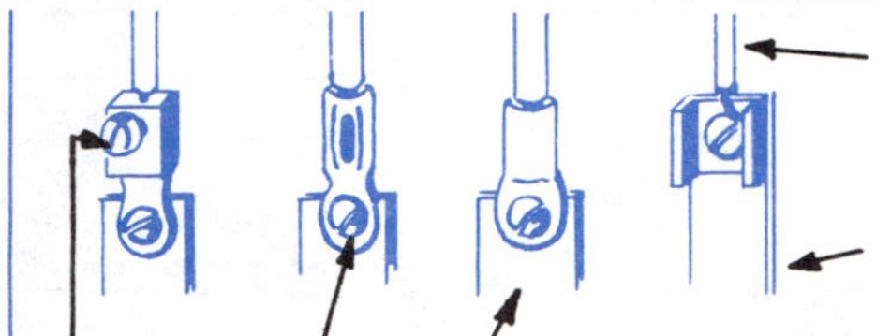

110-14 Conductors of dissimilar metals not to be intermixed unless terminal, lug, or connector identified for purpose.

Terminals for aluminum shall be identified for such use.

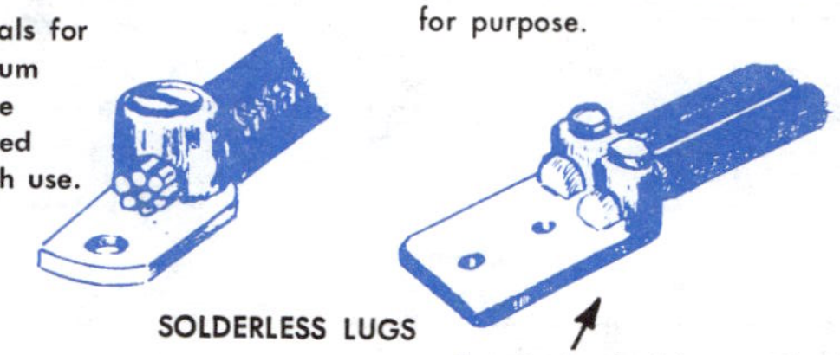

Terminals shall be so identified when used for more than one conductor or used to connect aluminum wire. (110-14 (a))

110-14(a) Connections to terminals shall insure good connection without damaging conductors. See 110-14(c) for 60°C and 75°C terminal ratings.

CONNECTIONS TO TERMINALS

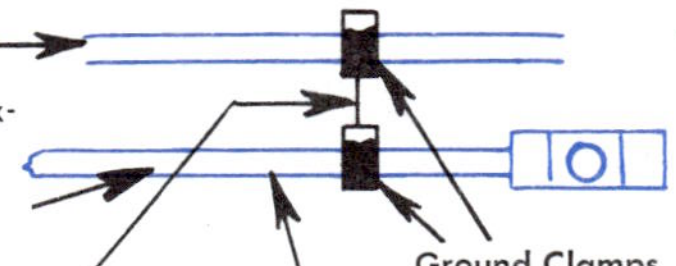

250-33, Ex. 1 For additions to existing installations, metal enclosures and raceways for open wiring, knob-and-tube, and NM cable without an EGC need not be grounded if less than 25′ and suitably isolated or guarded.

NEC 250-33, Ex. 2 Short sections of metal conduit used to protect cables from physical damage need not be grounded.

NEC 250-33 Basic rule-metal enclosures for conductors shall be grounded.

NEC Table 250-95 Equipment grounding conductor for 15-, 20- and 30-amp circuits shall have ampacity at least equal to overcurrent protection of circuit. For circuits over 30 amps, use Table 250-95.

GROUNDING -- METAL ENCLOSURES

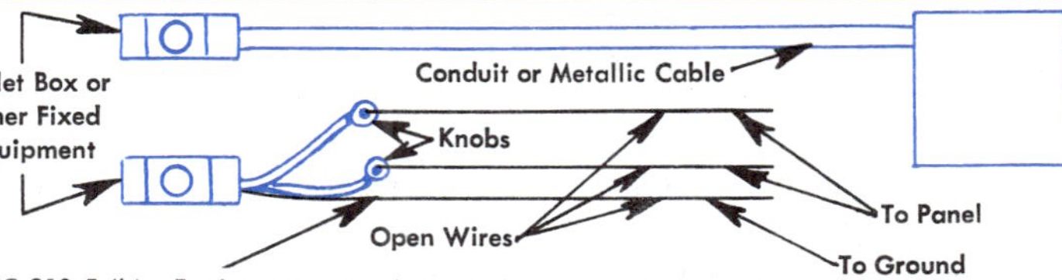

NEC 210-5 (b) Equipment grounding conductor either green, green with one or more yellow stripes, or bare.

250-57(a) Metal parts of equipment are considered to be grounded if connected to any of the EGCs in 250-91(b).

250-57(b) If a separate grounding conductor is used, this conductor bare, covered, or insulated with green or green with one or more yellow stripes on outer covering.

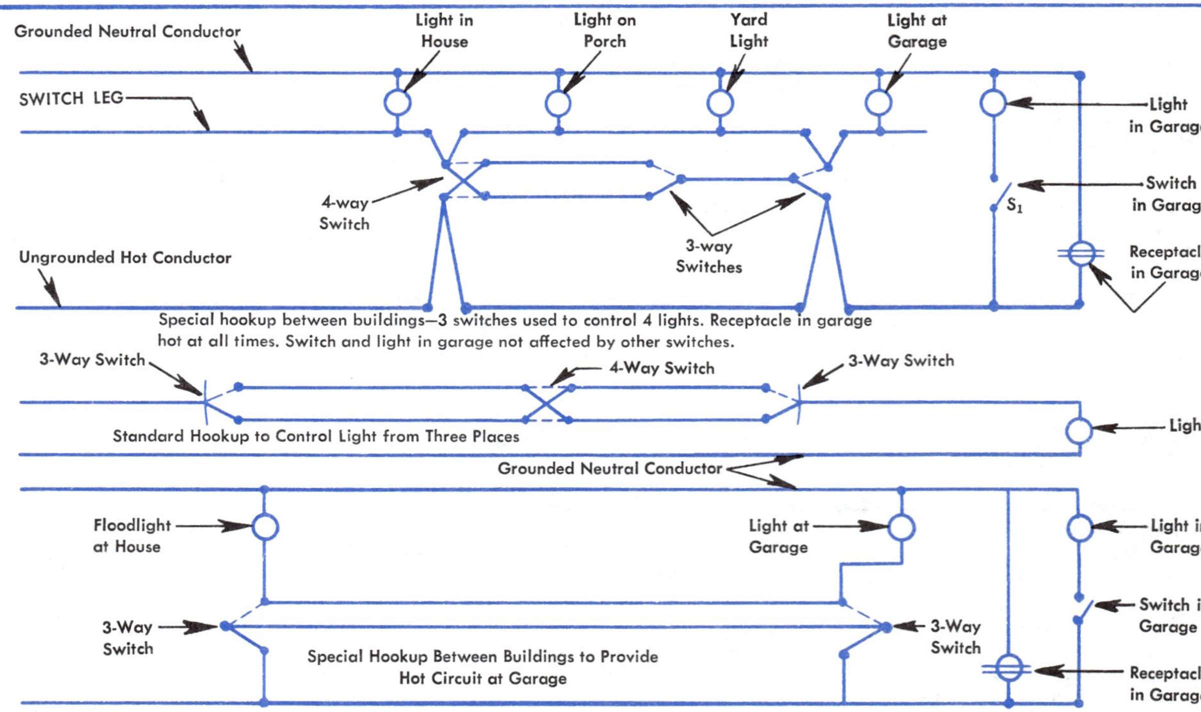

NEC 380-2 (a) Switching on circuits containing three-way and/or four-way switches shall be done in the ungrounded conductor. Wiring between switches and lights, if in metal enclosures, shall be run with both polarities in the same enclosure.

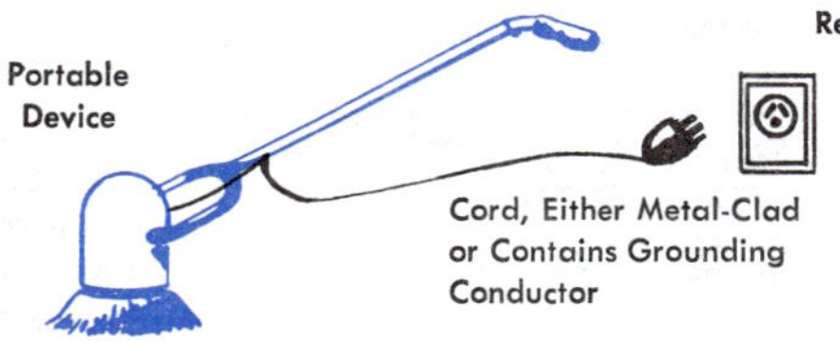

250-59 Noncurrent-carrying metal parts of cord- and plug-connected equipment which must be grounded may be grounded: (a) by means of metal enclosure of conductors or (b) by an EGC which is part of the cable or flexible cord, and which is either bare, or has a green or green with one or more yellow stripes on outer covering.

250-95, Ex. 1 Fixture wires and cords, per 240-4, are permitted to contain a #18 or larger Cu EGC, provided it is not smaller than circuit conductors.

NEC 250-45 (c) In residential occupancies the following cord- and plug connected equipment shall be grounded: clothes washers and dryers, refrigerators, freezers, air conditioners, dishwashers, electrical aquarium equipment, sump pumps, portable hand-held power tools and motor-operated appliances, such as saws, sanders, drills, lawn mowers, wet scrubbers, snow blowers, portable handlamps, and hedge clippers. Exception: Listed tools and appliances protected by a double insulation system and so marked need not be grounded.

NEC 250-59 (a), Ex and (b), Ex The grounding member of grounding type attachment plugs on the supply cord of hand-held portable tools or appliances may be of the self-restoring type.

NEC 410-58 (c) A grounding terminal or grounding device shall be used only for grounding purpose.

GROUNDING CORD-AND-PLUG CONNECTED EQUIPMENT

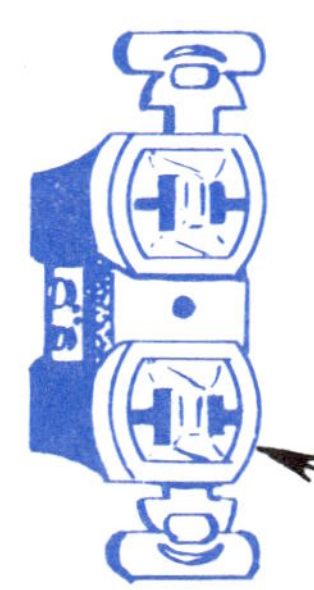

NONGROUNDING TYPE RECEPTACLE

NEC 410-56 (a) Receptacles for portable cords shall be rated not less than 15 amps for 125 volts or 250 volts. (See Ex for 10 amps, 250 volts nonresidential. Also see 210-21 (b).)

NEC 210-7(a)(b) Receptacles installed on 15 A and 20 A circuits shall be of grounding type. See Exception.

NEC 210-7 (b) (c) Receptacles having grounded contacts must be grounded. Branch circuits must include a grounding conductor unless the raceway provides grounding. (Fine Print Note 2): For extension to non-grounding existing circuits, an adjacent water pipe may be used for grounding. (See 250-50(a, b), Ex.)

NEC 210-7 (d) Grounding-type receptacles to be used as replacements for existing nongrounding types. Except where receptacle box does not have grounding means, a GFCI or nongrounding type receptacle is permitted.

NEC 250-74, Ex 2 Support hole in yoke, if designed and listed for purpose, may be used to bond receptacle to flush-type boxes.

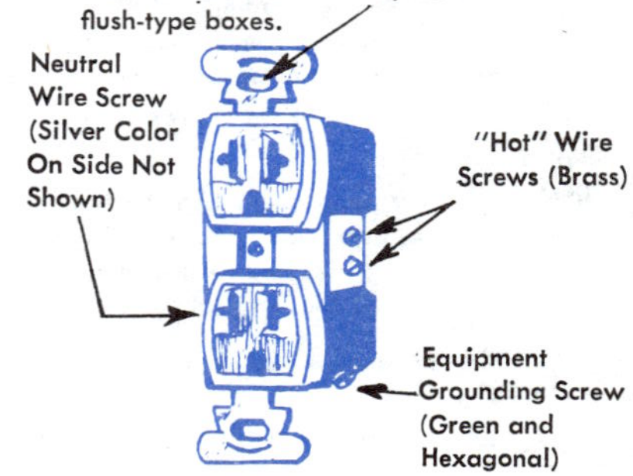

GROUNDING TYPE RECEPTACLE

RECEPTACLE GROUNDING

SWITCHES

(LEVITON)

NEC 380-14 (a) **AC general-use snap switch, suitable for use on AC only, may be used to control resistive and inductive loads not to exceed the ampere rating of the switch or the voltage of the circuit, or tungsten-filament loads of rating of switch at 120 volts, or motor loads not to exceed 80% of rating of switch at rated voltage. Snap switches rated 20 A or less shall be listed and marked CO/ALR when used with aluminum wire.**

NEC 380-8(a), Ex. 1, 2, 3 Switches and CBs shall be readily accessible and not over 6′-7″ from floor or platform except: (1) busway switches permitted at same level as busway; (2) switches located adjacent to motors, appliances, or other equipment; (3) hookstick operated isolating switches.

NEC 380-8(b) Voltage between adjacent snap switches in same box shall not exceed **300** volts. If voltage between adjacent switches exceeds **300** volts then permanent barriers must be installed.

NEC 380-14 (b-1, 2, 3) **AC-DC general-use snap switches suitable for AC or DC may be used to control resistive loads not exceeding the ampere rating of the switch or the voltage of the circuit, or inductive loads not exceeding 50% of the ampere rating of the switch at applied voltage, or when horsepower rated may control motor leads not exceeding voltage applied, or tungsten-filament loads not exceeding the ampere rating of the switch at applied voltage, when "T" rated. (See 430-83, 430-109, 430-110 for motors, and 600-6(b) for signs.)**

ELECTRIC-DISCHARGE LIGHTING

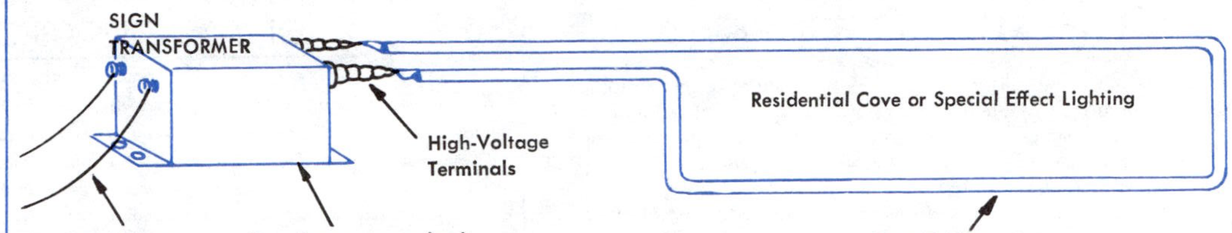

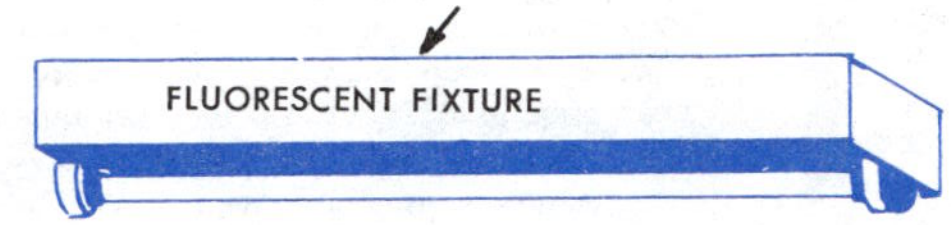

410-73(e) In general, all fluorescent fixtures installed indoors shall have ballasts with integral thermal protection. See Exceptions for (1) simple reactance ballasts, (2) exit fixtures, and (3) exit lighting.

NEC 410-75 Equipment having an open-circuit voltage of more than 300 V shall not be installed in dwelling occupancies. Equipment having an open-circuit voltage of more than 300 V shall not be installed in or on dwelling occupancies unless so designed that no live parts are exposed when lamps are being handled.

NEC 410-76(b) Fluorescent fixtures mounted on combustible low-density cellulose fiberboard required to have 1½″ air space or listed for direct mounting.

NEC 410-76(b), FPN Combustible low-density cellulose fiberboard includes sheets, panels and tiles formed of bonded plant fiber and has a density of 20 pounds or less per cubic foot. Does not include solid or laminated wood or material integrally treated with fire-retardant chemicals to limit flame spread to not over 25′ as per ANSI A2.5-1977 test methods.

MANUFACTURED WIRING SYSTEM--ARTICLE 604

NEC 604-1 applies to field-installed wiring using factory manufactured subassemblies for branch circuits, remote-control circuits, signaling and communication circuits in accessible areas. (Drop ceiling with lift-out panels considered accessible – See Definitions.)

604-4 Manufactured wiring systems are permitted to be installed in (1) accessible locations, (2) dry locations, (3) plenums, (4) spaces used for environmental air where listed and installed per 300-22, and (5) outdoor locations where listed for the purpose.

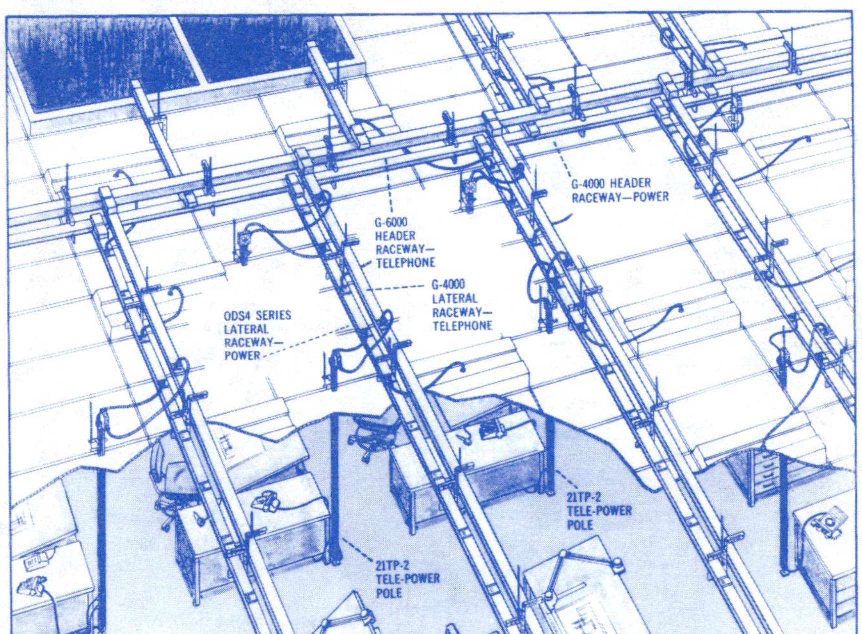

Illustration, courtesy; The Wiremold Company (Electrical Division)

604-6(a)(1)(2) Wiring systems shall be AC or MC cable with #10 or #12 Cu insulated conductors with a bare or insulated Cu EGC sized equivalent to ungrounded conductors. Conduit is also permitted if it is listed flexible metal conduit or listed liquidtight flexible conduit with #10 or #12 Cu insulated conductors with a bare or insulated copper EGC, sized equivalent to ungrounded conductors. See Exceptions.

NEC 604-7 **Unused outlets to be capped to close connector openings.**

NEC 604-6 (c) Component parts to be listed for appropriate system.

NEC 604-6 (b) **Receptacles and connectors to be locking type, uniquely polarized and identified for purpose and part of listed assembly for appropriate system.**

NEC 604-5 Not permitted where conductors or cables limited as per 333, 334.

NEC 604-5 Installations to conform to Articles 110, 200, 210, 220, 250, 300, 310, 333, 334, 350, 410, 545, 640, 700, 725, and 800 plus other applicable sections. This includes securing AC cable every 4½′ and with 12″ from boxes or fittings. See 333-7, Ex. 1–5.

NEC 342 NONMETALLIC EXTENSIONS--AERIAL CABLE

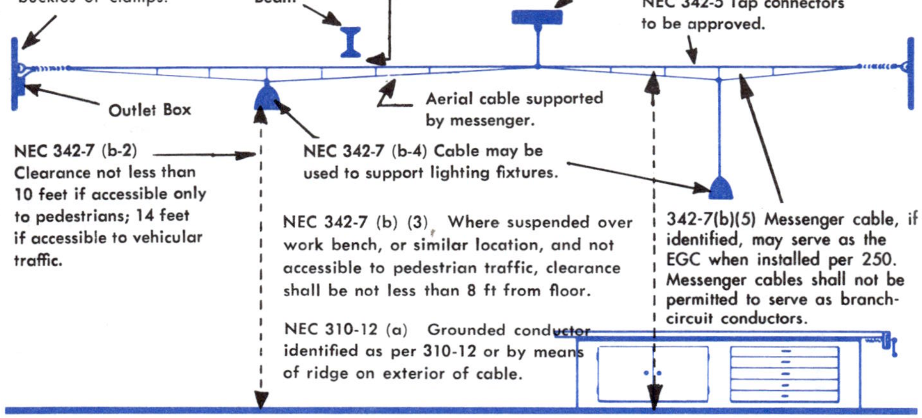

NEC 342-4(c) Not permitted to be installed where voltage between conductors exceeds 300 V.

NEC 342 NONMETALLIC SURFACE EXTENSIONS

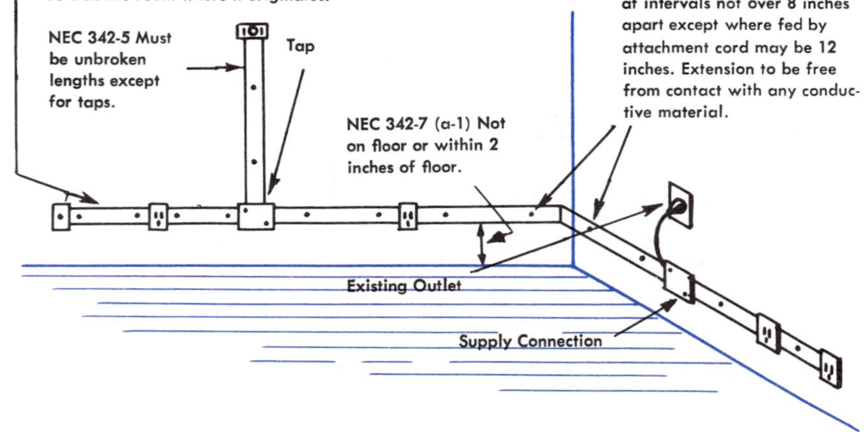

342-3 Nonmetallic surface extensions permitted: (a) extension from an existing outlet on a 15 A or 20 A circuit, (b) in an exposed dry location, (c) where building is occupied for residential or office purposes and not over 3 floors in height per 336-5(a)(1).

NEC 342-4 They are not permitted: (a) as aerial cable to substitute for one of the general wiring methods, (b) in unfinished basements, attics, or roof spaces, (c) if the voltage between conductors exceeds 150 volts, (d) if subject to corrosive vapors, or (e) if extended outside room where they originate.

PANELBOARD -- LIGHTING AND APPLIANCE

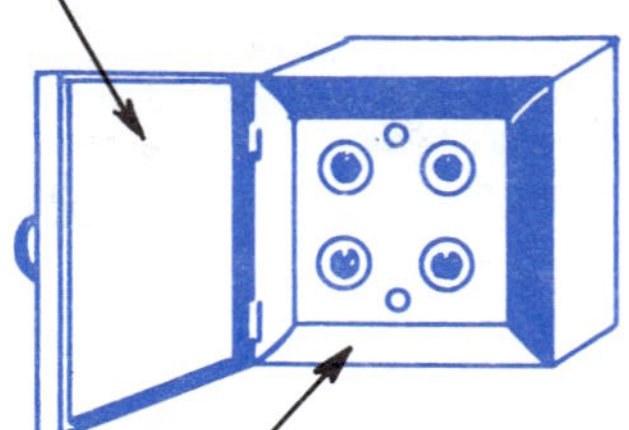

NEC 384-16 (a) Lighting and appliance panelboards shall be protected on supply side not to exceed rating of panelboard. (See Exceptions.)

240-30 Cabinets or cutout boxes shall be used to enclose all overcurrent devices. See Exceptions when (1) protection is afforded as part of an assembly, (2) rooms or enclosures accessible only to qualified people, (3) only operating handle is exposed.

NEC 240-51(b) Plug fuses of Edision-base type may be used as replacement in existing installations where no evidence of overfusing.

NEC RULES DEALING WITH BRANCH CIRCUITS

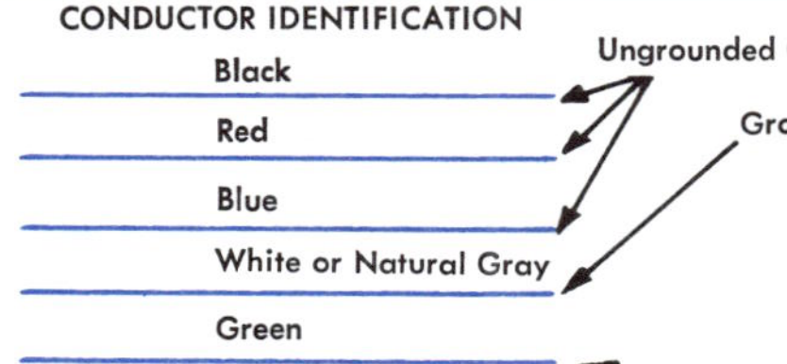

NEC 310-12 (c) Ungrounded conductors shall be identified by colors, or a combination of colors plus a distinguishing marking, other than white, natural gray, or green. These colors shall clearly distinguish the ungrounded conductor from the grounded or grounding conductor and shall not conflict with markings required by 310-11 (b-1).

NEC 210-5(a) The grounded conductor of a branch circuit shall be white or natural gray. Each other system in same raceway or enclosures shall have the grounded conductor (if required) identified by outer covering of white with colored stripe (not green).

NEC 310-12(a) Grounded conductors #6 or smaller shall be white or natural gray. For sizes over #6, see 200-6. Also see 210-5.

NEC 210-5 (b) Any conductor intended for grounding purposes shall be green, green with one or more yellow stripes, or bare.

210-4 (d) Specific color code requirements for ungrounded conductors are not provided, but if there is more than one nominal voltage system in a building, each systems' ungrounded conductor color code shall be permanently posted at each panelboard.

BRANCH CIRCUITS -- PERMITTED LOADS

NEC 210-23 Individual branch circuits may supply any loads.

NEC 210-21 (b-1) A single receptacle on an individual branch circuit shall have rating not less than that of the branch circuit.

NEC 210-23 Individual branch circuits may supply any loads. Branch circuits having two or more outlets may supply only loads as follows: (a) 15- and 20-amp circuits may supply lighting units and/or appliances, the rating of any cord-and-plug connected appliance not exceeding 80 percent of circuit, and total rating of fixed appliances not over 50 percent; (b) 30-amp circuits may supply heavy-duty lampholders in other than dwellings, or appliances anywhere, the rating of a cord-and-plug connected appliance not exceeding 24 amps; (c) 40- or 50-amp branch circuits may supply fixed cooking appliances in any occupancy or fixed lighting units with heavy-duty lampholders or infrared heating units or other utilization equip. in other than dwellings; (d) circuits over 50 amps to supply only non-lighting outlet loads.

SPACE HEATING PANELS (NEC 424, PART I)

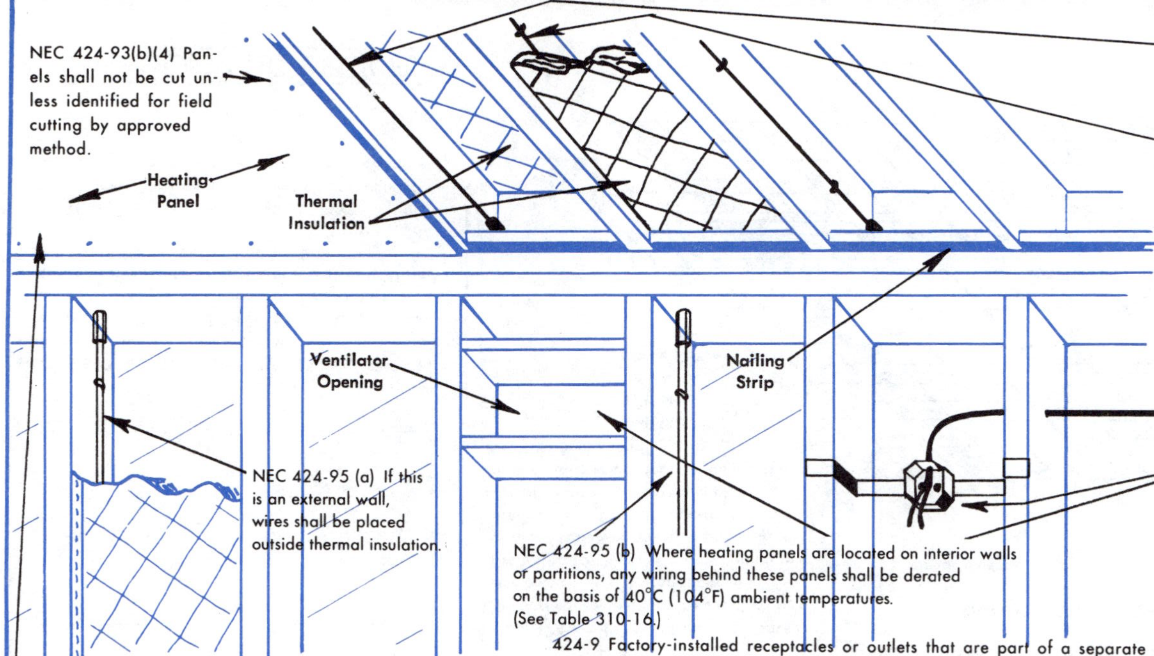

NEC 424-93(b)(4) Panels shall not be cut unless identified for field cutting by approved method.

NEC 424-94 Wiring installed above heated ceiling shall be considered as operating at an ambient temperature of 50 degrees C and shall be derated accordingly. Wiring shall be spaced at least 2 inches above heated ceiling.

NEC 424-94, Exception: Wires located above thermal insulation which is at least 2 inches thick need not be derated.

NEC 424-92(3) Each panel shall be marked with identifying name or symbol, catalog number, volts and watts, or volts and amps.

NEC 424-92(1) Panels to be marked in a location readily visible before finish is applied.

NEC 424-93 (a-3) Panels shall be separated at least 8 inches from outlet and junction boxes for surface fixtures and at least 2 inches from any part of recessed fixtures, ventilating or other room openings. Surface fixtures shall not be mounted under heating panels.

NEC 424-91 (a) Heating panel – complete assembly with junction box or flexible conduit. (b) Heating panel set – rigid or nonrigid assembly with nonheating leads or terminal junction suitable for connection to wiring system.

NEC 424-95 (a) If this is an external wall, wires shall be placed outside thermal insulation.

NEC 424-95 (b) Where heating panels are located on interior walls or partitions, any wiring behind these panels shall be derated on the basis of 40°C (104°F) ambient temperatures. (See Table 310-16.)

424-9 Factory-installed receptacles or outlets that are part of a separate listed assembly and are permanently installed in electric baseboard heaters are permitted in place of those required by 210-50 (b).

NEC 424-93 (a) (2) (b) Panels not permitted through or above partitions, walls, cupboards or similar parts of building.

SPACE HEATING CABLES (NEC 424, PART E)

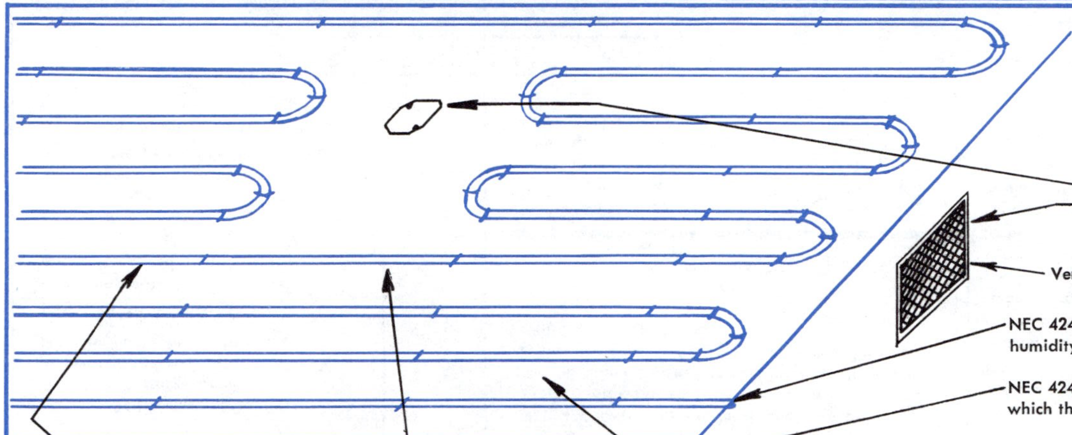

NEC 424-35 Each cable shall be marked with identifying name or symbol, catalog number, volts and watts, or volts and amps.

NEC 424-35 Each cable shall have permanent marking on each nonheating lead within 3 inches of terminal end. The lead wire shall be identified by: 120 V nominal, yellow; 208 V nominal, blue; 240 V nominal, red; 277 V nominal, brown; and 480 V nominal, orange.

NEC 424-39 Cables shall be separated at least 8 inches from outlet and junction boxes, surface fixtures and at least 2 inches from any part of recessed fixtures, ventilating, or other room openings. Surface fixtures shall not be mounted under heating cables.

NEC 424-38 (b) (c) Cables shall not be installed in closets except for low heat to control humidity, provided they are located above an unobstructed area to the floor.

NEC 424-38 (a) (b) Cables shall not extend through walls or extend beyond the room in which they originate. Exception: A single run may pass over partition.

NEC 424-41 (c) Heating cables shall be applied only to gypsum board, plaster lath, or other nonconducting surfaces. With metal lath or other conducting surfaces, a covering layer of plaster (brown or scratch coat) shall be first applied to the surface.

NEC 424-41 (e) The entire ceiling surface shall have a finish of thermally noninsulating material whose nominal thickness is ½ inch, or other noninsulating material identified as suitable for this use and installed as per specifications.

NEC 424-41 (h) Cables shall not contact metal or other conducting surfaces.

NEC 424-41(b) Adjacent runs for cable not exceeding $2\frac{3}{4}$ VA per foot shall not be installed less than $1\frac{1}{2}$" on centers.

NEC 424-41 (f) Cables shall be secured at intervals not exceeding 16 inches by means of staples, tape, plaster, or other approved means. Metal fasteners shall not be used with metal lath or other conducting surfaces. (See Exception.)

NEC 424-41 (a) Cables not permitted in walls except single runs to reach a dropped ceiling.

NEC 424-45 Cables shall be inspected and approved before covered or concealed.

NEC 110-2 Installation acceptable only when approved. (See Definitions for <u>approved</u>.)

SPACE HEATING CABLES

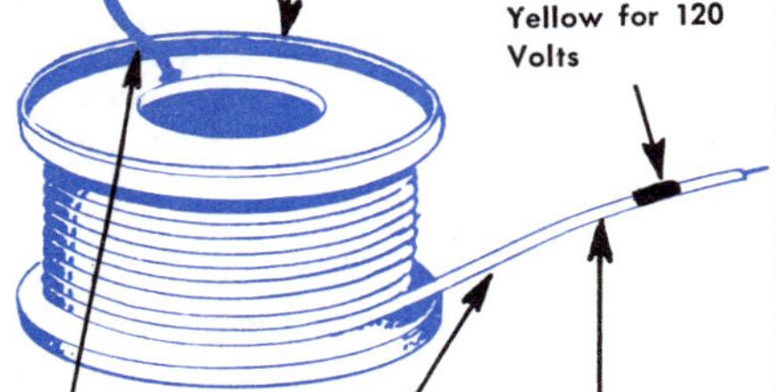

NEC 424-34 Heating cables shall have factory-assembled nonheating, general-use wire leads at least 7 feet in length.

NEC 424-35 Nonheating leads of heating cable shall be marked and color coded: 120 V yellow, 208 V blue, 240 V red, 277 V brown, and 480 V orange.

NEC 424-43 (a) Free nonheating leads from a junction box to a location within the ceiling shall be installed with approved wiring methods, such as: conductors in raceways, Type UF, NMC, MI cable, or other approved conductors.

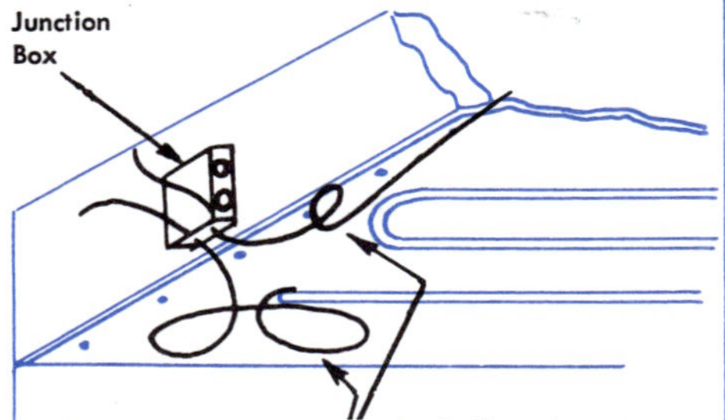

NEC 424-43 (c) Excess leads shall not be cut. Excess lead wires shall be embedded in plaster. Not less than 6 inches of free leads shall be left in junction box, and markings on leads shall be visible in junction box.

NEC 424-40 Embedded cables may be spliced where necessary, by approved means. In no case shall the length of the heating cable be altered.

FIXED OUTDOOR ELECTRIC DE-ICING AND SNOW MELTING EQUIPMENT

NEC 426-23 (a) Nonheating power supply leads to be in junction boxes with 6 inches of free conductor. (Also see **426-22** (e).)

NEC 426-20(a) Not over 120 VA per sq ft.

NEC 426-22 (a) (b) Nonheating leads shall have additional protection unless they have a metallic grounding sheath or braid.

426-4 Branch circuit conductors supplying fixed electric deicing and snow-melting equipment shall have an ampacity of at least 125% of total load of heaters.

NEC 426-27 (a) All noncurrent-carrying, exposed metal parts likely to be energized shall be grounded as per Article 250. (c) All non-current-carrying metal parts likely to be energized to be bonded and connected to panelboard with an equipment grounding conductor, sized as per Table 250-95.

NEC 426-20(e) Not permitted to bridge expansion joints unless expansion and contraction are provided for.

426-28 Branch circuits for outdoor fixed electric deicing and snow-melting equipment shall have ground fault protection of equipment.

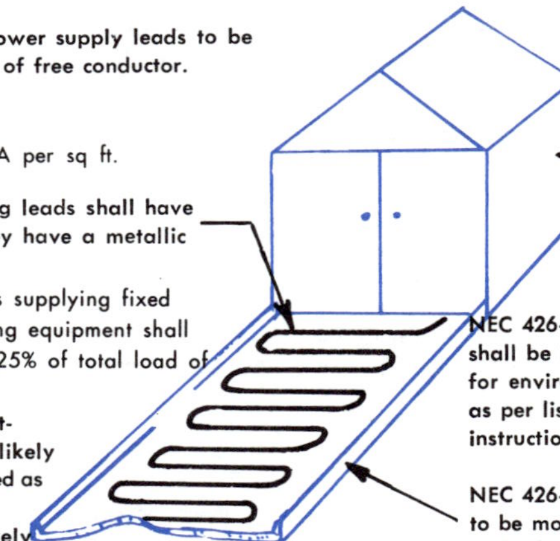

NEC 426-10 (1) (2) Material shall be identified as suitable for environment and installed as per listing and installation instructions.

NEC 426-20 (c) (1) (2) (3) Cable to be mounted on 2-inch pad with 1½-inch overpour, or within 3½-inch pad at least 1½-inch cover, or specially investigated cable.

THERMOSTATIC DEVICES

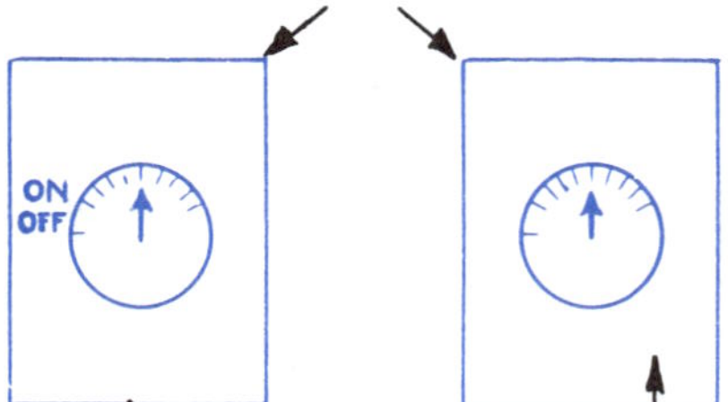

NEC 424-20 (a) Thermostatic devices permitted as both controller and disconnect means when located as required by **424-19**, and when all ungrounded conductors are opened when manually placed in the OFF position, and circuit cannot be automatically closed when manually placed in OFF position.

NEC 424-20(b) If these devices do not have "on" or "off" positions, they need not open all ungrounded conductors, and shall not be acceptable as disconnect means.

NEC RULES DEALING WITH PANELS OR CABLES IN CONCRETE OR POURED MASONRY FLOORS (424-44)

NEC 424-3(a)(b) Fixed electric space heating equipment may be supplied by any size branch circuit. Branch circuits limited to 15, 20, or 30 amps, where two or more outlets for fixed space heating are installed. Exception permits 50 amp circuit to infrared heaters in other than residential. Branch circuit sized per continuous load.

NEC 424-44(e) Leads shall be protected where they leave the floor by RMC, IMC, PVC, EMT, or by other approved means.

NEC 424-44 (f) Bushings or approved fittings shall be used in the floor slab where the leads emerge.

424-44(c) Not permitted to bridge expansion joints unless expansion and contraction are provided for.

NEC 424-44 (c) Cable shall be secured in place by nonmetallic frames or spreaders or other approved means while the concrete or other finish is applied.

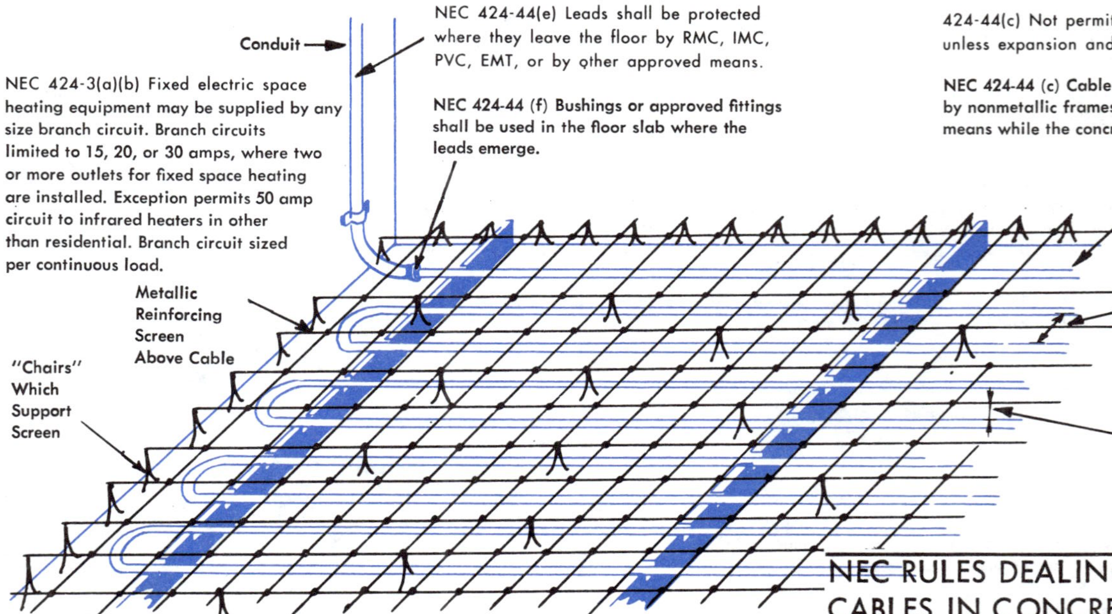

NEC 424-45 Installations to be inspected and approved before covered or concealed.

NEC 424-44(a)(b) Constant-wattage heating cables shall not exceed 16½ VA per linear foot of cable and shall not be less than 1" on centers.

NEC 424-44 (d) Between the heating cable and any metal in the slab, a spacing shall be maintained except for grounded metal-clad cable.

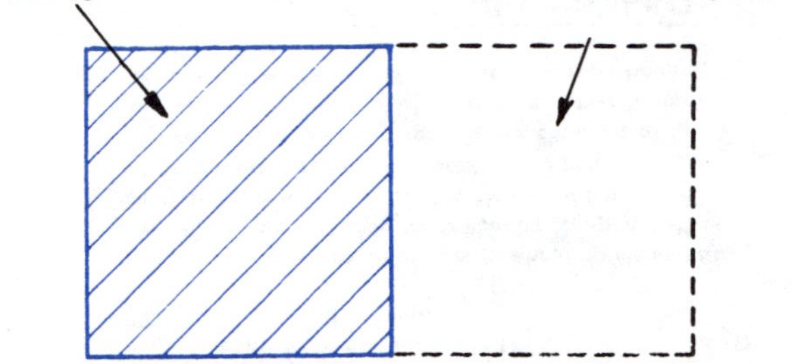

NEC 220-3(d)(1) When calculating loads in existing dwellings for previously unwired portions or new additions, either of which exceeds 500 sq ft, the VA-per-square-foot basis shall be used. For new circuits or extended circuits, the VA-per-square-foot, or the actual total load served per 220-3(c) shall be used.

NEC 220-3(d-2) In other than dwellings, adding new or extending existing circuits to be calculated as per 220-3(b, c)

CALCULATING LOADS -- BUILDING ADDITIONS

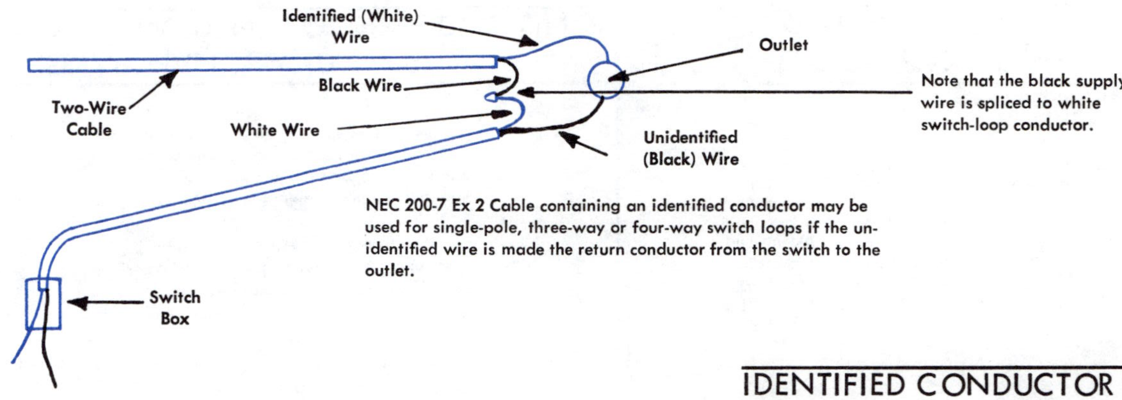

NEC 200-7 Ex 2 Cable containing an identified conductor may be used for single-pole, three-way or four-way switch loops if the unidentified wire is made the return conductor from the switch to the outlet.

IDENTIFIED CONDUCTOR USED AS SWITCH LOOP (200-7, EX 2)

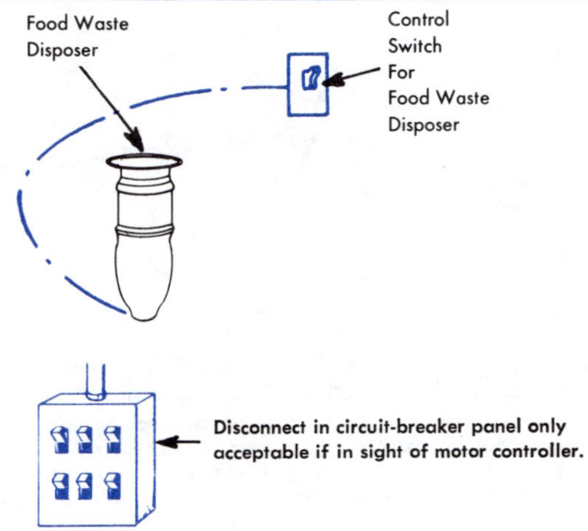

NEC 422-27 The disconnecting means for a motor-driven appliance of more than ⅛ hp shall be located within sight of the motor controller and shall comply with Part H of Article 430.

NEC 422-27, Ex. Where appliance unit switch disconnects all ungrounded conductors, and switch is marked with "off" position, then switch or circuit breaker permitted to be out of sight, only if conforming to 422-25(a, b, c, d).

DISCONNECT MEANS -- MOTOR APPLIANCE (OVER 1/8 H.P.) (422-27)

NEC 424-65 Duct heater controller equipment shall be accessible with disconnect at or within sight of controller. (Ex see 424-19 (a).)

NEC 424-58 Heaters installed in air ducts must be identified as suitable for such use.

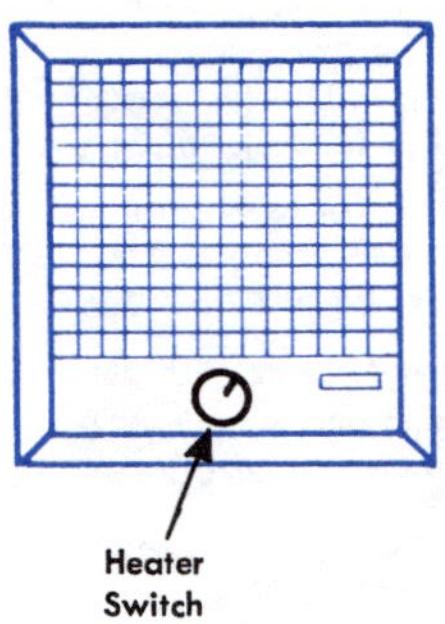

NEC 424-11 Fixed electric space heaters requiring supply conductors over 60°C shall be clearly and permanently marked.

NEC 424-13 Unless approved for direct contact with combustible material, fixed heating equipment must have required spacing.

NEC 424-12(a)(b) Fixed electric space heating equipment shall not be used: (a) where exposed to severe physical damage, unless adequately protected, nor (b) in damp or wet locations, unless approved for such locations.

ELECTRIC HEATERS -- INSTALLATION

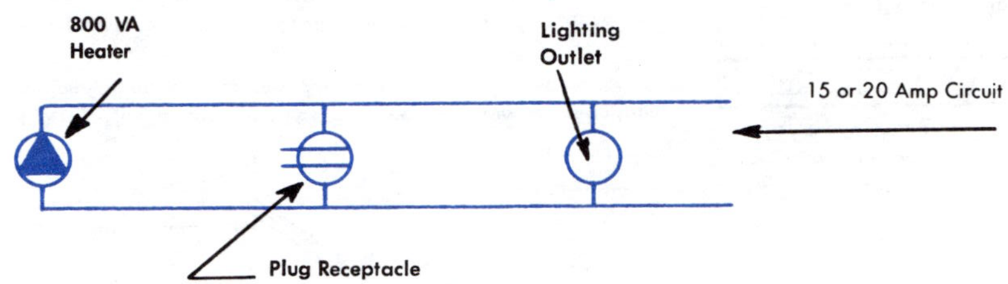

NEC 422-28(a) Appliances (other than motors) shall be considered as protected against overcurrent when supplied by branch circuits of 422-4, 422-5, and 422-28(e). Exception: For motors see NEC 430 and 440.

APPLIANCES -- BRANCH CIRCUIT PROTECTION (422-28)

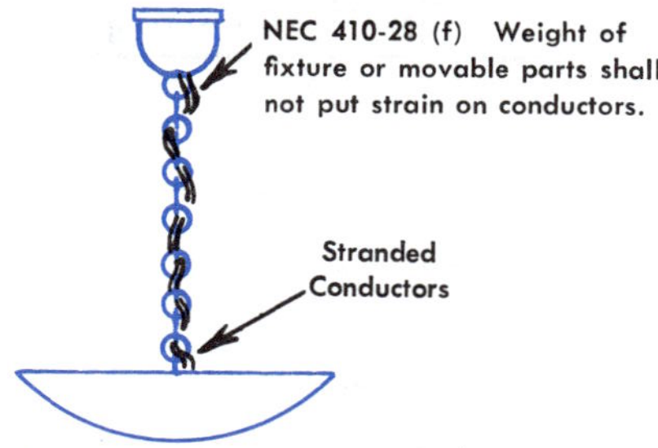

NEC 410-28 (e) Stranded conductors shall be used on chain fixtures and other movable parts.

NEC 410-17 Fixtures required to be grounded.

NEC 410-20 All fixtures with exposed metal parts shall provide means to connect equipment grounding conductor to the fixture.

410-4(d) No parts of pendants, cord-connected fixtures, hanging fixtures, lighting track, or ceiling fans shall be located within a 3′ horizontal and 8′ vertical zone from top of bathtub rim. For purposes of this section, the zone is all encompassing and shall include space directly over tub.

CHAIN SUPPORTED FIXTURES (410–26)

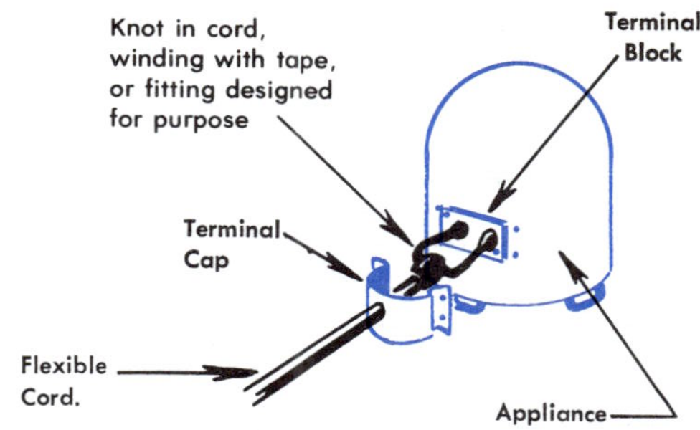

NEC 400-10 Flexible cords shall be so connected to devices and fittings that tension will not be transmitted to joints or terminal screws.

NEC 422-8(c)(1), 400-7(a)(6) Flexible cord permitted for connection to appliances to facilitate frequent interchange.

NEC 210-23(a) One cord-and plug appliance on a 15- or 20-amp ckt. not to exceed 80% of ckt. rating. Total fastened in place appliances not to exceed 50% of circuit rating if lights and other portable equip., or both are supplied.

CORD CONNECTED APPLIANCES

NEC 422-15(b) Infrared lamps over 300 VA shall not use screw-shell lampholders. See Exception.

NOT OVER 300 VA

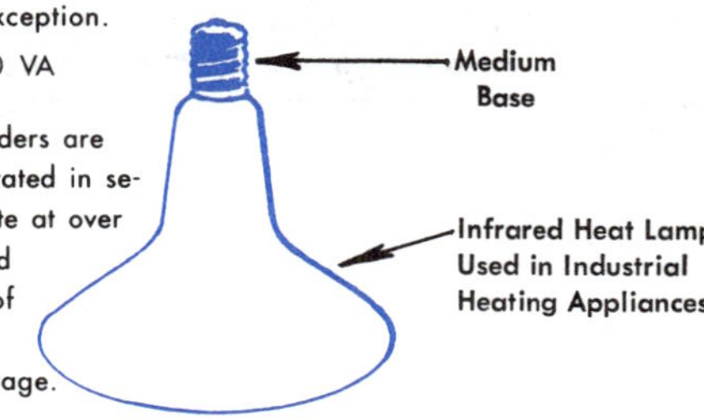

422-15 (c) Lampholders are permitted to be operated in series if circuits operate at over 150 V to ground and the voltage ratings of lampholders is not less than circuit voltage.

NEC 422-15(a) Infrared heating lamps rated at 300 VA or less may be used with lampholders of the medium-base unswitched type, or other types identified for the purpose.

INFRARED LAMPS -- INDUSTRIAL HEATING

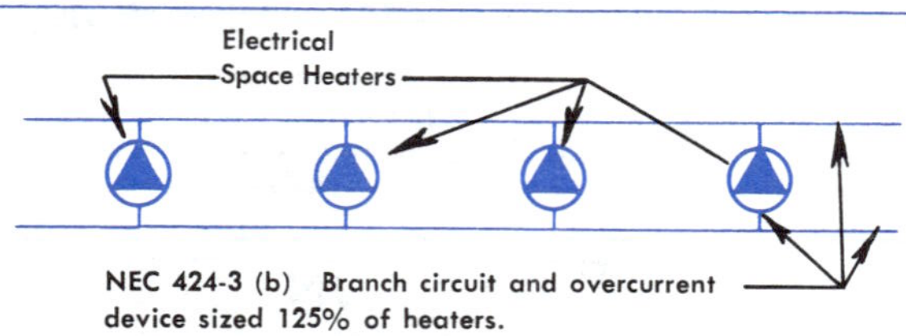

NEC 220-15 The computed load of a feeder supplying fixed electrical space heating equipment shall be the total load connected thereto unless duty-cycle features are present.

NOTE: NEC 220-15 does not apply when feeder capacity is calculated in accordance with optional methods in NEC 220-30, 220-31 for single dwelling or single apartment in multifamily dwelling or in NEC 220-32 for multifamily dwellings.

NEC 424-22 (b) When resistance-type heating elements of over 48 amps are used in space heating equipment, they shall be subdivided into smaller circuits of not over 48 amps and protected at not more than 60 amps. The rating of the overcurrent device shall be as per 424-3 (b) where subdivided load is less than 48 amps. See 424-72 (a) for resistance type boilers.

SPACE HEATING EQUIPMENT OVERCURRENT PROTECTION

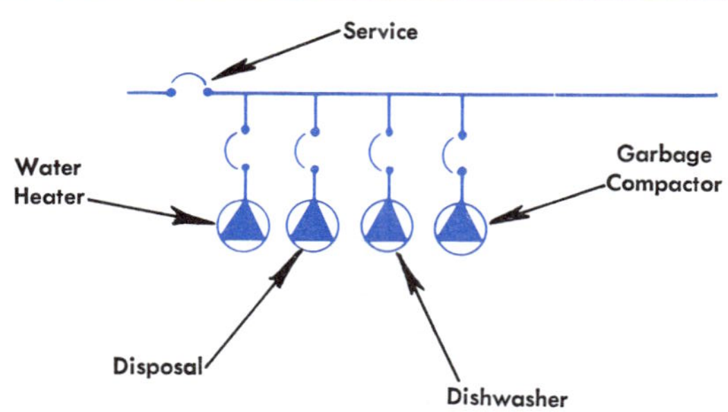

NEC 220-17 Where four or more fixed appliances, other than electrical ranges, clothes dryers, air conditioning equipment, or space heating equipment are connected to the same feeder in a one-family, two-family, or multifamily dwelling, a demand factor of 75% may be applied to the fixed appliance load.

NEC 220-21 In making load calculations, the smaller of two dissimilar loads may be omitted from the total where it is unlikely that both loads will be served simultaneously.

FIXED APPLIANCE AND NONCOINCIDENT LOADS -- DWELLINGS

NEC 220-19 Thirty 12 kW ranges supplied by a 3ϕ, 4-wire, 120/208-volt feeder, equal number of ranges on each feeder.

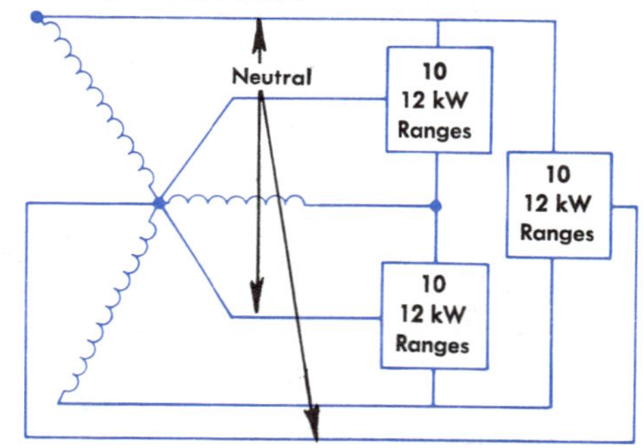

NEC 220-19 Where single-phase ranges are supplied by a 3-phase, 4-wire feeder, the load shall be computed (as per Table 220-19) on a demand basis of twice the maximum number between any two phases.

Method No. 1
10 12 kW ranges per phase × 2 = 20 ranges
Table 220-19 gives 35 kW.
35 kVA ÷ 2 = 17,500 × 3 = 52,500 VA
Method No. 2
10 × 2 = 20 ranges
Table 220-19 gives 35 kW
35 kVA ÷ 2 = 17,500 × 3 = 52,500 VA
52,500 ÷ (208 × 1.73) = 146 A

ELECTRIC RANGES -- 3Ø FEEDER DEMAND

GROUNDING -- RANGES AND SIMILAR APPLIANCES

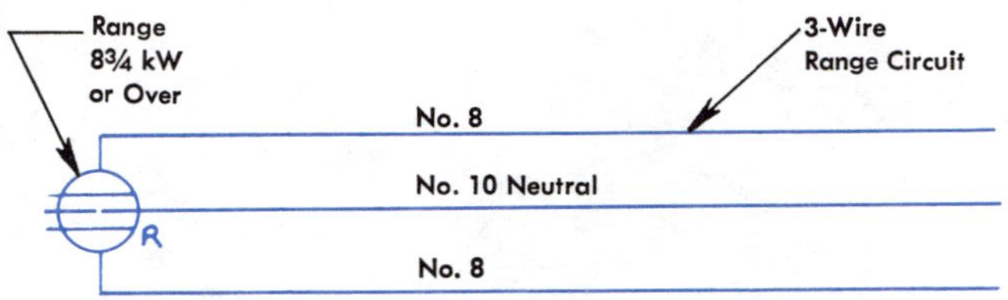

NEC 210-19(b) Conductors for ranges 8¾ kVA and larger shall not be rated less than 40 A.
NEC 210-19(b), Ex. 2 The neutral conductor for a household electric range may have an ampacity equal to 70% of that of the branch-circuit rating, but shall not be smaller than No. 10 in any case.

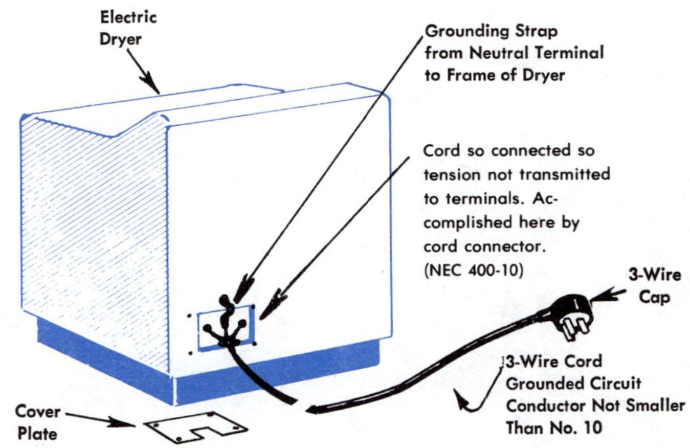

250-60 For new branch-circuit installations, frames of electric ranges, cook-tops, ovens, dryers, and their associated outlet or junction boxes shall be grounded the same as other electrical equipment. See 250-57 and 250-59. For existing branch-circuit installations, the grounded circuit conductor (neutral) may be used as the EGC, if all following conditions are met: (1) circuit is 120/240 V, 1ϕ, 3-wire, or 120/208 V derived from a 3ϕ, 4-wire system; (2) grounded conductor is #10 or larger; (3) grounded conductor is insulated or is uninsulated and part of an SE cable and is fed from service equipment; (4) grounded contacts of receptacles are bonded to equipment. This does not apply to mobile homes or travel trailers.

AIR CONDITIONING EQUIPMENT

NEC 110-3 (b) Equipment that is listed or labeled shall be installed or used as per instructions included in the listing or labeling. (For additional information, see UL standard for Safety, Central Cooling Air Conditioners, No. 465-Paragraph 56.25.)

Where proper fuse is used in panel, then non-fused disconnect or circuit breaker can be used. When nameplate states that unit is to be protected by "overcurrent device" or "circuit breaker or fuse", then a HACR (Heating, Air-Conditioning and Refrigeration) listed breaker is OK.

Where fuses are indicated in installation instructions and on nameplate, then fuses shall be used somewhere in circuit.

NEC 440-14 Disconnect means within sight and readily accessible from unit. Permitted on or within equip. See 430-107 for readily accessible. Ex. 2 cord- and plug-connected appliances.

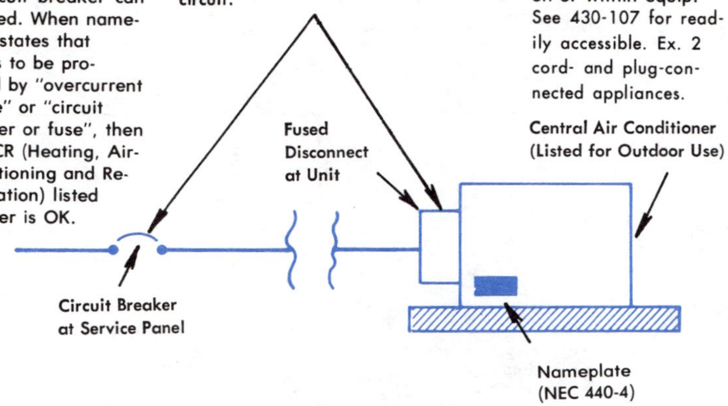

TYPICAL NAMEPLATE INFORMATION

	HP	VOLTS	PHASE	HZ	FLA	LRA
COMPRESSOR		230	1	60	15.3	71
FAN	¼	230	1	60	2.0	4.7

Fuse Size—Time Delay 25 Amp Max 35A ·

ALSO on nameplate "Listed Section of UL 465N, Central Cooling Air Conditioner." ALSO: Installation and operating instructions list fuses for overcurrent protection.

NEC 440-12 (a) Disconnect selected on basis of nameplate rated-load current, or branch-circuit selection current (whichever is greater), and locked-rotor current: (a-1) Ampere rating at least 115% of FLA where non-horsepower disconnect (such as circuit breaker) is used. (a-2) To determine size of hp rated disconnect, as required by 430-109, the hp corresponding to the FLA in Tables 430-148, 149, 150; the hp corresponding to the LRA in Table 430-151, where different hp ratings are obtained, the largest hp shall be used.

NEC 440-61 Grounding of room air conditioning units shall be in accordance with 250-42, 250-43, and 250-45.

NEC 250-42 (a) (b) (c) (d) (e) (f) Exposed noncurrent-carrying metal parts of fixed equipment shall be grounded if: (a) within 8 feet vertically or 5 feet horizontally of ground or grounded objects and possible contact by persons, (b) in damp or wet location and not isolated, (c) in electrical contact with metal, (d) in hazardous location, (e) if fed by metal enclosed wiring (except for 25 feet of isolated raceway as per 250-33, (f) See Exceptions if over 150 V to ground.

NEC 250-45 Exposed metal noncurrent-carrying parts of cord- and plug-connected equipment shall be grounded if: (a) in hazardous location, (b) See Exceptions if over 150 V to ground, (c) in residential occupancies: refrigerators, freezers, air conditioners, clothes washers, clothes dryers, dishwashers, sump pumps, and portable tools unless double insulated. See 250-45(d) for grounding of cord and plug equipment in other than residential occupancies.

NEC 440-62 (c) The total load of cord-and-plug room air conditioning equipment shall not exceed 50% of the rating of a branch circuit which also supplies lighting units or appliances.

440-62 (a) For determining branch-circuit requirements, a room air conditioner shall be considered a single motor when it is (1) cord- and plug-connected, and (2) rating is less than 40 A at 250 V, 1ϕ and (3) total rated-load current is provided on nameplate, and (4) branch-circuit, short-circuit, and ground-fault protective device is less than ampacity of branch-circuit conductors or receptacle rating.

NEC 440-63 A plug and receptacle may serve as disconnecting means for a single-phase air conditioner rated 250 volts or less when the manual controls are readily accessible and within 6 feet of floor, or a manually operable switch is readily accessible and within sight of unit.

NEC 440-14 Disconnect within sight and readily accessible for air conditioning or refrigerating equipment except for appliances connected by cord and plug.

NEC 440-62(b) The rating of a cord and plug connected room air conditioner shall not exceed 80% of its individual branch circuit rating where no other loads are supplied.

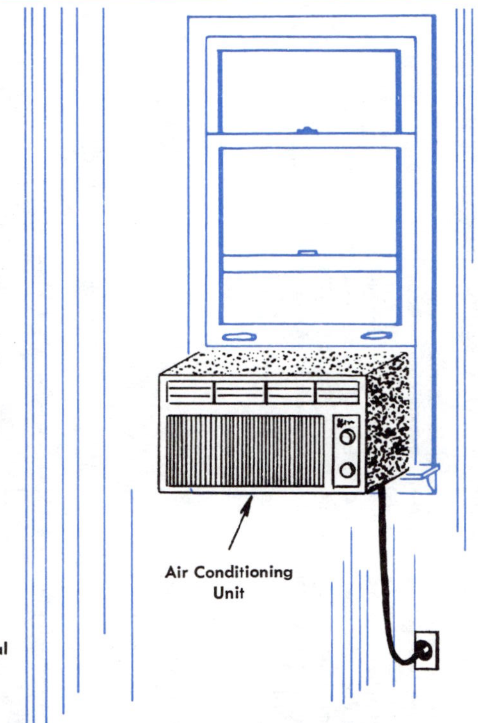

NEC 440-64 Supply cords for room air conditioners shall not exceed 10 ft at 120 volts or 6 ft at 208 or 240 volts.

NEC 440-60 Generally a room air conditioner must be in the window or wall of room being cooled, incorporate a hermetic-refrigerant motor-compressor, and operate at not over 250 V, 1ϕ. A 3ϕ unit rated over 250 V shall be directly wired with a recognized wiring method.

VOLTAGE DROP FORMULAS

The K factor (constant) is the resistance of a circular-mil-foot of wire (a wire 1′ long and having a cross-sectional area of 1 kcmil). The resistance increases with the temperature from 10.4 at 20°C (68°F) to 10.75 at 25°C (77°F). The American Electricians Handbook, The Standard Handbook for Electrical Engineers and other engineering data (also refer to Chapter 9, Table 8) lists the resistance of copper and aluminum conductors in ohms per 1000′ at 25°C (77°F). Based on this temperature, a K (constant) of 10.7 for copper and 17.4 for aluminum conductors is used.

K= 10.7
VD = voltage drop
L = length of circuit in feet (one way)
I = current
CM = circular mil area of conductor

$VD = \frac{2KLI}{CM}$ (single phase circuit)

$VD = \frac{KLI\ 1.73}{CM}$ or $VD = \frac{2KLI\ 0.866}{CM}$ (3-wire, 3-phase circuit or 4-wire, 3-phase wye circuit with balanced load)

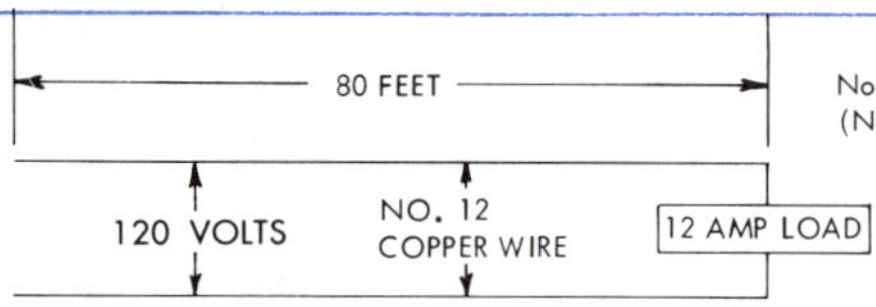

No. 12 wire has 6530 cir-mils (NEC Chapter 9, Table 8)

$VD = \frac{2KLI}{CM} = \frac{10.7 \times 160 \times 12}{6530} = 3.1$ volts $\%VD = 3.1 \div 115 = 2.7\%$

Check by using a resistance of 1.62 ohms per M/ft.

VD = IR R = 0.00162 Ω x 160 ft = 0.2592

VD = 12 x 0.2592 = 3.1 volts

THIS CIRCUIT HAS A VOLT DROP OF 3.1 VOLTS OR 2.7% AT 115 VOLTS WITH A 12 AMP LOAD.

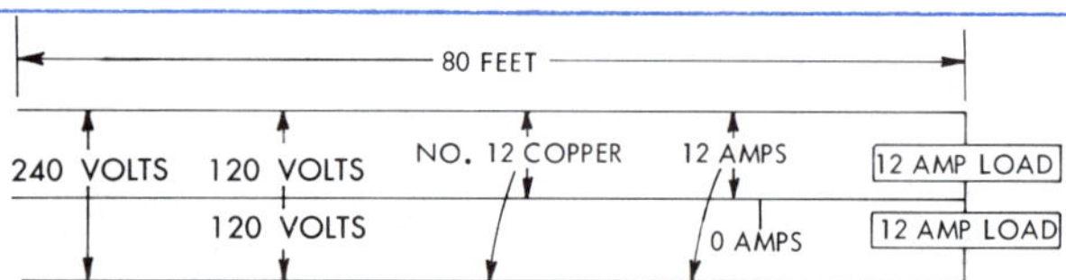

$VD = \frac{2KLI}{CM} = \frac{10.7 \times 160 \times 12}{6530} = 3.1$ VOLTS $\%VD = 3.1 \div 240 = 1.35\%$

THIS CIRCUIT HAS A VOLTAGE DROP OF 3.1 V OR 1.35% AT 120/240 V WITH A 12 A, 120/240 V BALANCED LOAD. THE NEUTRAL CARRIES NO CURRENT.

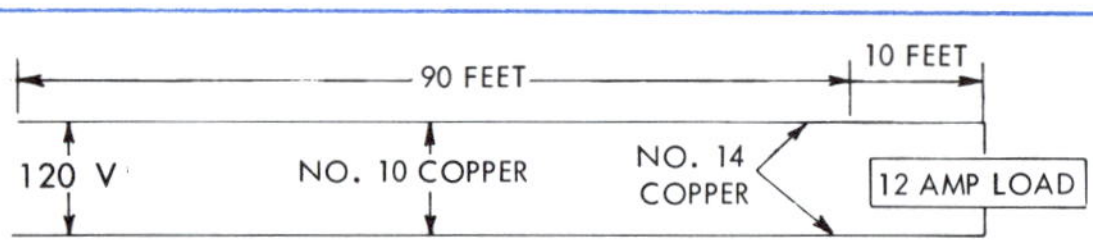

$VD = \frac{2KLI}{CM} = \frac{10.7 \times 180 \times 12}{10380}$

$= \frac{23112}{10380} = 2.2$ V

Check by using 1.018 ohms per M/ft.

VD = IR = 12 x .183 (.001018 x 180)
= 2.2V

$VD = \frac{2KLI}{CM} = \frac{10.7 \times 20 \times 12}{4110}$

$= \frac{2568}{4110} = .62$ VD

Check by using 2.57 ohms per M/ft.

VD = IR = 12 x .0514 (.00257 x 20)
= .62 VD

TOTAL VD = 2.2 + .62 = 2.82 VD
TOTAL % VD = 2.82 ÷ 120 V = 2.5% VD

Three-phase, 120/208 Volt (wye), 4-wire feeder with 45 amps per phase with balanced load.

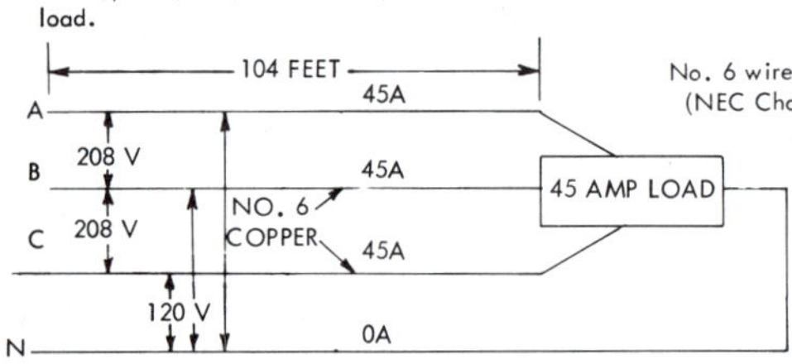

No. 6 wire has 26240 cir-mils (NEC Chapter 9, Table 8)

See page 186 this book for formulas of unbalanced load.

215-2, FPN No. 2 (1) For feeders, maximum allowable voltage drop should not exceed 3% at farthest outlet. (2) For branch circuits and feeders, overall voltage drop at farthest outlet should not exceed 5%.

TO FIND VOLTAGE DROP when the length of circuit, resistivity, and current are known, use the formula:

$$VD = \frac{2 \times L \times R \times I}{1000}$$

where
VD = voltage drop
L = length (in feet)
R = resistivity per Table 8, Chapter 9
I = current
1000 = 1000′ or less of wire

If the circuit is 104′, the resistivity is .491 ohms, and the current is 45 A, what is the voltage drop?

$$VD = \frac{2 \times L \times R \times I}{1000}$$

$$VD = \frac{2 \times 104 \times .491 \times 45}{1000}$$

$$VD = \frac{4595.76}{1000}$$

VD = **4.59 V**

TO FIND PERCENT VOLTAGE DROP divide the voltage drop by the volts in the circuit. Use the formula:

$$\%VD = \frac{VD}{V} \times 100$$

$$\%VD = \frac{4.59}{208} \times 100$$

%VD = **2.2%**

TO FIND CONDUCTOR SIZE when the voltage drop, length of circuit, and current are known:

$CM = \frac{2 \times K \times L \times I}{VD}$ (single-phase, 2-wire circuit)

$CM = \frac{2 \times K \times L \times I \times .866}{VD}$

TO FIND CURRENT when the voltage drop, length of circuit and conductor are known:

$I = \frac{VD \times CM}{2 \times K \times L}$ (single-phase, 2-wire circuit)

$I = \frac{VD \times CM}{K \times L \times 1.73}$ (three-phase circuit)

or $I = \frac{VD \times CM}{2 \times K \times L \times .866}$

TO FIND LENGTH OF CIRCUIT when the voltage drop, conductor size and current are known:

$L = \frac{VD \times CM}{2 \times K \times I}$ (single-phase circuit)

$L = \frac{VD \times CM}{K \times I \times 1.73}$ (three-phase circuit)

or $L = \frac{VD \times CM}{2 \times K \times I \times .866}$

TO FIND THE RESISTANCE of a conductor

L = length of conductor
R = resistance
K = constant (10.7 for copper, 17.4 for aluminum)

$R = \frac{KL}{CM}$; $L = \frac{R \times CM}{K}$; $CM = \frac{K \times L}{R}$

Trade Competency Test 5

Name ______________________ Date ______________

Section	Answer	
______	______	**1.** The minimum size of a copper equipment grounding conductor required for equipment connected to a 40 A circuit is No. _____. A. 8 C. 12 B. 10 D. 14
______	______	**2.** Splices that will be soldered shall first be made mechanically and _____ secure.
______	T F	**3.** If the resistance of a driven ground rod exceeds 25 Ω to ground, an additional rod shall be added.
______	______	**4.** The smallest permissible diameter of a bronze ground rod is _____″.
______	______	**5.** For a service run of rigid metal raceway interrupted by an expansion joint, the sections of rigid conduit shall be bonded by a(n) _____ jumper or other means.
______	______	**6.** The minimum branch circuit rating to a single electric range of over $8\frac{3}{4}$ kVA is _____ A.
______	T F	**7.** Stranded wires No. 10 and smaller may be connected to equipment by means of screws used with terminal plates having upturned lugs.
______	______	**8.** Double lockouts are required where rigid metal raceways in other than service runs enter a box or cabinet, if the raceway contains any wire of more than _____ V to ground.
______	T F	**9.** The grounding conductor in a cable assembly for cord-and-plug connected equipment need not be insulated.
______	______	**10.** The three wires of a single-phase, 120/240 V service are designated as *A*, *N*, and *B*. A 100 A, 240 V load is connected across *A* and *B*. A 50 A, 120 V load is connected across *A* and *N*. The neutral conductor shall be No. _____ TW copper. A. 4 C. 8 B. 6 D. 10
______	T F	**11.** A ground rod used to ground an electrical system shall be at least 8′ from a ground rod used to ground a lightning protection system.
______	______	**12.** The neutral feeder conductor shall be capable of carrying the maximum _____ load.
______	______	**13.** The minimum area of exposed surface offered by a plate electrode shall be _____ sq ft. A. 1 C. 3 B. 2 D. 4
______	______	**14.** Conductors run above heated ceilings and within thermal insulation shall be derated on the basis of _____°C ambient temperature.

_________ _________ **15.** Receptacles connected to a 30 A branch circuit supplying two or more outlets shall be rated not less than _________ A.

_________ _________ **16.** Noncurrent-carrying metal parts of cord-and-plug connected equipment may be grounded by a(n) _________.

_________ T F **17.** Iron wire shall not be employed for bonding jumpers.

_________ _________ **18.** A 15 A or 20 A receptacle can be installed on a _________ A general purpose circuit.

A. 20 | C. 30
B. 25 | D. none of the above

_________ _________ **19.** The maximum standard ampere rating of a fuse or inverse time circuit breaker shall be considered to be _________ A.

_________ T F **20.** In general, switch enclosures shall not be used for junction boxes.

_________ _________ **21.** If a change occurs in the size of the _________ conductor, a similar change may be made in the size of the grounded conductor.

_________ _________ **22.** Leads of 240 V heating cables are marked with _________.

_________ _________ **23.** The maximum permissable open-circuit voltage of electric discharge lighting equipment used in a dwelling is _________ V.

A. 500 | C. 1500
B. 1000 | D. 2000

_________ _________ **24.** _________ are permitted around service conduits entering concentric or eccentric knockouts.

_________ T F **25.** A single 15 A receptacle may be installed on an individual 20 A branch circuit.

_________ _________ **26.** A metal elbow for a service raceway installed at least _________″ underground does not have to be grounded.

_________ _________ **27.** In a 1-phase, 3-wire, AC system, the _________ shall be grounded.

_________ _________ **28.** Ungrounded conductors shall be clearly distinguishable from _________ and grounding conductors by color or other marking.

_________ _________ **29.** Ground rods are permitted to be buried if the trench is at least _________′ deep where rock bottom is encountered.

_________ _________ **30.** The minimum diameter of a steel ground rod shall be _________″.

A. 3⁄8 | C. 5⁄8
B. 1⁄2 | D. 3⁄4

_________ _________ **31.** Flexible cord shall not be used as a substitute for _________ wiring.

_________ _________ **32.** Unless special precautions are taken, exposed live parts operating at _________ V or more shall be guarded against accidental contact.

_________ _________ **33.** In dwelling units, equipment in which live parts are exposed while changing lamps shall have an open-circuit voltage not exceeding _________ V.

_________ _________ **34.** A single grounding electrode is permitted when the resistance to ground does not exceed _________ ohms.

A. 5 | C. 15
B. 10 | D. 25

_________ _________ **35.** Heating cables on ceilings shall be kept free from contact with metal or other _________ surfaces.

Trade Competency Test 6

Name ______________________ Date ______________________

Section	Answer		
______	T F	1.	The metal well casing enclosing a submersible pump shall be bonded to the EGC of the pump circuits.
______	______	2.	Fittings and connectors shall be designed and ______ for the specific wiring method with which they are used.
______	______	3.	Heating panels shall be separated from outlet boxes to be used for mounting fixtures by not less than ______″. A. 6 B. 8 C. 10 D. 12
______	T F	4.	Not less than 10′ of No. 4 bare copper conductor encased under the concrete footing in contact with the earth is an approved, made electrode.
______	______	5.	A dwelling unit with a computed load of 12 kVA shall be equipped with a(n) ______ A service.
______	T F	6.	Type NM nonmetallic-sheathed cable shall be permitted to be used in storage battery rooms.
______	______	7.	The earth shall not be used as the ______ equipment grounding means.
______	______	8.	The main bonding jumper in a panelboard may be a screw, a bus, or a(n) ______.
______	______	9.	A 6000 A overcurrent protection device requires that the equipment grounding conductor be rated at ______ kcmil for copper.
______	______	10.	Wiring located above electric heating panels shall be spaced not less than ______″ above the heated ceiling. A. 2 B. 3 C. 4 D. 5
______	T F	11.	Time switches need not be of the externally operable type.
______	______	12.	The minimum size grounded conductor required for a service with 600 kcmil per phase is No. ______ copper.
______	______	13.	A 15 A or 20 A branch circuit shall be permitted to supply ______ loads. A. lighting B. utilization equipment C. both A and B D. neither A nor B
______	______	14.	Equipment with open-circuit voltage exceeding ______ V shall not be installed in dwelling units.
______	______	15.	In general, flexible cords other than tinsel cords shall not be smaller than No. ______.
______	______	16.	A 30 A receptacle outlet supplying power to a cord-and-plug connected load shall not be loaded to more than ______ A.

______ ______ **17.** If more than one nominal voltage system exists in a building, each ungrounded conductor in each phase and system shall be ______.

A. black
B. orange
C. the same
D. identified

______ ______ **18.** Wiring that supplies equipment, located behind panels designed to allow ______, shall be adequately supported and kept off the ceiling tiles.

______ ______ **19.** The general rule requires a(n) ______ to be installed at an auxiliary building.

______ ______ **20.** The minimum burial depth of rigid metal conduit installed in solid rock is ______″ on all sides.

______ ______ **21.** A single receptacle installed on an individual branch circuit shall have a(n) ______ rating not less than that of the branch circuit.

______ ______ **22.** A 4160 V feeder circuit shall have the CB long-time trip rating selected at ______ times the ampacity of the conductors.

______ ______ **23.** Ground-fault circuit protection for personnel is required for all single-phase, 125 V, 15 A and 20 A receptacles installed in the ______ of a dwelling unit.

A. attic
B. bathroom
C. laundry
D. utility room

______ ______ **24.** All 600 V lighting fixtures installed outside on poles to illuminate a ballpark shall be set at a minimum height of ______′ where the supply circuits exceed 277 V to ground.

______ ______ **25.** The minimum burial depth for PVC conduit routed beneath a building is ______″.

______ ______ **26.** An insulated grounded conductor smaller than No. 4 shall be identified by a white or natural ______ outer finish.

______ ______ **27.** ______ conductors shall be used with chain fixtures.

______ ______ **28.** Disconnecting means for circuits to feeders and branch circuits shall be properly ______ to indicate the purpose.

______ T F **29.** Connections that depend solely on solder are not permissible for grounding equipment.

______ ______ **30.** Heavy-duty lampholders shall have a rating of not less than ______ W.

______ ______ **31.** Side- or back-wiring spaces are required in cabinets when, in addition to the supply circuit or a continuation thereof, there are devices connected to more than ______ conductors.

A. 6
B. 8
C. 10
D. 12

______ ______ **32.** Screw-shell lampholders shall not be used with infrared lamps of over ______ W except when identified as suitable for such use.

______ ______ **33.** Cables routed up a pole shall be protected at least ______′ above grade.

______ ______ **34.** The smaller of two dissimilar loads may be omitted in calculating the load on a feeder if it is unlikely that both loads will be connected ______.

______ T F **35.** Permanently installed 15 A and 20 A, 125 V receptacles that are part of the building wiring and used for construction personnel shall be GFCI protected.

Trade Competency Test 7

Name ______________________ Date ______________________

Section	Answer	
______	______	**1.** Steel cable trays are permitted to serve as EGC for circuits when ground-fault protection does not exceed ________ A.
______	T F	**2.** Wiring located less than 2″ above electrically heated ceilings and over 1′ of thermal insulation shall be derated.
______	______	**3.** The feeder load for one 10 kVA, one 16 kVA, and one 17 kVA household electric range is ________ kVA. A. 14.5 C. 15.6 B. 15 D. 16.1
______	______	**4.** The grounded conductor shall be ________ when service entrance cable is used in a feeder to supply a subpanel located in the same building.
______	______	**5.** Conduits, including fittings, entering a switchgear shall not rise more than ________″ above the bottom of the enclosure.
______	T F	**6.** Branch circuit conductors for fixed resistance space heaters shall be rated at 125% of the total load of heaters and motors.
______	______	**7.** The rating of any one cord-and-plug connected appliance used on a 30 A branch circuit shall not exceed ________ A. A. 21 C. 27 B. 24 D. 30
______	______	**8.** The maximum radius of a conduit bend enclosing THHN Cu conductors is ________″ for 2″ rigid metal conduit.
______	______	**9.** A 20 A circuit in a dwelling unit supplies only fixed resistance space heaters. This circuit may be loaded to a maximum value of ________ A. A. 12 C. 18 B. 16 D. 20
______	T F	**10.** Heating panels shall not be extended beyond partitions.
______	______	**11.** A feeder supplying a 5 kVA wall-mounted oven and a 7 kVA counter-mounted cooktop must have the capacity for ________ kVA. A. 7 C. 9.2 B. 7.8 D. 12
______	______	**12.** The general rule requires that raceways be installed ________ between junction boxes prior to the installation of conductors.
______	______	**13.** ________ means must be provided for disconnecting the disconnecting means of devices mounted on a busway out of reach.
______	______	**14.** A 3-pole circuit breaker counts as ________ devices in determining the count toward a lighting and appliance panelboard.
______	T F	**15.** Heating cable shall not be used with metal lath ceilings.
______	______	**16.** A 24″ ladder cable tray has a fill area of ________ sq in.

_________ _________ **17.** No. 8 THWN conductors supply a central air conditioner, consisting of a sealed (hermetic) motor-compressor. The branch circuit may be loaded to a maximum value of _________ A.

A. 30
B. 32
C. 36
D. 50

_________ _________ **18.** For the purpose of determining cable tray fill, the multiplier used to convert the diameter of 1000 kcmil, or larger, conductors to square inches is _________ Sd.

_________ _________ **19.** Conduits with No. 4 and larger conductors shall be provided with _________ for protection of conductors.

_________ _________ **20.** Type AC cable _________ in walls do not require support.

_________ _________ **21.** A 4160 V single-insulated conductor, installed underground in an electric duct, has an ampacity of _________ A.

_________ _________ **22.** The total rating of fastened-in-place appliances connected to a 20 A branch circuit that also supplies lighting units shall not exceed _________ A.

A. 10
B. 12
C. 16
D. 20

_________ _________ **23.** Single conductors in cable trays can be used as EGC's if they are No. _________ or larger.

_________ T F **24.** Heating cables in a concrete floor shall be placed on at least 2′ centers.

_________ _________ **25.** A 4″ rigid metal conduit installed with threaded couplings in a straight run shall be supported every _________′.

_________ _________ **26.** A 6″ rigid nonmetallic conduit shall be supported at _________′ intervals.

_________ _________ **27.** MC cables shall be supported at intervals not exceeding _________′.

_________ T F **28.** Embedded heating cables are permitted to be spliced only where necessary and if done by approved means.

_________ _________ **29.** Busways that do not exceed _________′ in length can be reduced in size without overcurrent protection in industrial establishments only.

_________ _________ **30.** Six current-carrying conductors routed through the same raceway shall be derated _________%.

_________ _________ **31.** The total load on overcurrent protection devices supplying continuous duty loads shall not exceed more than _________% of their rating.

_________ _________ **32.** A No. _________ is the smallest THWN neutral feeder conductor permitted to supply two 240 V, 12 kVA household ranges.

A. 4
B. 6
C. 8
D. 10

_________ _________ **33.** Noninsulated busbars installed in switchgears shall have a clearance of at least _________″ above the bottom of the enclosure.

_________ T F **34.** The nonheating leads of heating cables installed in concrete may be protected by rigid nonmetallic conduit.

_________ _________ **35.** The grounded circuit conductor can be used as an equipment grounding conductor and circuit conductor on the _________ side of the service equipment.

_________ T F **36.** A separable connector will meet Code requirements for a disconnecting means for a wall-mounted oven.

Trade Competency Test 8

Name ______________________ Date ______________

Section	Answer	
______	T F	**1.** In a 1000 sq ft addition to a dwelling unit, the "VA-per-square-foot" method is not used.
______	______	**2.** Heating elements that exceed ______ A in a heating unit shall be subdivided.
______	______	**3.** The neutral conductor of a 3-wire branch circuit to a household electric range shall not be smaller than No. ______ THW. A. 4 C. 8 B. 6 D. 10
______	______	**4.** Unbroken lengths of EMT ______′ or longer are not required to be supported where structural members do not allow fastening.
______	T F	**5.** The overcurrent protection device supplying power to a fixed water heater can be rated less than 125% of the unit's nameplate rating.
______	______	**6.** The grounded conductor in a subpanel shall be insulated from the metal enclosure unless the terminal bar is ______ for the purpose.
______	______	**7.** Frames of electric clothes dryers may be grounded by connection to the grounded circuit conductor only when this copper conductor is not smaller than ______ and it is an existing branch-circuit. A. No. 8 C. No. 12 B. No. 10 D. the other conductors
______	T F	**8.** Unit switches can be used as a disconnecting means for a fixed appliance where all conductors are disconnected simultaneously.
______	______	**9.** Lighting track shall be calculated at ______ VA for each 2′ or fraction thereof.
______	T F	**10.** Infrared lamps rated over 600 W shall not be used as screw-shell lampholders.
______	______	**11.** The nameplate on heating equipment shall be located to be ______ after installation.
______	______	**12.** Flexible cords identified for hard or extra-hard usage shall be permitted to be used for temporary wiring on ______ sites.
______	______	**13.** A heavy-duty, admedium lampholder shall have a rating of not less than ______ W. A. 150 C. 600 B. 300 D. 660
______	______	**14.** In general, all conductors, and when used the EGC, shall be routed in the ______ cable, raceway, trench, or conduit system.
______	______	**15.** Insulated or uninsulated busbars installed in switchgears shall be ______ mounted.

_________ T F **16.** In multifamily dwellings, the disconnecting means for a permanently connected appliance rated at not over 300 VA, is permitted to be the OCPD in the service equipment for the occupancy.

_________ _________ **17.** Heating elements of cables shall have a separation of at least _________″ from the edge of outlet boxes.

_________ _________ **18.** The voltage drop on a feeder circuit shall be held to _________%.

A. 2
B. 3
C. 5
D. 7

_________ T F **19.** The grounded conductor in a 277/480 V system can be identified white or gray.

_________ _________ **20.** A(n) _________ A or less overcurrent protection device with neutral loads shall be counted to determine if a panel is a lighting and appliance panelboard.

_________ _________ **21.** A circuit operating for _________ hours or more is considered a continuous duty load.

_________ _________ **22.** Receptacle outlets shall be installed above a kitchen countertop no more than _________″ apart.

A. 18
B. 24
C. 30
D. 36

_________ _________ **23.** The exposed blades of knife switches shall be de-energized when in the _________ position.

_________ T F **24.** A 20 A branch circuit supplying air conditioning equipment shall not supply lighting outlets.

_________ _________ **25.** Heavy-duty lighting track is identified for use with systems exceeding _________ A.

_________ T F **26.** A lockable branch circuit overcurrent protection device may serve as the disconnecting means for a motor-driven, permanently connected appliance rated not over ⅛ hp.

_________ _________ **27.** Angle connectors used on Greenfield shall not be _________.

_________ _________ **28.** Neon systems having an open circuit voltage exceeding _________ V shall not be installed in dwelling units.

_________ _________ **29.** A disposal installed in a dwelling unit shall be equipped with a cord that does not exceed _________″ in length.

_________ T F **30.** The total load of four or more fixed appliances in a dwelling unit can have an 80% demand factor applied.

_________ _________ **31.** Ceiling fans, mounted to a box identified for such use, shall not weigh more than _________ lb.

_________ _________ **32.** At least one receptacle outlet shall be installed above a show window for each _________′ or major fraction thereof.

A. 9
B. 12
C. 18
D. 24

Commercial Locations 3

WIRING PLANS

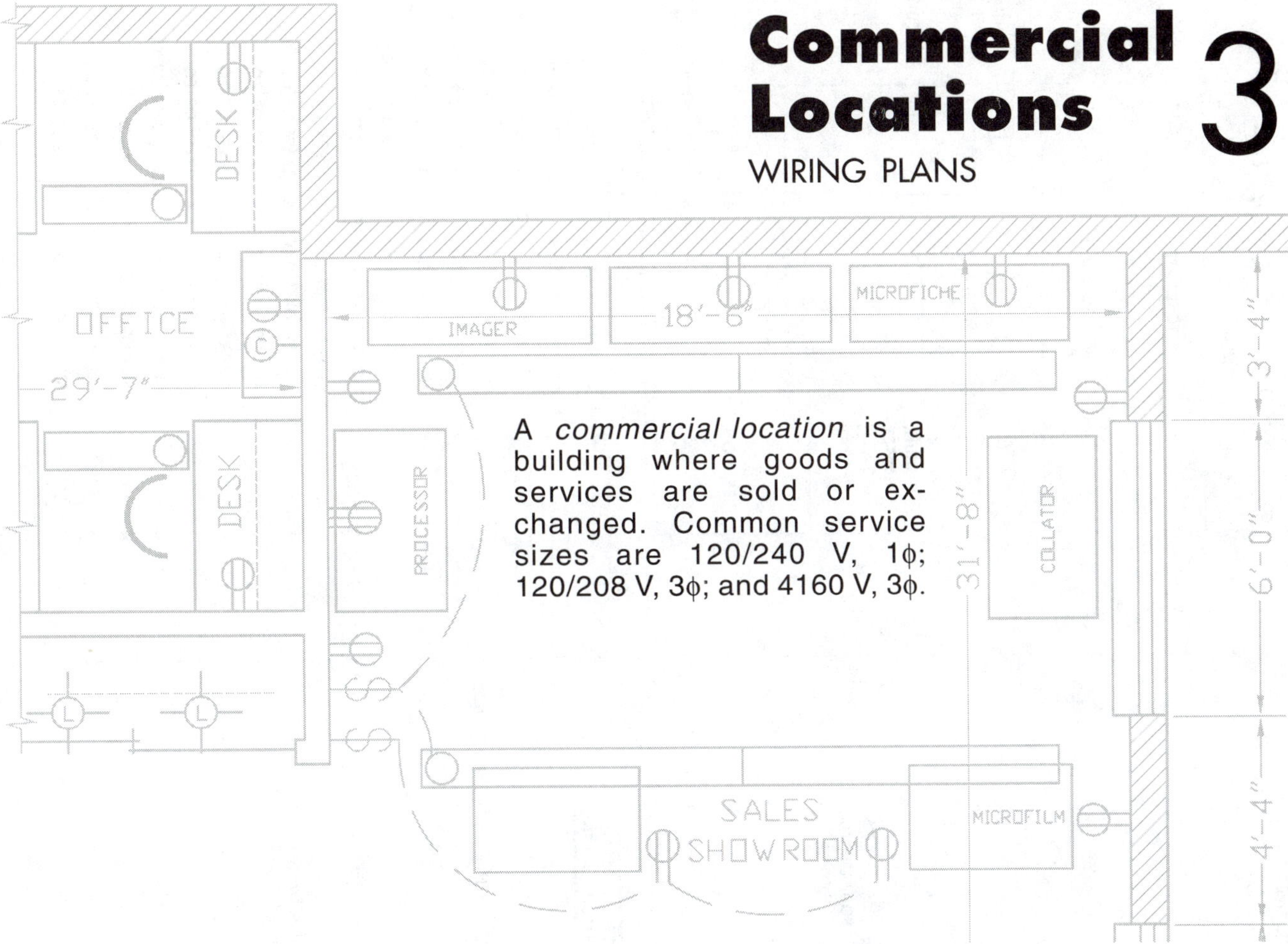

A *commercial location* is a building where goods and services are sold or exchanged. Common service sizes are 120/240 V, 1ϕ; 120/208 V, 3ϕ; and 4160 V, 3ϕ.

READING PRINTS

In preparing to handle a large installation, the electrical supervisor glances through the complete set of plans to obtain an overview of the entire project. The electrical plans are surveyed to determine the extent of the work, kinds of circuits and equipment, and locations of main elements. The electrical specifications are read, and the plans are referred to from time to time as necessary.

The electrical supervisor must refer to the plot, architectural, structural, framing, mechanical, and electrical plans to observe all the details essential to the work. Following are excerpts from electrical specifications for a store building, along with notes that might be taken from the architectural plans.

EXCERPTS FROM SPECIFICATIONS

Note: The Code references indicate where the main topic is found in the National Electrical Code. Always follow manufacturer's directions.

29-1 Main Service — 230

(a) The main service shall consist of a 4-wire, 3-phase, 120/208 V secondary system from utility company manhole on Sheppard Avenue to main switchboard.

(b) All underground conduit outside the building shall be buried at least 18″ below grade and painted with bitumastic paint.

(c) All service conductors shall be insulated, including the neutral conductor.

29-4 Conductors — 310

(a) Conductors shall be Type THW unless otherwise specified or required by applicable provisions of the NEC. All conductors to be copper unless otherwise noted.

(b) No conductors for lighting, plug receptacles, or power shall be smaller than No. 12 AWG copper.

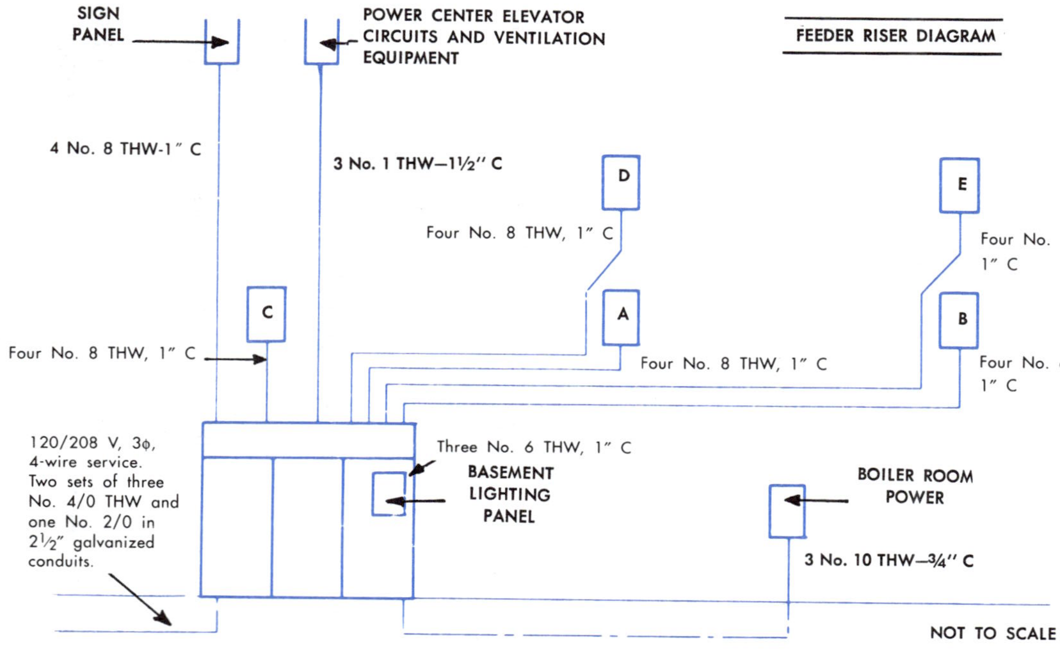
FEEDER RISER DIAGRAM
SIGN PANEL
POWER CENTER ELEVATOR CIRCUITS AND VENTILATION EQUIPMENT
4 No. 8 THW-1" C
3 No. 1 THW—1½" C
Four No. 8 THW, 1" C
Four No. 8 THW, 1" C
Four No. 8 THW, 1" C
Four No. 8 THW, 1" C
Four No. 6 THW, 1" C
C
D
A
E
B
120/208 V, 3φ, 4-wire service. Two sets of three No. 4/0 THW and one No. 2/0 in 2½" galvanized conduits.
Three No. 6 THW, 1" C
BASEMENT LIGHTING PANEL
BOILER ROOM POWER
3 No. 10 THW—¾" C
NOT TO SCALE

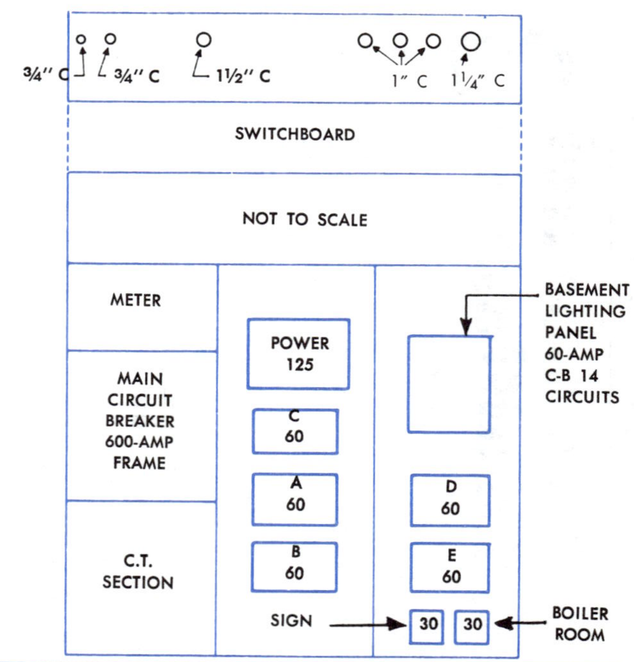
¾" C
¾" C
1½" C
1" C
1¼" C
SWITCHBOARD
NOT TO SCALE
METER
MAIN CIRCUIT BREAKER 600-AMP FRAME
C.T. SECTION
POWER 125
C 60
A 60
B 60
D 60
E 60
SIGN
30
30
BOILER ROOM
BASEMENT LIGHTING PANEL 60-AMP C-B 14 CIRCUITS

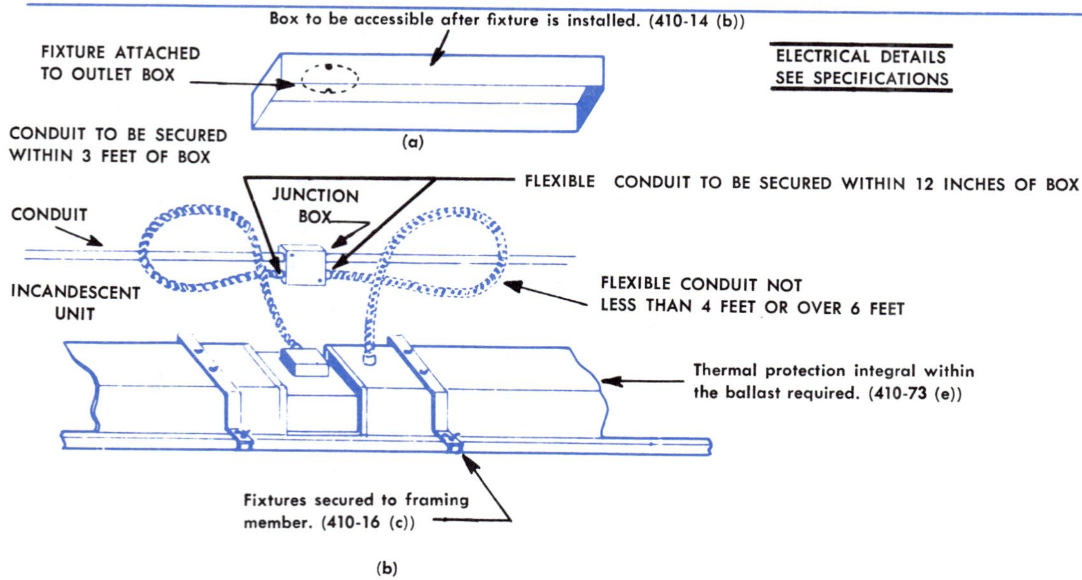
ELECTRICAL DETAILS
SEE SPECIFICATIONS
Box to be accessible after fixture is installed. (410-14 (b))
FIXTURE ATTACHED TO OUTLET BOX
CONDUIT TO BE SECURED WITHIN 3 FEET OF BOX
(a)
JUNCTION BOX
FLEXIBLE CONDUIT TO BE SECURED WITHIN 12 INCHES OF BOX
CONDUIT
INCANDESCENT UNIT
FLEXIBLE CONDUIT NOT LESS THAN 4 FEET OR OVER 6 FEET
Thermal protection integral within the ballast required. (410-73 (e))
Fixtures secured to framing member. (410-16 (c))
(b)

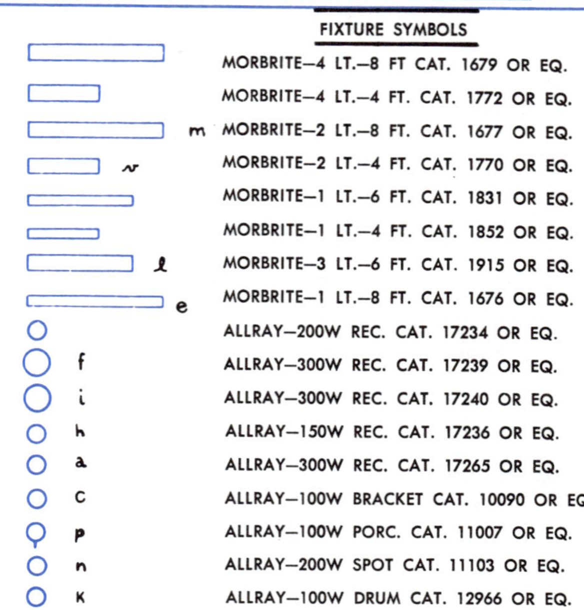
FIXTURE SYMBOLS
MORBRITE—4 LT.—8 FT CAT. 1679 OR EQ.
MORBRITE—4 LT.—4 FT. CAT. 1772 OR EQ.
m MORBRITE—2 LT.—8 FT. CAT. 1677 OR EQ.
MORBRITE—2 LT.—4 FT. CAT. 1770 OR EQ.
MORBRITE—1 LT.—6 FT. CAT. 1831 OR EQ.
MORBRITE—1 LT.—4 FT. CAT. 1852 OR EQ.
ℓ MORBRITE—3 LT.—6 FT. CAT. 1915 OR EQ.
e MORBRITE—1 LT.—8 FT. CAT. 1676 OR EQ.
ALLRAY—200W REC. CAT. 17234 OR EQ.
f ALLRAY—300W REC. CAT. 17239 OR EQ.
i ALLRAY—300W REC. CAT. 17240 OR EQ.
h ALLRAY—150W REC. CAT. 17236 OR EQ.
a ALLRAY—300W REC. CAT. 17265 OR EQ.
c ALLRAY—100W BRACKET CAT. 10090 OR EQ.
p ALLRAY—100W PORC. CAT. 11007 OR EQ.
n ALLRAY—200W SPOT CAT. 11103 OR EQ.
k ALLRAY—100W DRUM CAT. 12966 OR EQ.

(c) No block and tackle or other mechanical means shall be used to draw conductors smaller than No. 2 into raceways. Powdered soapstone will be the only lubricant permitted.

(d) Feeder voltage drop from service location to any distribution panel shall not exceed 1%.

29-5 Wiring Methods and Materials — 300 to 384

(c) Approximate locations of all outlets are marked on the drawings. The contractor shall check all such measurements and, in case of doubt, consult the architect as to exact locations so that boxes will be properly centered with acoustical tile, interior trim, paneling, etc.

29-6 Equipment for General Use — 400 to 480

(b) Lock-type wall switches shall be installed at points where a switch is marked Sk on drawing.

(m) Two special plug receptacles on the east wall of the Alterations Department shall be equipped with signal lights. These outlets. . .

29-7 Telephone and Loud Speaker Systems — 640, 800, and 810

(a) Telephone outlets shall be located approximately as shown. . .

29-8 Miscellaneous Equipment — 422

(j) Toilet exhaust fans shall be Nuway, 130 VA, Type C2, or. . .

ELECTRICAL PLANS

Before undertaking a detailed analysis of the store building installation, notice that four drawings are included. The Basement, First Floor, and Second Floor floor plans are included with a special drawing that shows a riser diagram, switchboard elevation, electrical details, and a list of fixture symbols. Discussion could be started with the Basement plan, then carried on to the First Floor and Second Floor plans in turn. Or, the First Floor might be dealt with, then Second Floor, and finally Basement.

A different order is followed, however, starting with the Second Floor, working downward to the First Floor, and lastly the Basement. This procedure simplifies explanation. The Second Floor is the least complicated area and makes a good starting point. The First Floor plan is more involved. The Basement plan is examined last because it covers service requirements.

SECOND FLOOR PLAN

Two lighting panels are used for this floor, panel E at the north wall, and panel D at the south wall. Conduit runs between lighting outlets and panelboards are indicated by solid lines showing them to be concealed in the hung ceiling. Plug receptacle runs are indicated by broken lines showing that they are to be run in the poured concrete slab. Telephone runs are indicated by lines marked --- *T* --- *T* ---. Speaker system lines are marked --- S --- S --- . Telephone outlets are indicated by standard symbols. Speaker outlets are indicated by the symbol S.

Conduit riser indications are shown in the south wall. At the left or west side, the first such riser is marked DOWN to show that a conduit run from the stair outlet leads to the floor below. The next two risers are marked UP, and since no conduit runs lead to them from the second floor, it is apparent that they originate below and continue upward toward the roof.

The final riser is marked DOWN. It is run of conduit leading downward from a lighting outlet at the head of the stairs. Two risers in the east wall of the passenger elevator shaft are both shown as DOWN. They are run of conduit leading downward from a lighting outlet at the head of the stairs.

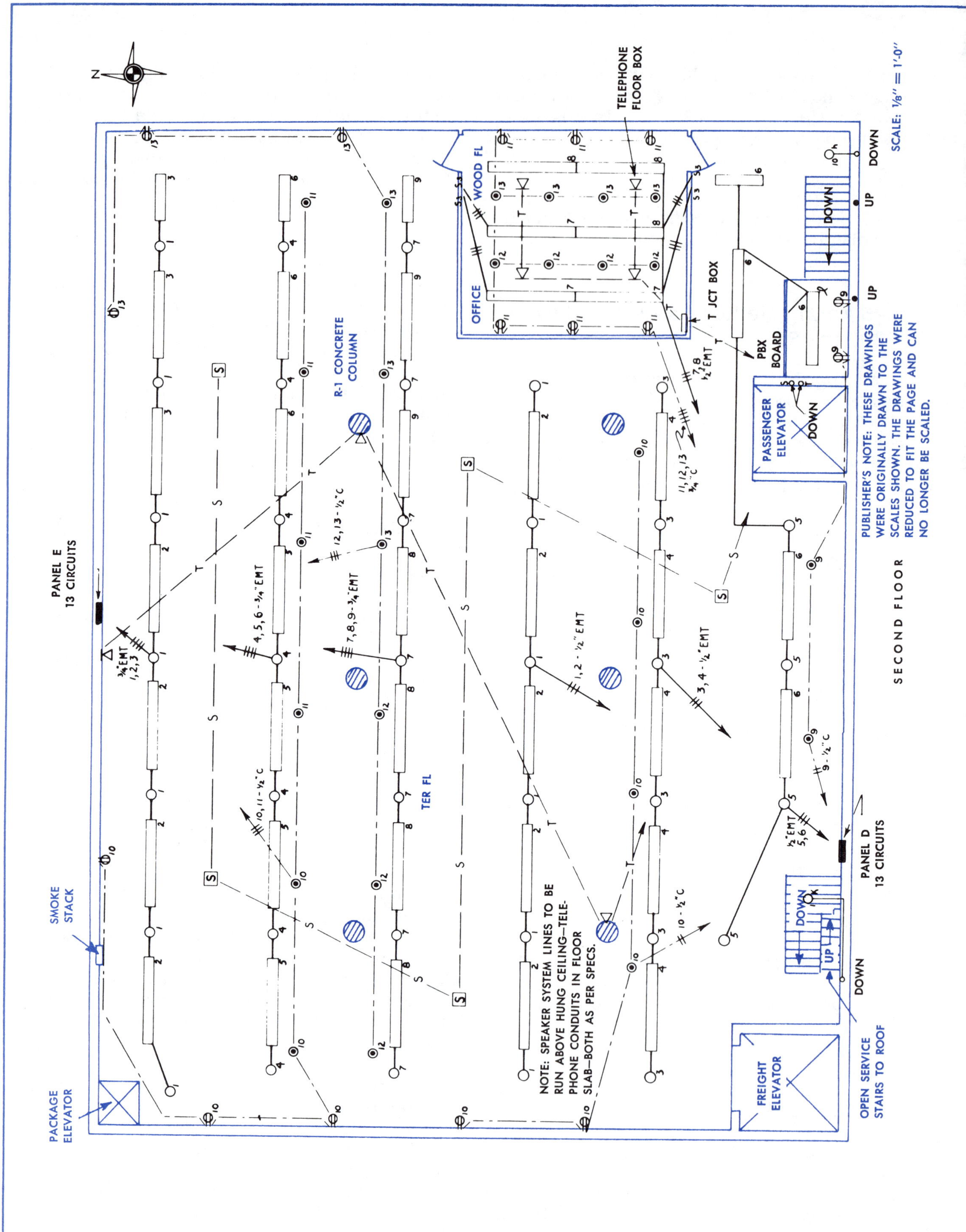
SECOND FLOOR
SCALE: 1/8" = 1'-0"
PUBLISHER'S NOTE: THESE DRAWINGS WERE ORIGINALLY DRAWN TO THE SCALES SHOWN. THE DRAWINGS WERE REDUCED TO FIT THE PAGE AND CAN NO LONGER BE SCALED.
NOTE: SPEAKER SYSTEM LINES TO BE RUN ABOVE HUNG CEILING—TELEPHONE CONDUITS IN FLOOR SLAB—BOTH AS PER SPECS.
PANEL E
13 CIRCUITS
PANEL D
13 CIRCUITS
TELEPHONE FLOOR BOX
T JCT BOX
PBX BOARD
OFFICE
WOOD FL
TER FL
R-1 CONCRETE COLUMN
PASSENGER ELEVATOR
FREIGHT ELEVATOR
PACKAGE ELEVATOR
SMOKE STACK
OPEN SERVICE STAIRS TO ROOF
UP
DOWN
3/4"EMT 1,2,3
4,5,6-3/4"EMT
7,8,9-3/4"EMT
12,13-1/2"C
10,11-1/2"C
1,2-1/2"EMT
3,4-1/2"EMT
11,12,13 3/4"C
7,8 1/2"EMT
9-1/2"C
1/2"EMT 5,6
10-1/2"C

Two risers in the east wall of the passenger elevator shaft are both shown as DOWN, one marked **T** to indicate a telephone run, the other **S** to indicate a speaker run. The PBX telephone board and the sound amplifier are in the small room to the right of the elevator shaft.

In determining the load on the VA-per-square-foot basis, 220-3(b) requires that the floor area be computed from outside dimensions of the building. The measurements are 72′ × 96′, which equals 6912 sq ft. From this amount, it is permissible to deduct 520 sq ft, which is the area of the office space and the area of the two elevators. The elevator shaft area and the office space do not come under the heading of general lighting in Table 220-3(b).

Table 220-3(b) lists the unit load for a store as 3 VA per sq ft. NEC 220-10(b) states that this value shall be increased by 25% where the load on a branch circuit is continuous (load is expected to continue for 3 hours or more). The unit load in the present case must be 1.25 × 3 VA, or 3.75 VA per sq ft. The minimum provision for the lighting load is 6392 sq ft × 3.75 VA, or 23,970 VA. Panels *D* and *E* must have a sufficient number of lighting circuits to furnish this amount of general illumination.

The Specifications provide that all lighting circuits shall be run with No. 12 Type THW copper wire, protected by 20 A circuit-breakers. NEC 210-3 states that branch circuits shall be classified according to the rating of the overcurrent protection devices. Because the load is continuous, these 20 A circuits, each of which can supply 2400 VA of power (20 A × 120 V), can be loaded to only 80% of this value, or 1920 VA per 384-16(c).

Calculations for Panel *E*

The fixture list, fixture specifications, and the print show the load on Circuit 1 as seven 200 VA incandescent units, totaling 1400 VA. Continuing to Circuit 2, 210-22(b) requires that the load on a lighting fixture that has a ballast shall be calculated according to the ampere rating of the ballast and not the VA of the lamps. The current used in the two ballasts that are part of each fixture on the circuit is 3.2 A or 384 VA.

The total VA for Circuit 2, which supplies four such units, is 1536 VA. Circuit 3 has two of these units and a third with a ballast rating of 1.7 A. The load is 972 VA for all units in Circuit 3. The load for this group of three circuits is 1400 VA + 1536 VA + 972 VA = 3908 VA.

Loads on the two groups 4, 5, 6 and 7, 8, 9 are the same as the above figures so that the total lighting power to be supplied by Panel *E* amounts to 3 × 3908 VA, or 11,724 VA. Table 220-11 lists a demand factor of 100% for lighting loads of this nature.

In addition to the lighting outlets, a number of wall receptacles and floor outlets are supplied from the panel. The Specifications require that No. 12 Type THW wire be used for these circuits. NEC 220-3(c)(6) designates a load of 180 VA or $1\frac{1}{2}$ A (180 divided by 120) for each of such other outlets. The eighteen plug receptacles connected to this panel represent a current input of 27 A × 120 V, or 3240 VA.

The load imposed on the feeder is 11,724 VA + 3240 VA = 14,964 VA. At this point, determine if the feeder will carry its required proportion of minimum lighting load. The floor plan shows that circuits from Panel *E* provide general illumination for an area of 36′ × 96′ = 3456 sq ft.

Since the unit load cannot be less than 3.75 VA per sq ft, the amount of general lighting burden allotted to Panel *E* cannot be less than 3456 × 3.75 VA, or 12,960 VA. This value must be used in the feeder calculations because it is larger than the actual connected load of lighting circuits.

The minimum value of lighting VA does not have to be used, but must be present in circuit and feeder capacity. The corrected load on the feeder then is 12,960 VA + 3240 VA = 16,200 VA.

With a balanced, 4-wire, 3-phase, 120/208 V system, the current per conductor may be determined by dividing VA by 360 (1.73 × 208 V = 360). The current in the feeder to Panel *E* equals 16,200 divided by 360 V, or 45 A. NEC Table 310-16 lists the nearest size of Type THW copper

PUBLISHER'S NOTE: THESE DRAWINGS WERE ORIGINALLY DRAWN TO THE SCALES SHOWN. THE DRAWINGS WERE REDUCED TO FIT THE PAGE AND CAN NO LONGER BE SCALED.

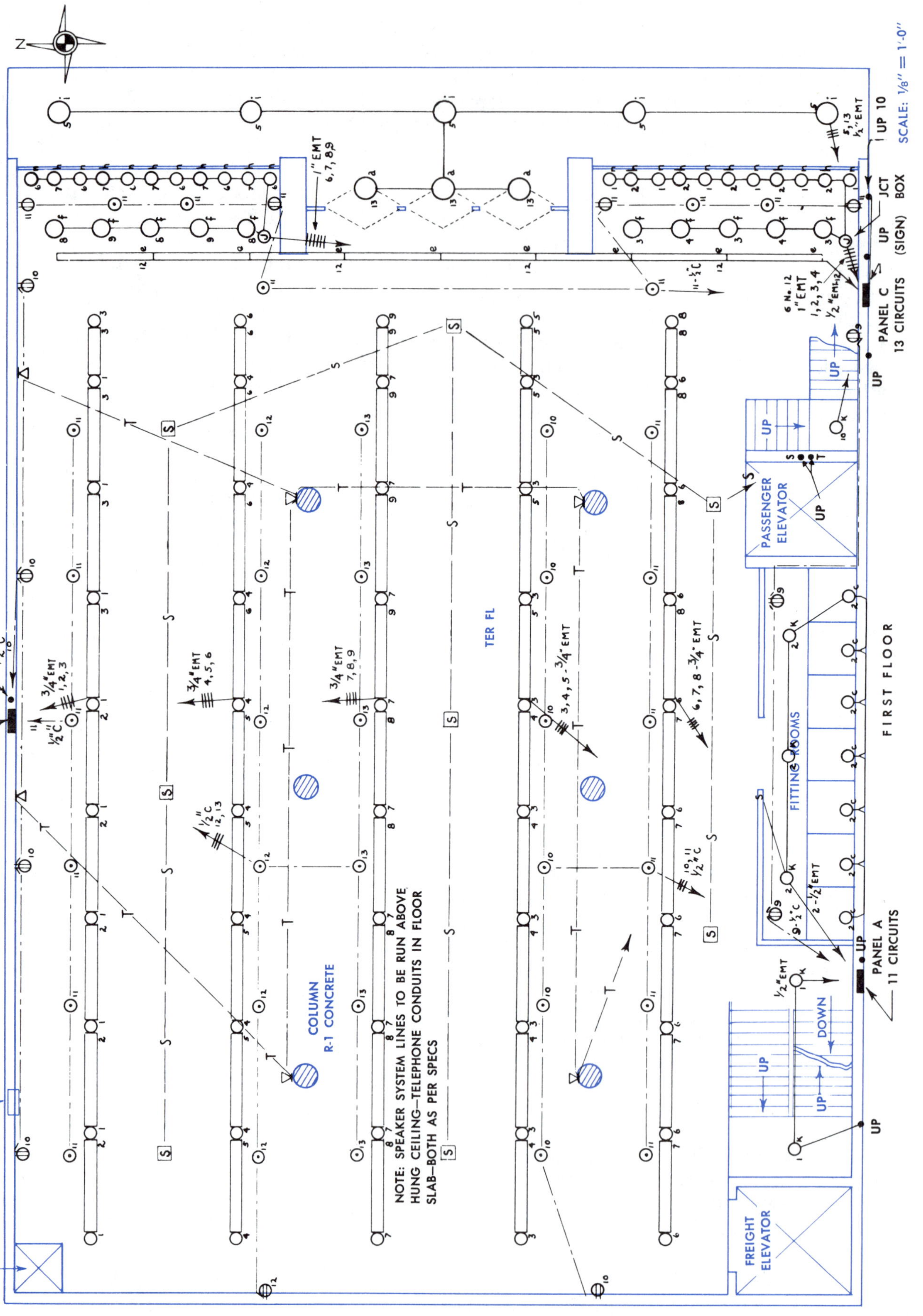

conductor to carry 45 A as No. 8, which is rated at 50 A if there are not more than three current-carrying conductors in a raceway.

Note 8(a) of Notes to Ampacity Tables of 0 to 2000 Volts states that where the number of conductors in a raceway exceeds three, the allowable ampacity of each conductor shall be limited to 80% of the values given in the table. Since there are four conductors, it seems that the capacity of the No. 8 wire is reduced to 80% of 50 A, or 40 A. Note 10, however, states that a neutral conductor that carries only the unbalanced current from other conductors shall not be counted in determining ampacities where there are more than three conductors in a raceway.

Such a neutral conductor is included in the feeder under consideration, and the carrying capacity of the No. 8 wire will not be reduced in value. Nevertheless, the rule may be altered later on because the neutral conductor of a 4-wire, 3-phase wye system that supplies circuits for electric discharge lamps carries some third harmonic current even when the load is perfectly balanced.

American Electricians Handbook and other engineering data (See Table 8, Chapter 9) lists the resistance of No. 6 copper as .491 ohms per 1000′. When floor heights and distances on the architectural plans are measured, allowing a sufficient length of conductor for connections at either end of the run, the length of single conductor is found to be 104′ from Panel *E* to the switchboard in the basement.

The preferred method for calculating voltage drop for a 3-phase circuit is by using the formula:

$$VD = \frac{2 \times R \times L \times I \times .866}{1000}$$

where

VD = voltage drop

R = resistivity per Table 8, Chapter 9

L = length (in feet)

I = current

1000 = 1000′ or less of wire

$$VD = \frac{2 \times R \times L \times I \times .866}{1000}$$

$$VD = \frac{2 \times .491 \times 104 \times 45 \times .866}{1000}$$

$$VD = \mathbf{3.98\ V}$$

$$\%VD = \frac{VD}{V} \times 100$$

$$\%VD = \frac{3.98}{208} \times 100$$

$$\%VD = \mathbf{1.91\%}$$

A FPN following 215-2(b) recommends that the voltage drop for feeders should be limited to 3% of the phase voltage at the farthest outlet. The recommendation for voltage drop for both feeders and branch circuits is 5%. In this calculation, the recommended voltage drop should not exceed 6.24 V. The 3.98 V is well below the 3% recommendation and is acceptable. The feeder to Panel *E* will consist of four No. 6 THW conductors in 1″ conduit.

Calculations for Panel *D*

The loads on the circuits are

1-1200 VA	2-1920 VA	3-1200 VA
4-1920 VA	5- 800 VA	6-1152 VA
7-1152 VA	8-1152 VA	9- 720 VA
10-1080 VA	11-1080 VA	12- 720 VA
13- 720 VA		

The circuits devoted to general lighting are 1, 2, 3, 4, 5, and 6. The general lighting circuits require a total of 8192 VA.

Measurement shows the floor area supplied with general illumination from this panel to be 2496 sq ft. The minimum VA required is 2496 × 3.75 VA, or 9360 VA. This value, being somewhat higher than actual load on the circuits, will be used in the feeder calculations.

The office has an area of 440 sq ft, requiring 1650 VA of illumination at the 3.75 VA rate. Table 220-3(b) lists office buildings at a minimum unit load of 3.5 VA per sq ft. For a continuous load, this minimum would have to be increased to 4.4 VA (1.25 × 3.5 VA).

Although this provision does not apply strictly to the one small office space here, it does indicate that office buildings and office areas need somewhat more lighting intensity than sales areas. Using the value given by Table 220-3(b), this area will require 440 × 4.4 VA, or 1936 VA. The lighting load connected to Circuits 7 and 8 amounts to 2200 VA. Because the connected load of 2200 VA × 1.25 = 2750 VA is larger than that required by

Table 220-3(b) and 220-10(b), the higher value of 2750 VA will be used.

Plug receptacle Circuits 9, 10, 11, 12, 13 take 4320 VA. The total load on the feeder is 9360 VA + 2750 VA + 4320 VA, or 16,430 VA, the current being 46 A (16,430 VA divided by 360), for which the minimum size of Type THW conductors is No. 8.

The length of single conductor between Panel *D* and the switchboard is approximately 83′. The voltage drop for a No. 8 stranded conductor is too high. Use of No. 4 conductor will result in an acceptable drop. Table 3A, Chapter 9 shows that four No. 4 Type THW copper conductors may be installed in $1\frac{1}{4}''$ conduit.

$$VD = \frac{2 \times R \times L \times I \times .866}{1000}$$

$$VD = \frac{2 \times .308 \times 83 \times 46 \times .866}{1000}$$

$$VD = \mathbf{2.04\ V}$$

$$\%VD = \frac{VD}{V} \times 100$$

$$\%\ VD = \frac{2.04}{208} \times 100$$

$$\%VD = \mathbf{.98\%\ VD}$$

FIRST FLOOR PLAN

Three lighting panels are shown on this floor, Panel *B* in the north wall, and Panels *A* and *C* in the south wall. Panel *C* is primarily confined to supplying outlets in the entrance, canopy, and show window areas. Panels *A* and *B* are confined to the general store illumination. The wiring plan on this floor is similar to that employed on the floor above.

A single riser is indicated in the north wall. The feeder conduit is associated with Panel *E* on the second floor. The riser near the left end of the south wall, marked UP, is the same one shown on the second floor plan, marked DOWN.

The riser next to Panel *A* is a feeder conduit to Panel *D*, above. The next two risers, also marked UP, were mentioned before. The third riser, marked UP 10, is the other end of the conduit supplying the lighting unit at the head of the stairs on the second floor. The two risers in the east wall of the passenger elevator shaft have been explained.

Calculations for Panel *B*

The loads on the circuits are

1-1800 VA	2-1920 VA	3-1556 VA
4-1800 VA	5-1920 VA	6-1556 VA
7-1800 VA	8-1920 VA	9-1556 VA
10- 720 VA	11-1080 VA	12-1260 VA
13-1080 VA		

The circuits devoted to general lighting are, 1, 2, 3, 4, 5, 6, 7, 8 and 9. The general lighting circuits use 15,828 VA.

Measurement shows the floor area lighted by these circuits to be 3344 sq ft. The minimum acceptable VA for this area is 3344 × 3.75 VA, or 12,540 VA. Since the connected load is greater than this amount, the higher value will be used in feeder calculations.

Plug receptacle Circuits 10, 11, 12, and 13 require 4140 VA. The load on the feeder is 15,828 VA + 4140 VA = 19,968 VA. The current is approximately 56 A. The minimum size of Type THW conductor is No. 6.

The length of single conductor between Panel *B* and the switchboard is approximately 92′. The voltage drop in a No. 6 conductor equals 4.48 V.

$$VD = \frac{2 \times R \times L \times I \times .866}{1000}$$

$$VD = \frac{2 \times .491 \times 92 \times 56 \times .866}{1000}$$

$$VD = \mathbf{4.48\ V}$$

$$\%VD = \frac{VD}{V} \times 100$$

$$\%VD = \frac{4.38}{208} \times 100$$

$$\%VD = \mathbf{2.1\%}$$

The feeder to Panel *B* will consist of four No. 6 Type THW copper conductors in 1″ conduit.

Calculations for Panel *A*

The loads on the circuits are

1- 300 VA	2-1000 VA	3-1800 VA
4-1920 VA	5-1556 VA	6-1800 VA
7-1920 VA	8-1556 VA	9- 540 VA
10-1260 VA	11-1080 VA.	

The circuits concerned with general lighting are 1, 2, 3, 4, 5, 6, 7, and 8. The general lighting circuits have a connected load of 11,852 VA.

Measurement shows the floor area lighted by these circuits as 3080 sq ft, for which the minimum VA must be not less than 3080 × 3.75 VA, or 11,550 VA. The connected load, being somewhat higher than this value, will be used in feeder calculations.

Plug receptacle Circuits 9, 10, and 11, take 2880 VA. The total load on the feeder is 11,852 VA + 2880 VA, or 14,732 VA. The current is approximately 41 A and calls for a minimum size of No. 8 Type THW conductor.

The length of single conductor from Panel *A* to the switchboard is about 73′. Voltage drop in a No. 8 conductor is 4 V.

$$VD = \frac{2 \times R \times L \times I \times .866}{1000}$$

$$VD = \frac{2 \times .778 \times 73 \times 41 \times .866}{1000}$$

$$VD = \mathbf{4\ V}$$

$$\%VD = \frac{VD}{V} \times 100$$

$$\%VD = \frac{4}{208} \times 100$$

$$\%VD = \mathbf{1.9\%}$$

The feeder to Panel *A* consists of four No. 8 Type THW copper conductors in 1″ conduit.

Calculations for Panel *C*

Except for front stair lighting, Circuit 10, and cove lighting Circuit 12, this panelboard feeds only show window, entrance, and canopy areas. The loads on the circuits are

1-1200 VA	2-1000 VA	3-1500 VA
4-1000 VA	5-1500 VA	6-1200 VA
7-1000 VA	8-1500 VA	9-1000 VA
10- 300 VA	11-1800 VA	12- 864 VA
13- 900 VA		

The circuits for lighting the show windows, Circuits 1, 2, 3, 4, 5, 6, 7, 8, and 9, take 10,900 VA.

NEC 220-12 provides that a minimum of 200 VA per linear foot be assigned for show window lighting. The two show windows are each 22′ long, so that a minimum of 8800 VA must be used in the feeder calculations. Here the greater value of 10,900 VA connected load will be employed.

The total load on Panel *C* is equal to 14,764 VA, the current being approximately 41 A, the smallest size of Type THW conductor No. 8. Since Panel *C* is directly over the switchboard, voltage drop in the feeder conductors will be negligible, and four No. 8 THW copper conductors in 1″ conduit is satisfactory.

Important Practical Consideration

Since the Specifications state that the basement ceiling is unfinished, and since floors are of poured, reinforced concrete construction, it is necessary to set basement lighting outlet boxes and conduit on the first floor deck at the same time that first floor plug receptacle boxes are spotted there. Plans and Specifications should always be checked with a view to detect these situations.

BASEMENT PLAN

The basement lighting panel is included in the main switchboard. A power panel, located in the boiler room, furnishes current to pump motors there and to a motor that operates the package elevator unit.

The basement contains a shipping and receiving space, an alteration department, a large stock-

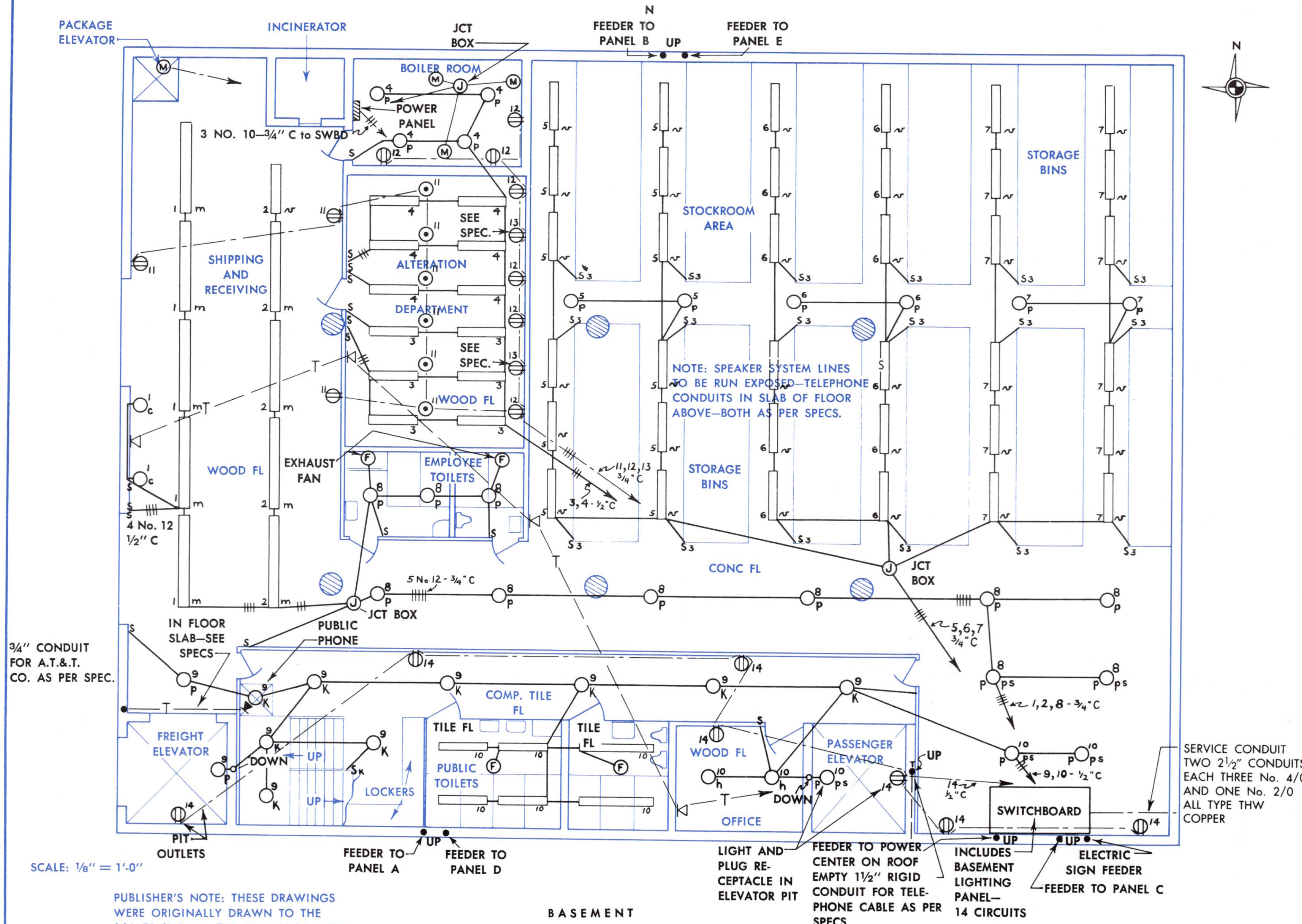

SCALE: 1/8″ = 1′-0″

BASEMENT

PUBLISHER'S NOTE: THESE DRAWINGS WERE ORIGINALLY DRAWN TO THE SCALES SHOWN. THE DRAWINGS WERE REDUCED TO FIT THE PAGE AND CAN NO LONGER BE SCALED.

room area, employees' toilets, public toilets, and a small office. Both employees' and public toilets have exhaust fans connected directly to respective lighting circuits so they are in operation at all times while light switches are turned ON. The Specifications furnish details regarding the fans, stating that they consume 130 VA each. Plug receptacle Circuit 13, in the alteration department, has two special outlets described in the Specifications.

A pair of risers in the north wall are feeders to Panels *B* and *E*. In the south wall, risers at the left are associated with Panels *D* and *A*. Toward the other end, a conduit rises to the power center on the roof. A feeder conduit to Panel *C* is next in line, followed by one to an electric sign on the roof. The DOWN marking in the west wall of the freight elevator shaft refers to a conduit leading to the pit light, and the identical marking in the passenger elevator wall shows a run to its pit light.

Calculations for Basement Panel

Loads on circuits are

1-1200 VA	2- 870 VA	3-1224 VA
4-1624 VA	5-1220 VA	6-1220 VA
7-1220 VA	8-1360 VA	9-1100 VA
10-1464 VA	11-1620 VA	12-1260 VA
13-1500 VA	14-1260 VA	

The total is 18142 VA. The current is approximately 50 A and a No. 6 conductor will be large enough to supply the panel.

Service Calculation

The service load may be obtained by adding together all the feeder requirements, including boiler room power, sign, and roof power. A load of 4800 VA is to be allowed for boiler room power, 8640 VA for the sign, and 36,000 VA for roof power. Starting with Panel E and continuing in the order of calculations, the total load is 16,200 VA + 16,430 VA + 19,968 VA + 14,732 VA + 14,764 VA + 18,142 VA + 8640 VA + 4800 VA + 36,000 VA, or 149,676 VA. The current is equal to 149,676 VA divided by 360, or 416 A.

Note: When making calculations for balanced 4-wire, 3-phase, 120/208 V systems, divide total VA by 360 in order to obtain current per phase wire.

Table 310-16 shows that a 600 kcmil Type THW conductor will carry 420 A. Two smaller conductors are more easily handled than a single large one. NEC 310-4 permits conductors of No. 1/0 and larger to be run in parallel, provided they are of the same length, conductor material, circular-mil area, insulation type, and are terminated in the same manner. Where parallel conductors are run in separate raceways or cables, the raceways or cables shall have the same physical characteristics.

Two parallel conductors on each leg of the circuit would each carry one-half of the total current, or 208 A. With all conductors in the same conduit, a total of six current-carrying wires, the rating of each must be reduced to 80% of the value indicated in Table 310-16. A size No. 4/0 Type THW conductor has an ampacity of 230 A where there are not more than three current-carrying conductors in a raceway. For six conductors, each must have a normal ampacity equal to 208 A divided by .8, or 260 A. A 300 kcmil Type THW would be required in this case. But if two sets of No. 4/0 conductors are installed in separate conduits, the full ampacity of each conductor is restored, and there is considerable saving of copper.

NEC 220-22 states that the neutral feeder must be large enough to carry the maximum unbalanced load. The *maximum unbalanced load* is that load that can be connected between the neutral and any ungrounded conductor, which means that loads that have no neutral component, such as power loads confined to the 3- phase wires, must be subtracted from total load in order to determine this value of current. Total load in the present instance is 416 A and the 3-phase power load 114 A, (boiler-room power plus roof power). The maximum unbalanced load on the neutral is 416 A – 114 A or 302 A.

NEC 220-22 provides that a demand factor of 70% may be applied to that portion of the neutral load in excess of 200 A. It appears that the neutral load could be reduced to 200 A + (.7 × 102 A) or 272 A. NEC 220-22 states that no reduction of neutral capacity shall be allowed for that part of the load that consists of electric discharge lighting. Since electric discharge lighting in the present example accounts for at least 150 A, there is no excess over 200 A to which a 70% factor might be applied. The neutral load, then, must be calculated on the basis of 302 A.

NEC 300-3(b) requires that a grounded conductor be run in the same raceway with each set of phase conductors. Either of two neutral wires in this case will carry one-half of the unbalanced current (302 A), or 151 A. The nearest size listed in the table is No. 2/0 Type THW, which has an ampacity of 175 A.

NEC 230-30 permits an uninsulated copper grounded conductor to be installed underground in duct or conduit. It may appear that a larger ampacity might be assigned to an uninsulated conductor, so that a size smaller than No. 2/0 THW could be used. But Note 5 of Notes to Ampacity Tables of 0 to 2000 Volts states that where bare conductors are used with insulated conductors their allowable ampacity shall be limited to that permitted for an insulated conductor of the same size. The present Specifications require that an insulated neutral conductor be used. The service consists of two sets of three No. 4/0 Type THW conductors and one set of No. 2/0 Type THW conductors in parallel.

Service Conduit Size

When all conductors in a raceway are of the same size, Tables 3A, 3B, and 3C, Chapter 9 determine the size of conduit required. Each table includes all sizes of conductors with all of the different types of conductor insulation included in one of the three tables. Table 3A includes Types such as RH, TW, and THW. Table 3B includes types such as THWN, THHN, and XHHW. Table 3C includes Types RHW and RHH. For different sizes of wire in a conduit, calculations are based upon Table 1 and Tables 4 through 8, Chapter 9.

Table 4 gives dimensions and percent of area of conduit and tubing. The other tables give dimensions for the various types of conductor coverings including rubber, thermoplastic, lead-covered, asbestos-varnished-cambric, cross-linked polyethylene, and bare.

In the store service, each conduit has three No. 4/0 Type THW conductors and one No. 2/0 Type THW conductor. Table 5 gives the cross-sectional area of a No. 4/0 Type THW conductor as .3904 sq in. and a No. 2/0 Type THW conductor as .2781 sq in. The area of all four conductors equals (3 × .3904 sq in.) + .2781 sq in., or 1.4493 sq in. Table 1 allows 40% of conduit cross-sectional area to be used when there are three or more conductors of different sizes. Cross-sectional area of the conduit, therefore, must be equal to 1.4493 sq in. divided by .4, or 3.6233 sq in. According to Table 4, the required size of conduit is 2½″.

480 V Lighting System

A method of lighting that has become increasingly popular in commercial and industrial establishments is the four-wire, three-phase, 277/480 V electric discharge system. Lighting units are connected between neutral conductor and phase wires, imposing 277 V on ballast assemblies. Switching may be at the panelboard if circuit breakers are listed and marked SWD per 240-83(d). Where wall switches are used and the voltage between adjacent switches exceeds 300 VA, a permanent barrier is required in outlet boxes between switches per 380-8(b).

Aluminum Wire and Conduit

In recent years, aluminum wire and conduit have competed with copper wire and steel conduit. Since aluminum conduit is made in the same standard trade sizes as steel, there are no Code problems. Aluminum conductors, however, vary with respect to copper in both allowable ampacity and ohmic resistance because ampacity for a given size is less than that of copper and resistance is higher. The sizes of wire and conduit required will be calculated when Type THW aluminum feeder and service conductors are substituted for Type THW copper conductors in the store building example.

The formula for calculating voltage drop for aluminum conductor is the same as the formula used for copper conductors. The resistivity (R) for copper and aluminum conductors is given in Table 8, Chapter 9.

The formula for calculating voltage drop (VD), %VD, and number and size of copper and aluminum conductors for Panels *A–E* is

$$VD = \frac{2 \times R \times L \times I \times .866}{1000}$$

where

VD = voltage drop

R = resistivity per Table 8, Chapter 9

L = length (in feet)

I = current

1000 = 1000′ or less of wire

COPPER					
Panel	**R**	**L**	**I**	**Conductors**	**Conduit**
A	.778	73	41	4 No. 8 THW	1″
B	.491	92	56	4 No. 6 THW	1″
C	.778	73	41	4 No. 8 THW	1″
D	.308	83	46	4 No. 4 THW	1¼″
E	.491	104	45	4 No. 6 THW	1″

ALUMINUM					
Panel	**R**	**L**	**I**	**Conductors**	**Conduit**
A	.808	73	41	4 No. 6 THW	1″
B	.508	92	56	4 No. 4 THW	1¼″
C	.808	73	41	4 No. 6 THW	1″
D	.808	83	46	4 No. 2 THW	1¼″
E	.508	104	45	4 No. 4 THW	1″

SERVICES

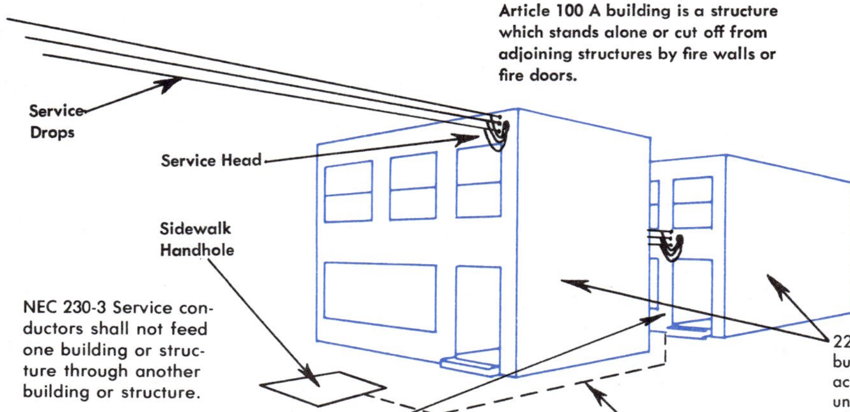

Article 100 A building is a structure which stands alone or cut off from adjoining structures by fire walls or fire doors.

NEC 230-3 Service conductors shall not feed one building or structure through another building or structure.

225-8(b) Conductors supplying each building shall be provided with readily accessible means for disconnecting ungrounded conductors from source of supply. The disconnecting means for garages, etc., for residential may be 3- and 4-way snap switches.

NEC 250-24(a), Ex. 1, 2 Where grounded system serves two or more buildings from the same service, each building shall have netural to its own grounding electrode except if second building has single-branch circuit and no equipment that requires grounding or an equipment grounding conductor is run to second building.

NEC 230-6 Conductors installed under at least two inches of concrete beneath a building or enclosed by concrete or brick at least 2 inches thick or in a transformer vault shall be considered outside a building.

SERVICE-ENTRANCE CONDUCTORS

NEC 230-54 (f) Drip loops required to prevent moisture and are used to connect service-entrance wire to service-drop wires below level of service head.

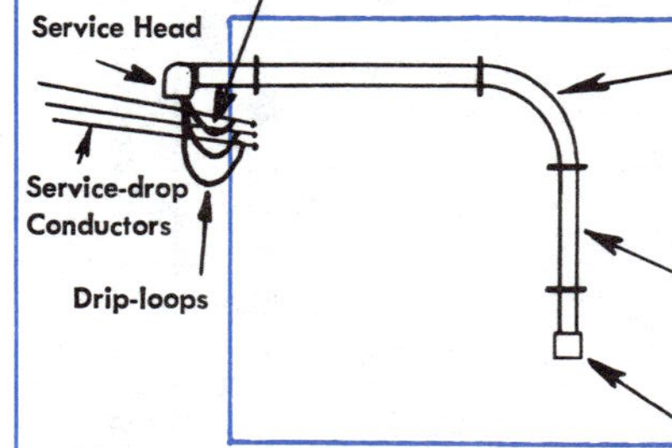

Service-entrance. (Service-entrance conductors in service-entrance conduit)

NEC 230-46 Service-entrance conductors shall be without splice. (See Ex. 1–6.)

Service-entrance conductors feed service-entrance equipment.

230-43 Wiring methods for service conductors are limited to: (1) open wiring on insulators, (2) IGS Cable, (3) Rigid Metal Conduit, (4) IMC, (5) EMT, (6) ENT, (7) SE Cable, (8) Wireways, (9) Busways, (10) Auxiliary Gutters, (11) Rigid Nonmetallic Conduit, (12) Cablebus, (13) MC Cable, (14) MI Cable, (15) Flexible Metal or Liquidtight Flexible Metal Conduit, 6′ maximum length, with equipment bonding jumper, (16) Liquidtight Flexible Nonmetallic Conduit.

SERVICE EQUIPMENT

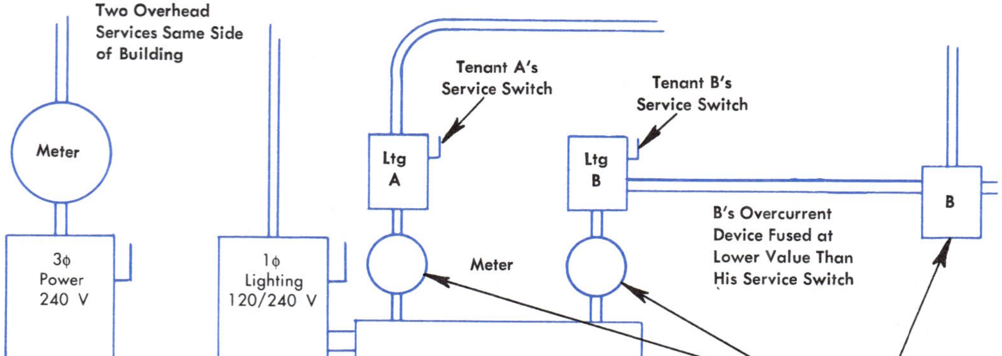

NEC 230-91(b) In a multiple-occupancy building, each occupant shall have access to overcurrent devices except as permitted by 240-24(b), Ex. 1.

NEC 230-93 If necessary to prevent tampering, the service equipment for a special load may be locked.

NEC 230-92 If service equipment locked or inaccessible, overcurrent devices of lower rating must be installed in an accessible location.

NEC 230-72(a)(b) If supplied by two or more services on the same side of the building, the service equipment shall be grouped and marked for identification. Service for fire pumps permitted to be in another location.

SERVICE DISCONNECTING MEANS

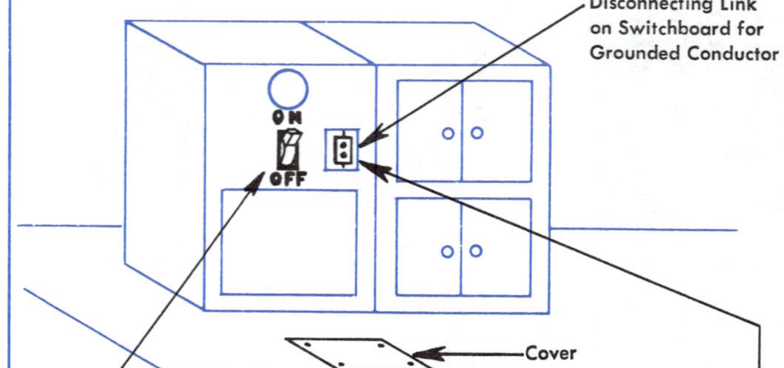

NEC 230-76 The service disconnecting means shall be operable by hand and shall consist of either a manually operable switch or circuit breaker or an electrically operated switch or circuit breaker also operable by hand.

NEC 230-75 If the switch or circuit breaker does not interrupt the grounded conductor, other means of disconnection must be provided in the cabinet or switchboard.

NEC 384-3 (c) Switchboard or panelboards used as service equipment shall have means within the service disconnect section for connecting the supply neutral to the metal frame. (Size as per 250-79(d).)

GROUNDING ELECTRODE CONDUCTOR

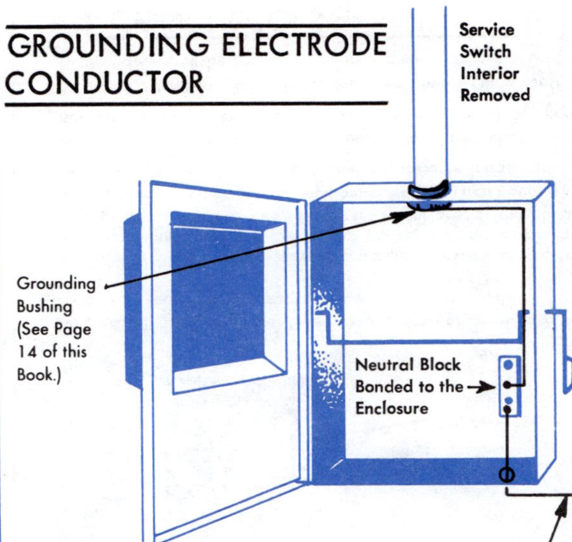

NEC 250-79(a) Main and equipment bonding jumpers shall be of copper or other corrosion-resistant material and shall be sized as per 250-79(d)(e).

NEC 250-91 (a) **The grounding electrode conductor shall be of copper, aluminum, or copper-clad aluminum, and selected to be resistant to any corrosive conditions or be suitably protected against corrosion. Shall be solid or stranded, insulated, covered, or bare and continuous without splice or joint except: (1) busbars; (2) where more than one service (as per 230-40, Ex. 2), the GEC, sized per 250-94, serving the largest service entrance conductors, may be tapped for GECs for other enclosures. This tap permitted to be sized per 250-94 for largest phase conductor serving respective separate enclosures, and then only permitted where connection extends within enclosure.**

NEC 250-92(a) Grounding electrode conductor No. 4 and larger may be attached to the surface, also a No. 6 if not exposed to physical injury. Aluminum grounding conductors may not be run in direct contact with masonry or earth or where subject to corrosive conditions. If run outside, aluminum grounding conductors shall not be installed within 18 inches of the earth.

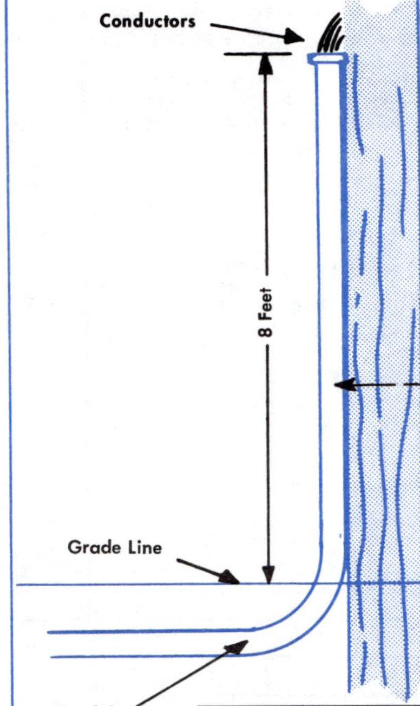

NEC 230-49 **If underground service conductors are carried up a pole, they shall be protected by suitable means to a height of at least 8 feet above ground. (230-49 refers to 300-5.)**

NEC 300-5(d) Conductors emerging from the ground to be suitably protected to a height of 8 ft. and at least 18 inches below ground. Where subject to physical damage, rigid metal conduit, IMC, Schedule 80 PVC or equivalent acceptable.

CONDUCTOR PROTECTION-POLE

NEC 300-15(b), Ex. 4 Switch or receptacle permitted without a box for conventional on-site frame construction for use with romex if device has brackets that securely fasten the device to a structural member of the walls or ceiling.

NEC 336-18, Ex. 2 Nonmetallic-sheathed cable permitted to feed a device without a box if the device is identified for the use and cable is secured within 12 inches of the device and loop of at least 6 inches of free cable available on the interior side of the wall to permit replacement.

NEC 336-25 Switches and receptacles with integral enclosures permitted with romex if identified for such use per 300-15(b), Ex. 4.

NEC 545-10 A receptacle or switch with an integral enclosure and mounting means permitted in manufactured buildings if tested and listed for such use.

NEC 550-10(i), Ex. Devices with integral enclosures provided with special wall or ceiling mounting **brackets are permitted in manufactured buildings without a box if devices and brackets are tested and approved for wall and ceiling construction systems used.**

NEC 551-47(j), Ex. Where approved devices are used without boxes in recreational vehicles the extra loop of cable shall be considered as an integral part of the device.

Integral Receptacle

Mounting Bracket

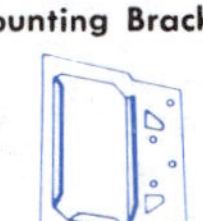

INTEGRAL DEVICES

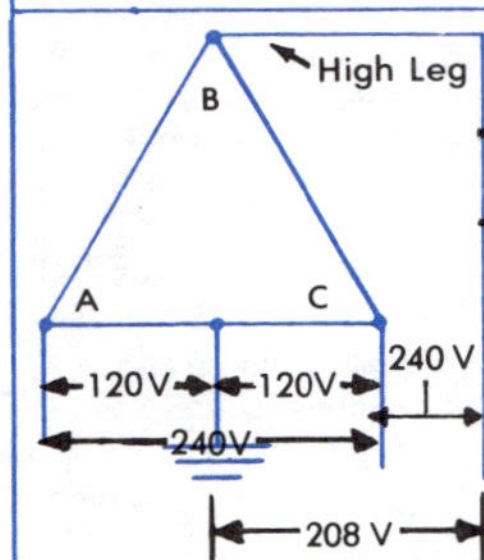

This wire to be orange or identified by tagging or other effective means

NEC 215-8 The 4-wire delta phase conductor having the highest voltage to ground (so-called high-leg) shall be orange in color or identified by tagging or other effective means where the neutral is present.

NEC 384-3 (f) High-leg connected to B (center) phase in switchboard or panelboard.

HIGH-LEG -- IDENTIFICATION

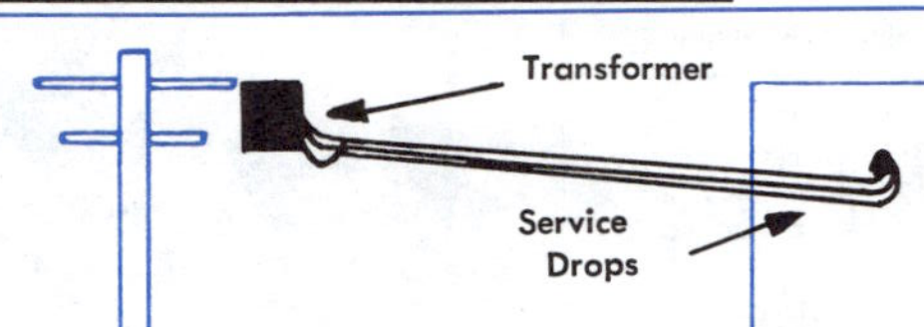

NEC 230-2 A building or structure shall have only one service. (See Exceptions.) If more than one, a permanent plaque or directory at each, indicating other services and area served by each.

NUMBER OF SERVICES

NEC 230-8 Underground raceway shall be sealed.

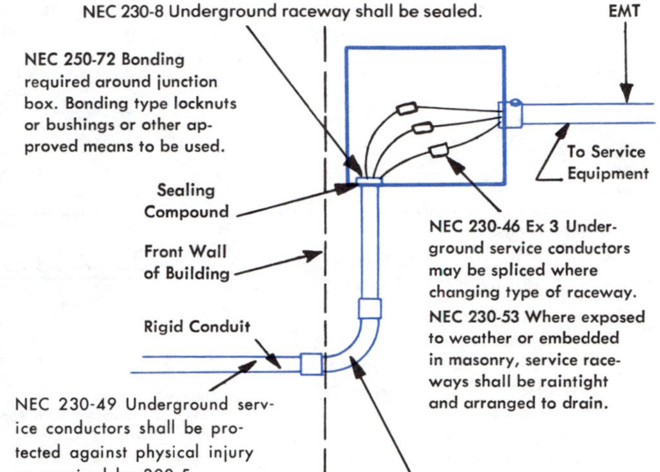

NEC 250-72 Bonding required around junction box. Bonding type locknuts or bushings or other approved means to be used.

NEC 230-46 Ex 3 Underground service conductors may be spliced where changing type of raceway.

NEC 230-53 Where exposed to weather or embedded in masonry, service raceways shall be raintight and arranged to drain.

NEC 230-49 Underground service conductors shall be protected against physical injury as required by 300-5.

NEC 300-5 (c) If raceway is under building and contains underground cable it shall extend beyond outside walls of building.

SERVICE-UNDERGROUND

FLAT CONDUCTOR CABLE -- TYPE FCC-- ARTICLE 328

NEC 328-15 Transition from one system to FCC system made with transition assembly identified for this use.

Transition Assembly (See (328-2 Definitions.)

NEC 328-11 All connections to be made with connectors identified for use and all bare cable ends to use listed insulating ends. All cable connections and ends to be sealed against dampness and liquid spillage.

Transition Assembly

NEC 328-16 All system components firmly anchored by adhesive or mechanical anchoring system identified for this use.

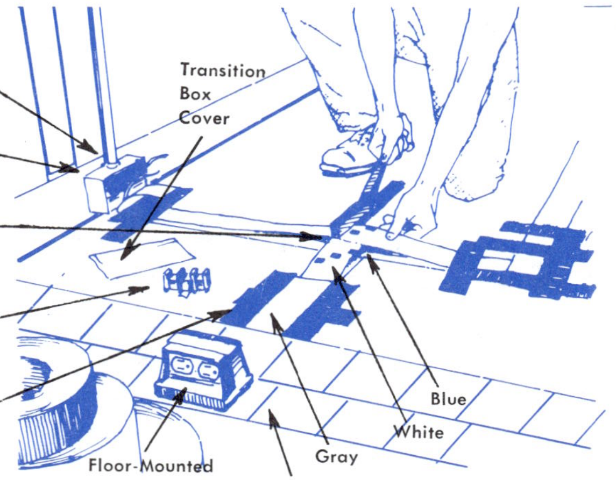

NEC 328-10 Carpet squares not larger than 36 in. by 36 in.

NEC 328-10 Type FCC cable to be covered with carpet squares that have release-type adhesive.

NEC 328-6 Branch circuits: (a) Voltage not over 300 volts between ungrounded conductors or 150 volts to ground; (b) Not over 20 amps for general-purpose circuit or 30 amps for individual circuit.

NEC 328-5 Not permitted: (1) outdoors or wet locations; (2) corrosive vapor areas; (3) hazardous areas; (4) residential, school and hospital buildings.

NEC 328-3 System to comply with 210, 220, 240, 250, 300.

NEC 328-1 Type FCC cable is a field-installed wiring system using associated accessories for branch circuits and installed under carpet squares as per Article 328.

MULTIOUTLET ASSEMBLY

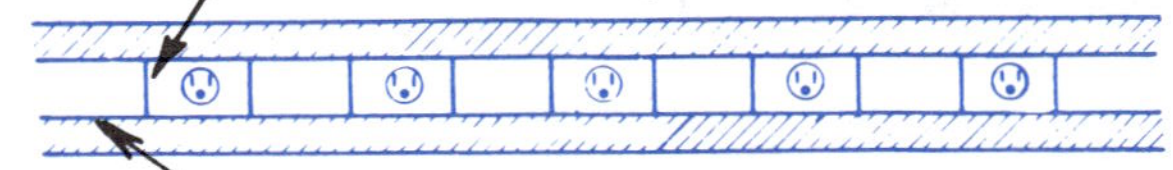

220-3(c)(7), Ex. 1 Each 5′ or fraction thereof of fixed multioutlet assembly shall be considered as one outlet of 180 VA capacity, except where a number of appliances may be connected simultaneously, then each 1′ or fraction will be considered one outlet. NOTE: This exception does not apply to dwelling units or guest rooms of motels or hotels.

NEC 353-2 Multioutlet assembly may be used in dry locations. It shall not be used: (1) where concealed except that back and sides of metallic assembly may be surrounded by building finish or baseboard; (2) where exposed to severe physical damage; (3) where voltage between conductors is 300 V or more unless metal is not less than .040″ thick; (4) where subject to corrosive vapors; (5) in hoistways; nor (6) in any hazardous location.

NEC 353-3 May be used through dry partition where no outlet is concealed and provisions are made for removing the cap or cover.

HEAVY-DUTY LAMPHOLDERS -- RATING

NEC 720-5 For circuits of less than 50 V, standard lampholders rated not less than 660 VA are required.

NEC 210-21(a) Lampholders connected to circuits having a rating greater than 20 A shall be of the heavy-duty type and shall be rated not less than 660 VA, if of admedium type, and not less than 750 VA if any other type. See page 182 for lampholders.

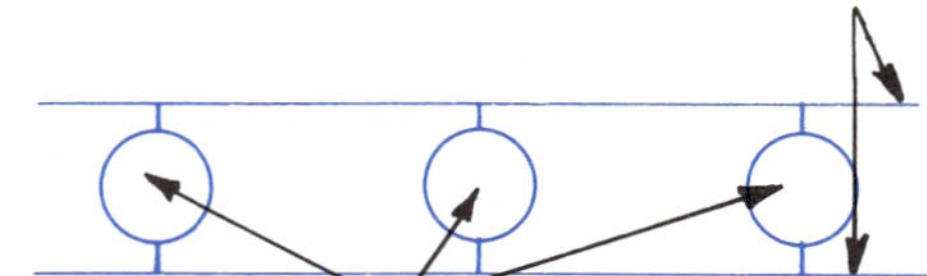

CONDUCTORS -- OVERCURRENT PROTECTION

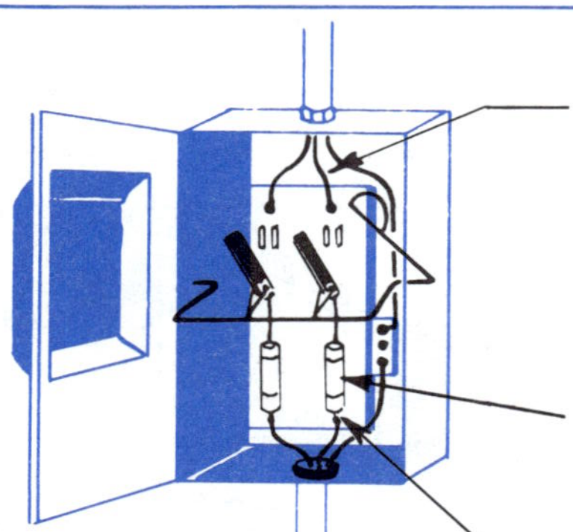

NEC 240-21 (General Rule) Overcurrent protection required where conductors receive their supply.

240-3(b) If allowable ampacity of conductor does not correspond with a standard rating of overcurrent device, the next larger rating may be used, provided that the current is not greater than 800 A and conductor is not part of multioutlet branch circuit supplying receptacles for cord and plug-connected portable loads.

NEC 210-19 (a) **Branch-circuit conductors shall have an ampacity not less than branch circuit and not less than maximum load served.**

NEC 240-3 (General Rule) Conductors other than flexible cords and fixture wires shall be protected per their ampacities per 310-15.

RECEPTACLE RATINGS

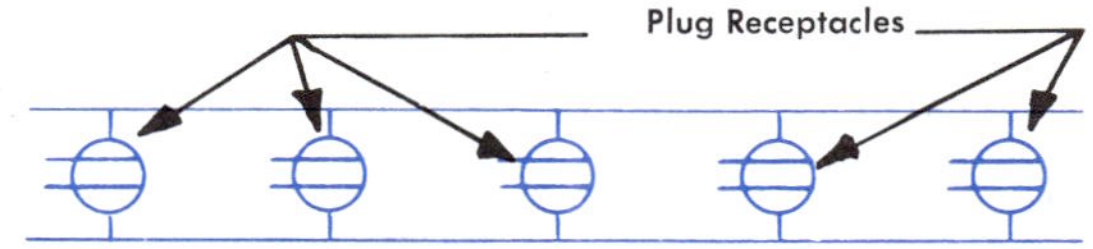

NEC 210-21(b)(1)(2) Plug receptacles shall have a rating not less than load served; when connected to circuits having two or more outlets, they shall conform to following:

15-amp circuit—max. rating 15 amps
20-amp circuit—rating 15 or 20 amps
30-amp circuit—rating 30 amps
40-amp circuit—rating 40 or 50 amps
50-amp circuit—rating 50 amps

Receptacles installed on 15-amp and 20-amp circuits must be of the grounding type. (210-7)

NEC 210-21 (b) (1) A single receptacle installed on an individual branch circuit shall have a rating not less than branch circuit.
NOTE: 15 A and 20 A receptacles shall not supply a cord- and plug-connected load where total load exceeds 80% of rating of receptacle. Table 210-21(b)(2)

CIRCUIT BREAKERS

NEC 240-83(d) Where CBs are used to switch 120 V and 277 V fluorescent fixtures, they shall be marked SWD and listed for such switching duty.

NEC 380-11 CBs used as a switch must conform to 240-81 and 240-83.

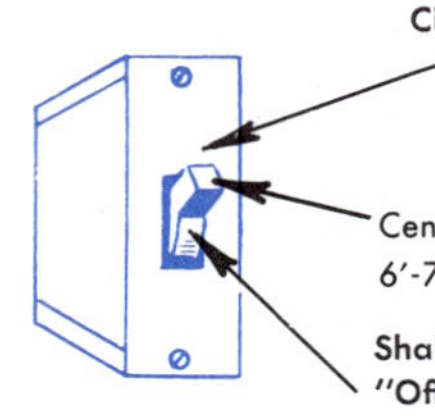

240-3(b) If allowable ampacity of a conductor does not correspond with a standard rating of overcurrent device, the next larger rating may be used, provided that the current is not greater than 800 A.

NEC 240-82 A CB shall be of such design that alteration of its operating characteristics will be difficult.

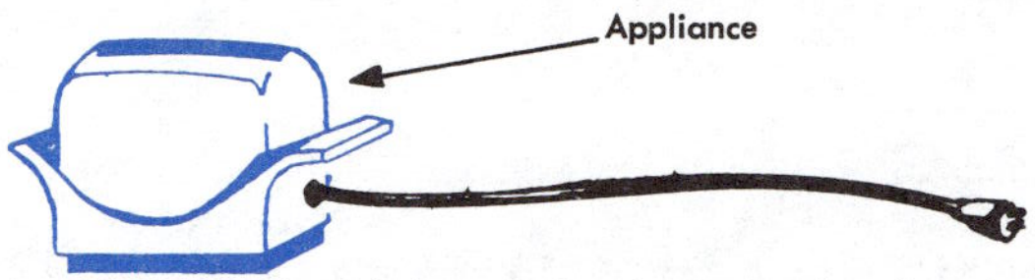

240-4 Flexible cords (including tinsel) to be protected at their ampacities per Table 400-5(A).

Ex. 1. Cord listed with specific appliance or portable lamp is permitted to be connected to a standard branch circuit as follows:
Tinsel on No. 18 or larger cord, 20-amp circuit.
No. 16 or larger cord, 30-amp circuit.
Cord rated 20-amp or over for 40- or 50-amp circuit.
Ex. 2. Fixture wire connected to 120-volt or more standard branch circuit permitted to be protected as follows:
No. 18 (up to 50 ft.) or No. 16 (up to 100 ft.), 20 amp circuit.
No. 14 or larger, 20- or 30-amp circuit.
No. 12 or larger, 40- or 50-amp circuit.
EX. 3 Listed extension cordsets with No. 16 or larger conductors permitted on 20-amp branch circuit.

CORDS AND FIXTURE WIRE -- OVERCURRENT PROTECTION

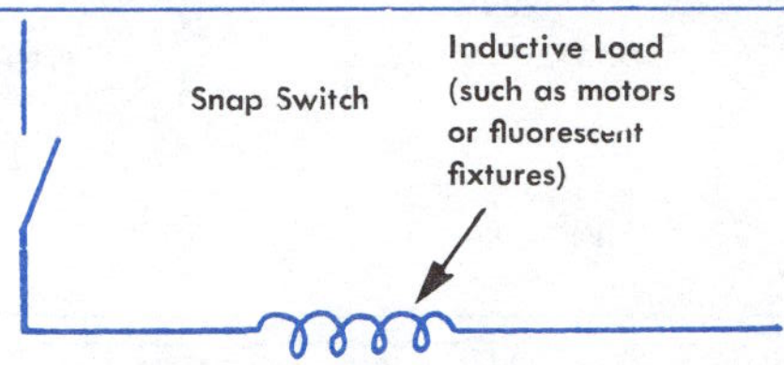

NEC 380-14 (a–d) AC general-use snap switches controlling inductive loads shall not exceed rating of load at applied voltage. They shall have rating not exceeding 50% of ampere rating of switch at applied voltage. Where rated 20 A or less and connected to aluminum wire, they shall be marked CO/ALR and be listed for purpose.

SNAP SWITCHES -- RATING

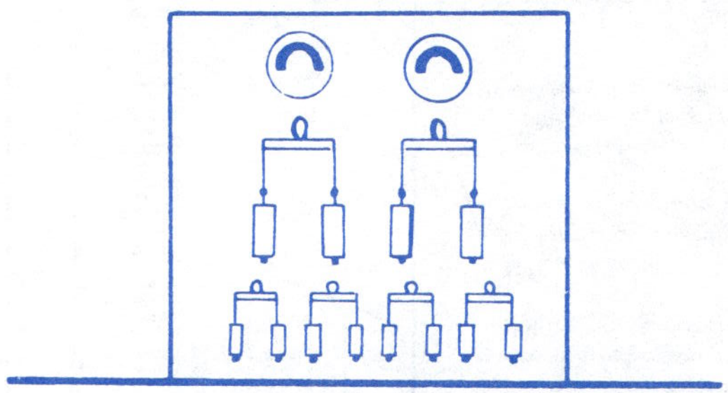

NEC 384-5 Switchboards with exposed live parts shall be located in permanently dry locations accessible only to qualified persons.

SWITCHBOARDS -- LOCATION

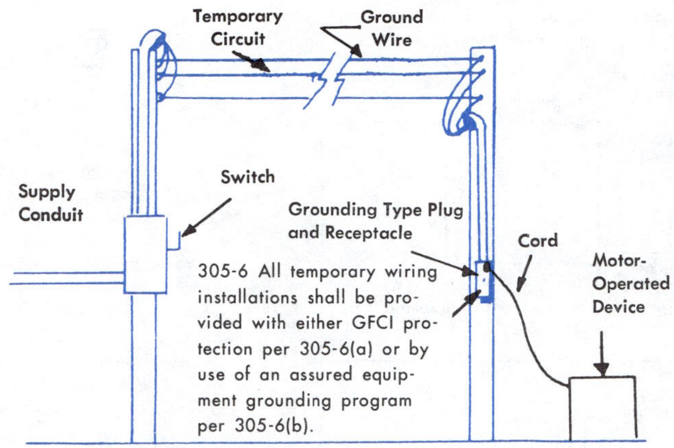

305-6 All temporary wiring installations shall be provided with either GFCI protection per 305-6(a) or by use of an assured equipment grounding program per 305-6(b).

305-4(a)–(h) Services shall conform to 230; feeders and branch circuits to be protected at rated ampacity; receptacles to be of grounding type and connected to grounding conductor; receptacles not permitted on lighting branch circuits; suitable disconnecting switches or plug connectors required to disconnect all ungrounded conductors of each temporary circuit; lamps to be protected by suitable fixture or lampholder with guard; box not required for splices or junction connections when using cable assemblies, cords or cables. See 110-14(b) and 400-9.

TEMPORARY WIRING

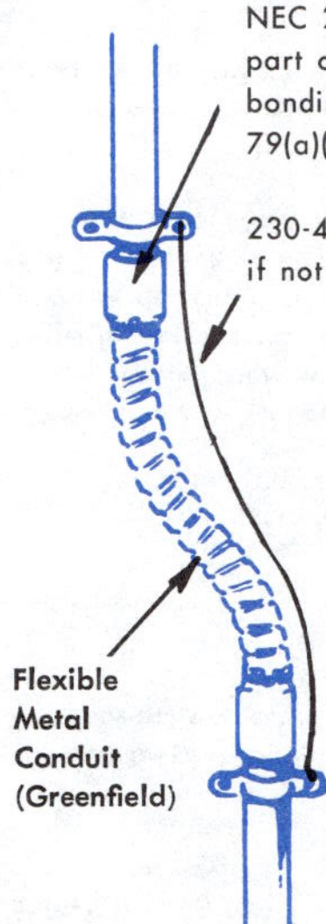

NEC 230-43(15) Flexible Metal Conduit permitted as part of service raceway if not over 6′ long with bonding jumper routed with flex per 250-79(a)(c)(d)(f). Also see 350-2.

230-43(15) Greenfield permitted for service conduit if not over 6′ and equipment bonding jumper.

350-14, Ex. Flexible Metal Conduit permitted as grounding means where conduit and fittings are listed for purpose per 250-9(b) but total length shall not exceed 6′ and not over a 20 A circuit. Bonding jumper may be installed on outside of conduit per 250-79(f) and sized per Table 250-95.

350-10 Minimum trade size for Flexible Metal Conduit is ½″. See Exceptions 1–5.

350-16 Maximum of 4 quarter bends in one run of conduit.

FOR USE RESTRICTIONS, SEE 350-5

350-18 Flexible conduit shall be secured within 12″ of box or fitting and not over 4½′ in remainder or run except: (1) where fished; (2) not over 3′ at terminals for flexibility; (3) up to 6′ at fixtures per 410-67(c).

FLEXIBLE METAL CONDUIT

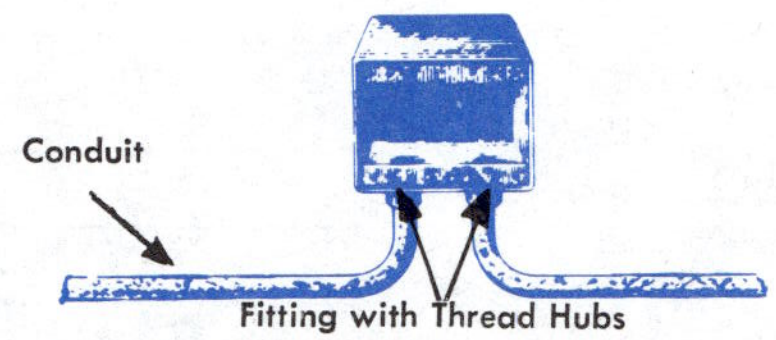

NEC 370-23(d) Threaded boxes or fittings not over 100 cu in. in size which contain no devices are considered fastened if two or more conduits are threaded into box on two or more sides and are supported within 3′ of the box.

370-23(e) Threaded boxes or fittings with devices not over 100 cu in. are considered supported if two or more conduits are threaded into box and are firmly secured within 18″ of the box.

THREADED BOX SUPPORTS

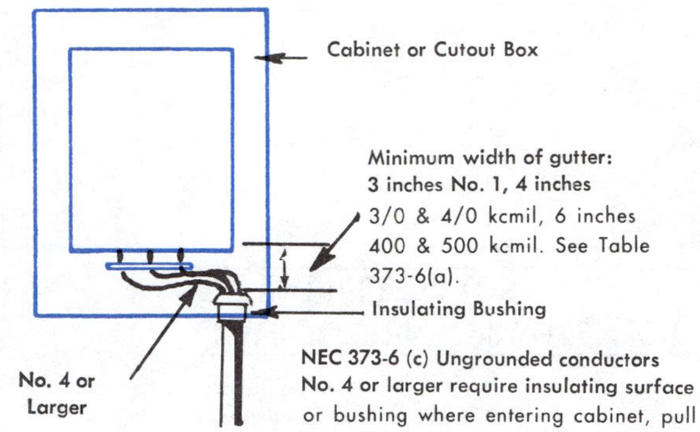

NEC 373-6 (c) Ungrounded conductors No. 4 or larger require insulating surface or bushing where entering cabinet, pull box, junction box, or auxiliary gutter.

DEFLECTION OF CONDUCTORS

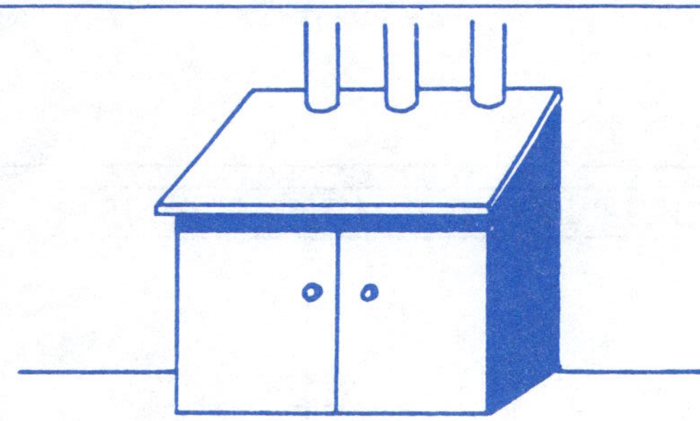

NEC 384-6 A switchboard in a wet location or outside a building shall have a weatherproof enclosure to comply with 373-2(a).

SWITCHBOARD -- OUTSIDE

LIGHTING AND APPLIANCE PANELBOARD

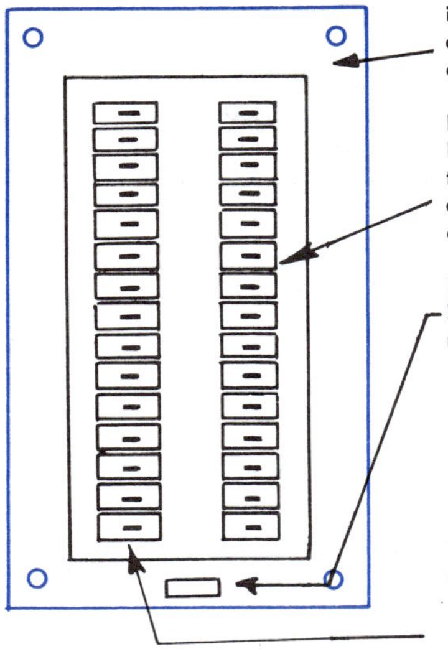

NEC 384-14 A lighting and appliance switchboard is one having more than 10% of its overcurrent devices rated 30 amperes or less for which neutral connections are provided.

NEC 384-15 Not over 42 overcurrent devices of lighting and appliance panelboard, other than the mains, permitted in one cabinet. A two-pole circuit breaker counts as two overcurrent devices, a three-pole breaker as three.

NEC 384-16 (a) Ex 1 Panelboard may be protected by feeder overcurrent device if it does not exceed rating of panelboard.

NEC 384-16 (a) Ex 2 For existing installations individual protection not required in individual residential occupancy.

NEC 384-16 (a) (c) Lighting and appliance switchboards shall be protected individually by not more than two circuit breakers or sets of fuses unless the feeder provides adequate protection. Load on an overcurrent device shall not exceed 80% of its rating where load will continue for three hours or more, unless device is listed for 100% continuous operation. See Exception.

NEC 384-19 Except as permitted for services, panelboard fuses shall be installed on load side of switches.

SNAP SWITCHES IN PANELBOARDS

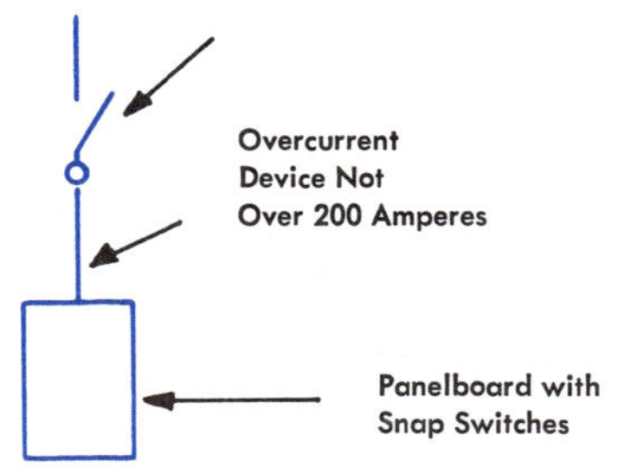

NEC 384-16 (b) Panelboards with snap switches rated 30 amps or less shall have overcurrent protection not in excess of 200 amps.

OVERCURRENT PROTECTION OF PANELBOARDS

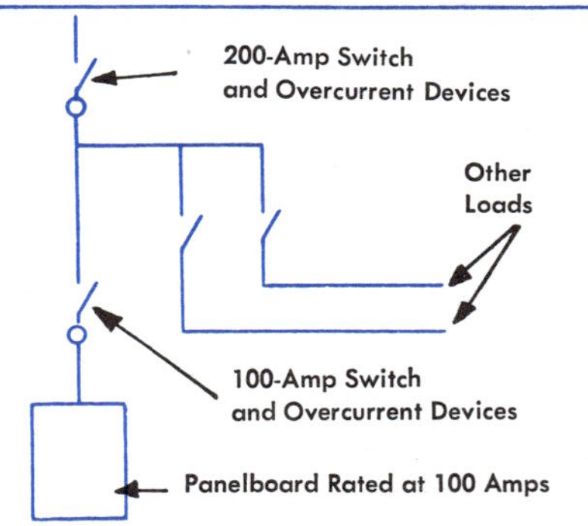

NEC 384-16 (a) A lighting and appliance panelboard shall be protected by not more than two sets of fuses or circuit breakers having a rating not greater than that of the panelboard.

FLEXIBLE CORDS

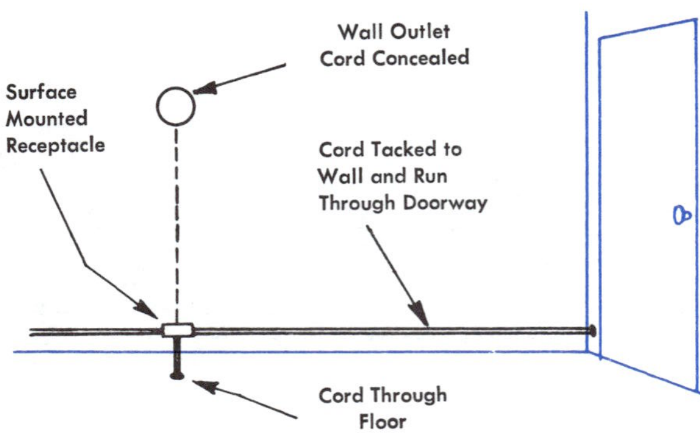

NEC 400-8 Flexible cords and cables shall NOT be used: (1) as a substitute for fixed wiring; (2) thru holes in walls, floors, or ceilings; (3) thru doors, windows, or similar openings; (4) attached to building surfaces; (5) concealed by building walls, floors, or ceilings; (6) in raceways. See Exception.

NEC 400-7 Flexible cords and cables, permitted uses: (1) pendants; (2) fixtures; (3) portable lamps or appliances; (4) elevator cables; (5) cranes and hoists; (6) stationary equipment to facilitate frequent interchange; (7) prevent transmission of vibration or noise; (8) where mechanical connections and fastening means of appliance are designed to permit removal for maintenance and repair; (9) data processing cables as per 645-5; (10) connection of moving parts; (11) temporary wiring as per 305-4(b)(c).

NEC 400-9 Flexible cords shall be used only in continuous lengths, not spliced or tapped except junior hard-service and hard-service flexible cord #14 and larger if spliced per 110-14(b).

NEC 400-11 Flexible cord used in show windows and show cases shall, in general, be S, SO, SE, SEO, SOO, SJ, SJE, SJO, SJEO, ST, STO, STOO, SJT, SJTO, SJTOO, or AFS.

NEC 400-12 In general, flexible cords shall not be smaller than No. 18. (See Tables 400-4, 400-5.)

FLOOR BOXES

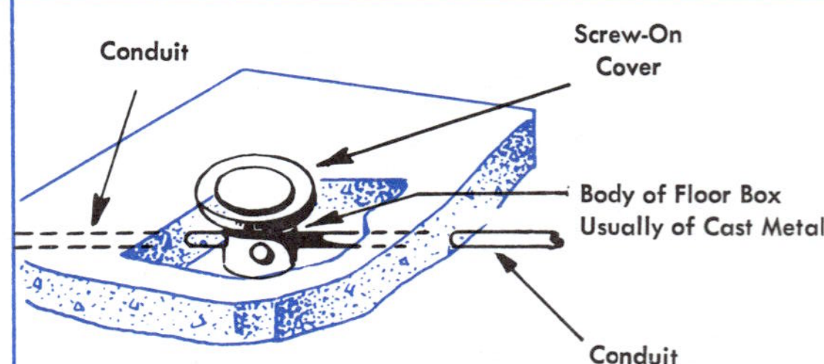

370-27(b) Receptacles located in floors shall be installed in floor boxes listed specifically for this application. See Exception for show windows.

SHOW WINDOW WIRING

NEC 220-12 Unit load of 200 VA per linear foot required.

NEC 600-5(a) In commercial buildings and occupancies accessible to pedestrians, a separate 20 A circuit to outside outlet for sign required. Not required for entry to inside hall or corridor.

NEC 210-62 At least one receptacle outlet required above show window for each 12 ft or major fraction thereof, measured horizontally.

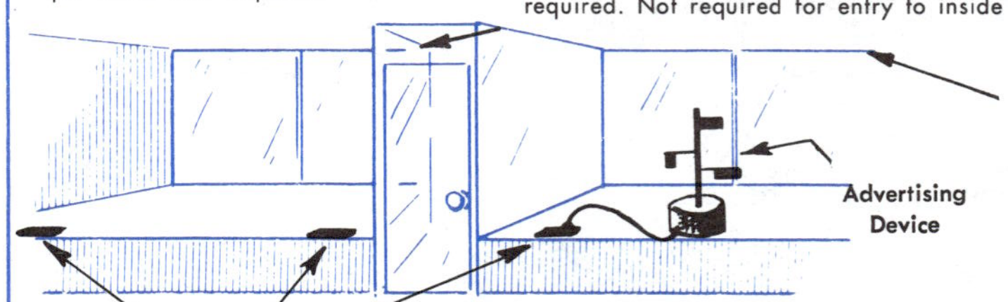

370-27(b), Ex. Receptacles in elevated floors of show windows and similar locations may be installed in standard type receptacle boxes where enforcing authority judges them to be free from physical damage, dirt, and moisture.

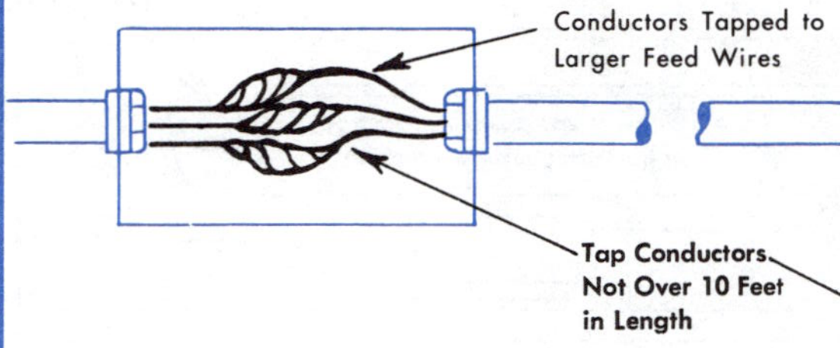

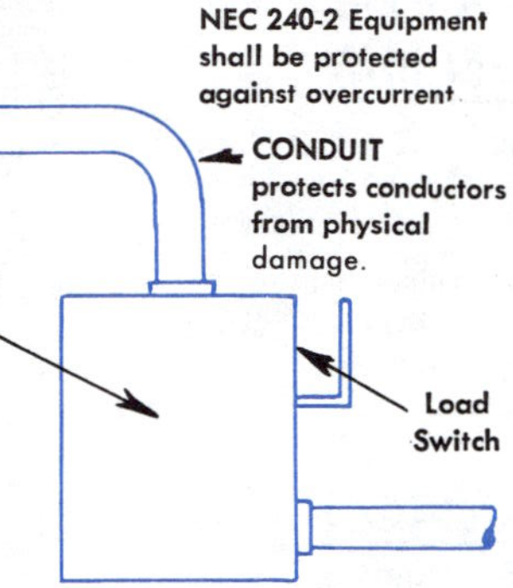

240-21(a) In general, each ungrounded conductor shall have an overcurrent device connected in series with it at point at which it receives its supply,

240-21(b) Feeder 10′ Tap Rule. Overcurrent protection not required where a feeder or transformer secondary conductor receives supply, if tap conductor is not over 10′ long, ampacity of tap conductor is not less than load served or rating of device supplied, and not less than overcurrent device tap conductor feeds. Also tap conductors not permitted beyond switchboard, panelboard, disconnecting means, or control devices they supply, and are protected from damage.

NEC 240-23 (General Rule) Where a change occurs in the size of the ungrounded conductor, a similar change may be made in the size of the grounded conductor.

FEEDER 10′ TAP RULE

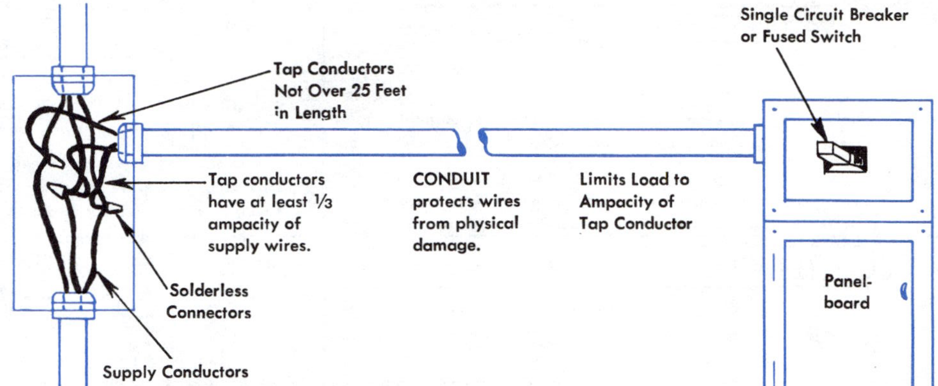

FEEDER TAP NOT OVER 25 FEET LONG

240-21(c) A smaller conductor may be tapped to a larger one without overcurrent protection at that point if smaller conductor has an ampacity at least one-third that of rating of overcurrent device from which it is supplied, and provided that tap is suitably protected from physical damage, is not over 25′ long, and terminates in a single CB or set of fuses that limits current in smaller conductor to a suitable value.

FEEDER TAP OVER 25 FEET LONG

240-21(e) High bay (over 35′ high at walls) manufacturing buildings, maintained and serviced by qualified persons, permit feeder taps not over 25′ horizontally and not over 100′ total length where the tap conductor: (a) has at least one-third ampacity of feeder overcurrent protection; (b) terminates in single overcurrent device that limits load to ampacity of tap conductors; (c) suitably protected from physical damage and in raceways; (d) continuous and contains no splice; (e) No. 6 copper or No. 4 aluminum or larger; (f) does not penetrate walls, floors, or ceilings, (g) tap not less than 30′ from floor.

FEEDER TAP RULES

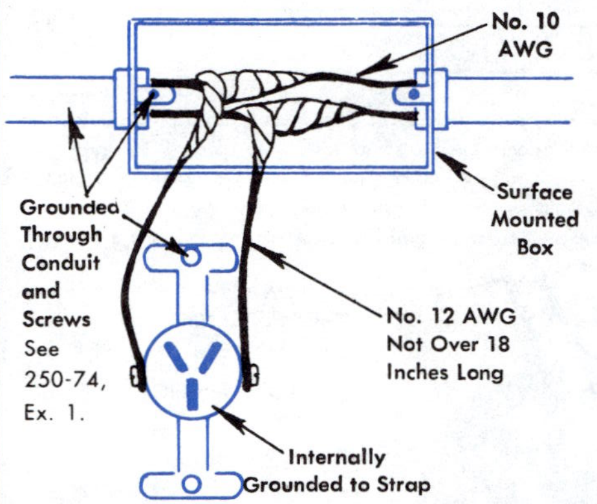

NEC 210-19 (c) Ex 1 Branch-circuit tap conductors (for other than cooking loads as per 210-2) shall be at least No. 14 and shall be rated not less than 15 amps when tapped from less than 40-amp circuit and not less than 20 amps when from 40- or 50-amp circuit when loads consist of: (a) lampholder or fixture where tap is not over 18″ long; (b) fixture has tap conductors per 410-67; (c) individual outlets, other than those for receptacles, less than 18″; (d) industrial, infrared lamp heating appliances; (e) nonheating leads of de-icing and snow-melting cables and mats. For fixture wires and cords, see 240-4.

BRANCH-CIRCUIT TAP RULES

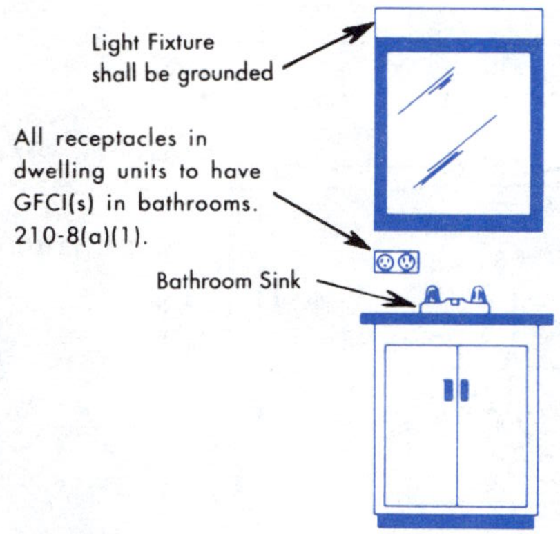

NEC 250-42 (General Rule)
Noncurrent-carrying parts of fixed equipment shall be grounded if so located as to be potentially dangerous to persons or capable of creating or of increasing a hazardous condition.

EQUIPMENT GROUNDING-- FIXED EQUIPMENT

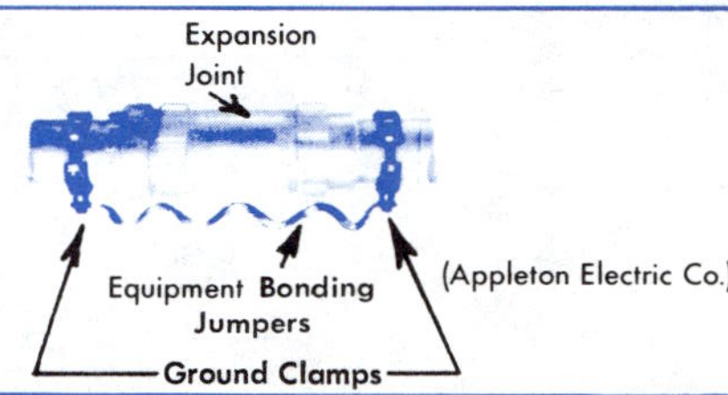

(Appleton Electric Co.)

250-77 Expansion fittings and telescoping conduits shall be made electrically continuous by bonding jumpers or other means.

250-79(f) Bonding jumper permitted on outside of conduit if not 6′ long.

BONDING JUMPERS

NEC 250-114 Where more than one grounding conductor enters a box, the electrical connection shall be such that the removal of a device will not interrupt the grounding continuity.

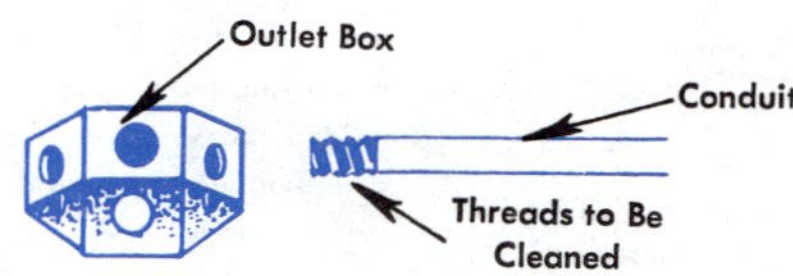

NEC 250-75 To insure electrical continuity any nonconducting coating shall be removed from conduit or cable armor before fastening to metal enclosure.

NEC 250-114(a) The connection shall be made between a grounding conductor and a metallic box, by means of a grounding screw used for no other purpose, or by a listed grounding device.

250-76 Metal conduits or cable systems containing conductors of more than 250 V to ground, shall use one or more of the methods specified in 250-72(b) through (e) to assure electrical continuity.

GROUNDING-BOXES

BOXES -- GROUNDING

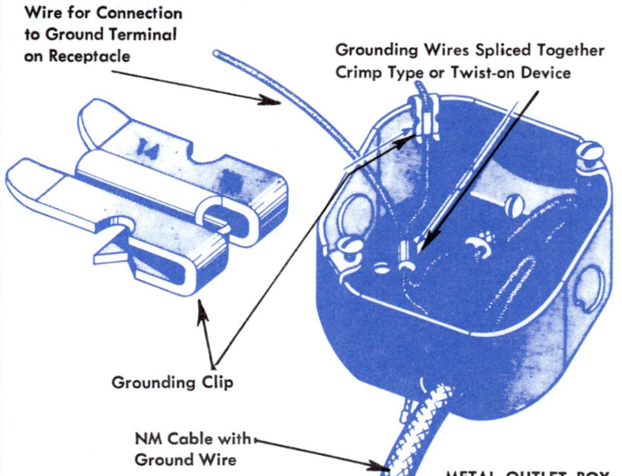

250-113 EGCs shall be connected by listed pressure connectors, clamps, exothermic welding, or other listed means. Sheet metal screws shall not connect EGC to enclosure. Connections or fittings that depend solely on solder are not permitted.
NEC 250-114 Where more than one grounding conductor enters box, they shall be so connected that removal of fixture or receptacle will not interrupt the grounding continuity.

CONDUCTORS -- VERTICAL SUPPORTS

NEC 300-19(a) Conductors in vertical raceways shall be supported if the vertical rise exceeds the values in Table 300-19(a).

Not exceeding:
100 ft. for No. 18 through No. 8 copper aluminum
100 ft. for No. 6 through No. 1/0 copper
200 ft. for No. 6 through No. 1/0 aluminum
50 ft. for 350,000 kcmil to 500,000 kcmil copper
120 ft. for 350,000 kcmil to 500,000 kcmil aluminum

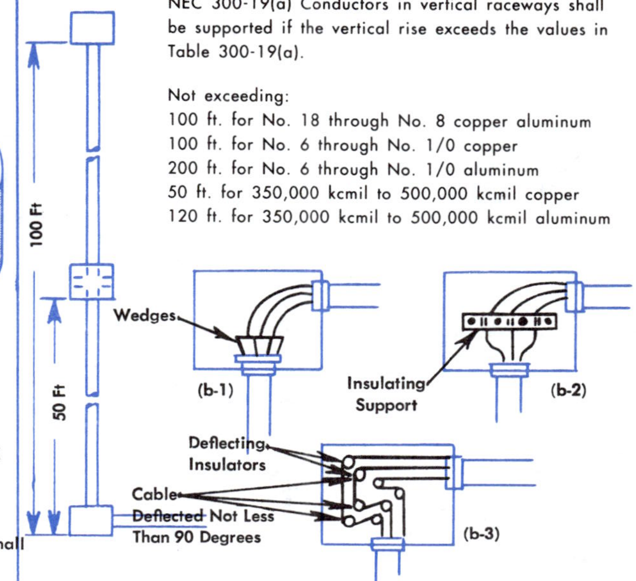

300-19(b) Conductors shall be supported by one of the following methods: (1) wedges, (2) insulating supports, (3) deflecting insulators. See 300-19(b)(3) for method of installation of conductors and supports and 300-19(b)(4) equivalent method.

AIR PLENUMS -- WIRING METHODS

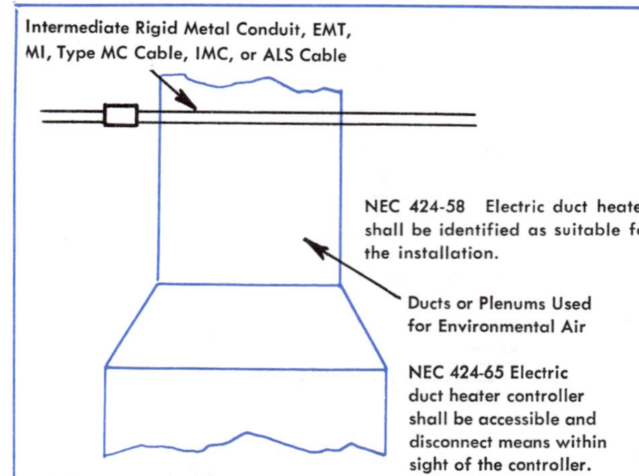

NEC 424-58 Electric duct heaters shall be identified as suitable for the installation.

NEC 424-65 Electric duct heater controller shall be accessible and disconnect means within sight of the controller.

NEC 300-22 (b) Where necessary to run wiring system thru air handling ducts or plenums, types MI or MC cables, EMT, flexible metallic tubing, IMC or rigid metal conduit may be used. Physically adjustable equipment and devices in ducts may be wired with flexible metal conduit or liquid-tight flexible metal conduit in lengths not exceeding 4 ft.

NEC 300-22(a) No wiring of any kind shall be installed in ducts used to transport dust, loose stock, flammable vapors, or commercial cooking.

NEC 300-22(c) Wiring methods in hollow spaces above lowered ceilings, that are also used for supply or return plenums for environmental air, are limited to Types MI, MC, AC cables, and other factory-assembled multiconductor control or power cable specifically listed for the use. Other cables and conductors to be in EMT, IMC, RMC, flexible metallic tubing, or conduit, or where accessible, surface metal raceway, metal wireway, or solid bottom metal cable tray with solid metal covers. Ex. 1–Liquidtight flexible metal conduit in single lengths not over 6′ permitted. Ex. 4–Listed prefabricated cable assemblies of manufactured metallic wiring systems where listed for this use permitted. Ex. 5–Stud or joist space in dwelling units where used as air handling permits normal wiring to pass through if perpendicular to space. See 800-53(a) for communication cables. Also see 725-2(b), 760-3(b), and 820-53(a).

RACEWAYS EXPOSED TO DIFFERENT TEMPERATURES

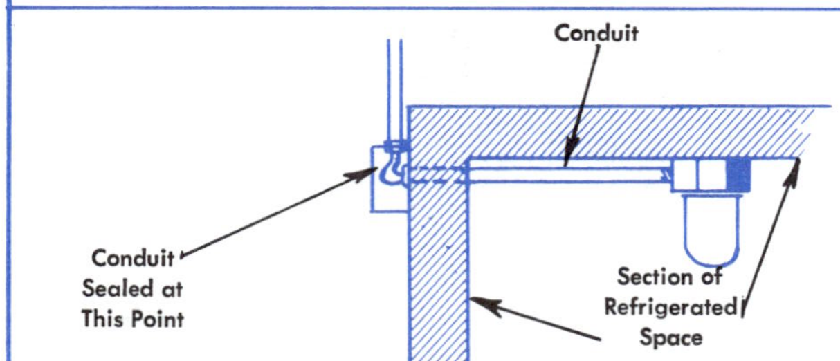

NEC 300-7(a) To prevent circulation of air through interior conduit runs whose ends are exposed to widely different temperatures, conduit shall be sealed off.
NEC 300-7 (b) **Expansion joints shall be provided where necessary to compensate for thermal movement.**

CONTINUITY -- RACEWAYS AND BOXES

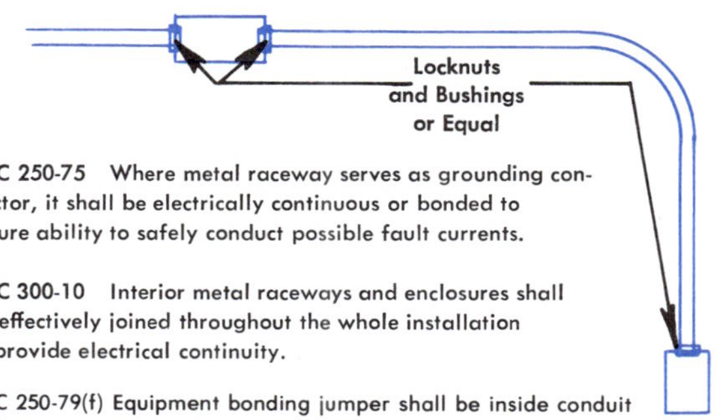

NEC 250-75 Where metal raceway serves as grounding conductor, it shall be electrically continuous or bonded to assure ability to safely conduct possible fault currents.

NEC 300-10 Interior metal raceways and enclosures shall be effectively joined throughout the whole installation to provide electrical continuity.

NEC 250-79(f) Equipment bonding jumper shall be inside conduit with circuit wires except if not over 6 feet, then may be outside conduit if routed with conduit.

WIRING WITHIN FLUORESCENT FIXTURES

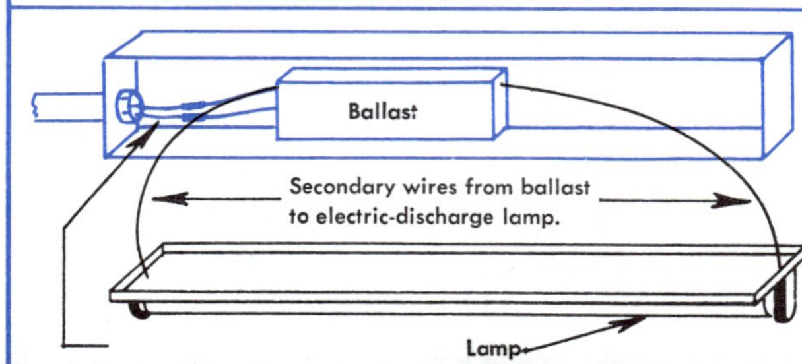

NEC 300-3(c)(2), Ex. 1 Secondary wiring to electric-discharge lamps of 1000 volts or less may occupy the same enclosure as branch-circuit conductors, if insulated for the secondary voltage of the ballast.

NEC 410-31 Branch-circuit conductors within 3″ of a ballast shall have a temperature rating of at least 90°C, such as RHH, THHN, THHW, FEP, FEPB, SA, XHHW, and THW.

DATA PROCESSING SYSTEMS

NEC 300-22(d) Wiring systems used with data processing systems and located within air handling spaces created by raised floors shall be permitted per 645.

NEC 645-5(b)(c) Specifically approved computer cable or flexible cords permitted to connect or interconnect data processing equipment. See 645-5(d) for specific rules for wiring under raised floors.

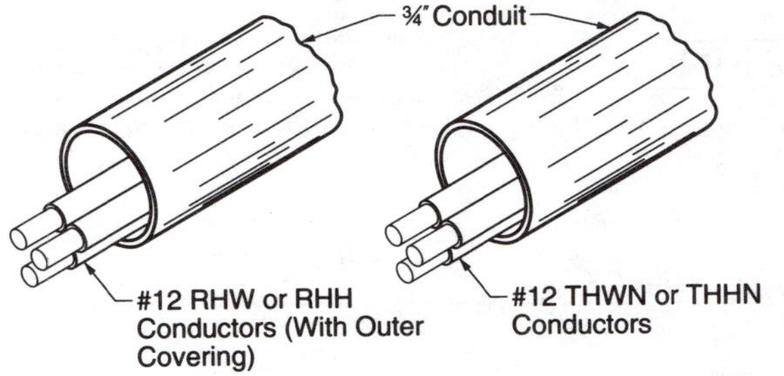

Ch 9, Table 1 permits conduit or tubing with more than two conductors to be filled to 40%. App. C, Table C1, lists six #12 RHW or RHH conductors for 3/4" EMT.

Ch 9, Table 1 permits conduit or tubing with more than two conductors to be filled to 40%. App. C, Table C1, lists 16 #12 THWN or THHN conductors for 3/4" EMT. Where there are three or more current-carrying conductors in a raceway, ampacity shall be reduced per Note 8 (a) of Notes to Ampacity Tables of 0 to 2000 Volts.

CONDUCTORS--NUMBER IN CONDUIT

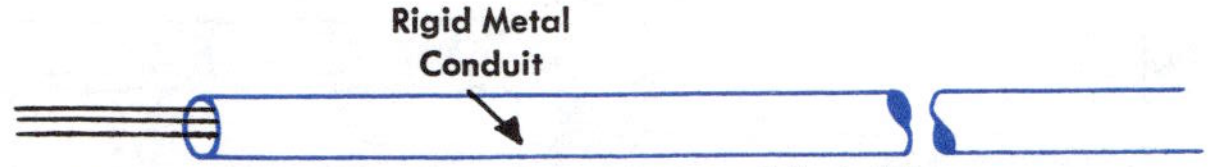

NEC 346-6 The number of conductors in a conduit shall be as per the percentage of fill permitted in Table 1, Chapter 9 of NEC.

NEC Chapter 9, Notes to Tables:

Note 1. See App. C for specific conduit or tubing for permitted conduit fill for same size conductors.
Note 2. Table 1 does not apply to short sections of conduit or tubing but to complete raceway system.
Note 3. EGCs are included in conduit fill calculations. Actual conductor dimensions shall be used to calculate fill.
Note 4. Conduits or tubing nipples, not over 24", are permitted. Derating required by Note 8 (a) to Ampacity Tables does not apply. Fill is permitted to be at 60%.
Note 5. Actual dimensions of conductors, such as multiconductor cables, shall be used if not included in Ch 9.
Note 6. Use Table 5 for dimensions of conductors of different size installed in same raceway. Use 4 for specific conduit or tubing area.
Note 7. When calculating fill for conductors of same size, the next higher number shall be used when 0.8 and larger.
Note 8. Dimensions for bare conductors are per Table 8, Ch 9.
Note 9. Multiconductor cables shall be treated as a single conductor for conduit fill.
Note 10. For conduit fill for three conductors, if ratio of raceway's inside diameter and cable or conductor's outside diameter is between 2.8 and 3.2, jamming can occur. Use next larger raceway.
See worked examples in lesson text.

BOXES -- STRAIGHT PULLS

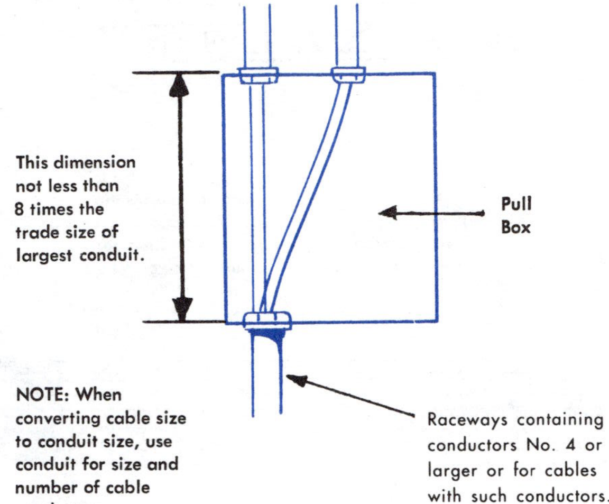

NOTE: When converting cable size to conduit size, use conduit for size and number of cable conductors.

370-28(a)(1) In straight pulls, length of box shall be not less than 8 times trade diameter of largest raceway.

BOXES -- ANGLE PULLS

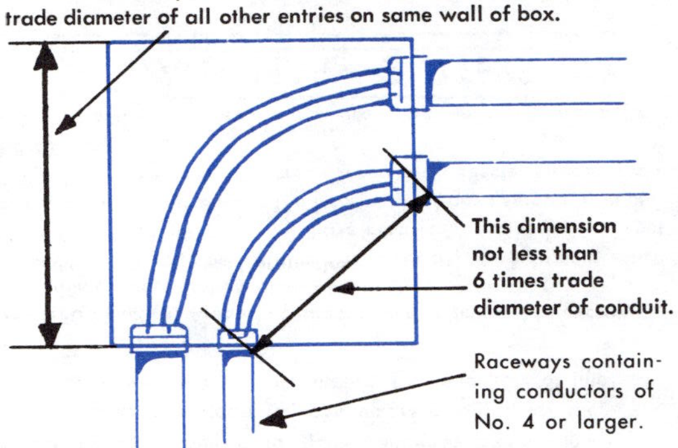

370-28(a) The following relates to boxes and conduit bodies used as pull or junction boxes, containing wire No. 4 or larger, and cables No. 4 or larger.

370-28(a)(2) For angle and "U" pulls, distance between raceway entry and opposite side of box at least 6 times trade size of largest conduit plus trade size of all entries on same wall of box. Distance between raceways for same conductor not less than 6 times trade size conduit. For cables, the equivalent conduit size required for conductors is to be used.

Exception: Where conduit enters wall of box opposite removable cover, this distance as per Column for one conductor per terminal in Table 373-6 (a).

AUXILIARY GUTTERS

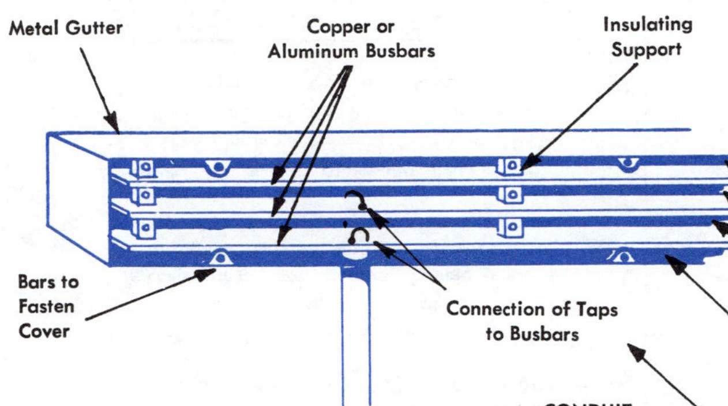

Auxiliary gutters used to supplement wiring spaces at meter and distribution centers and similar points of wiring systems.

NEC 374-1 Auxiliary gutters may enclose only conductors or busbars. They shall not enclose switches, overcurrent devices, appliances, or similar equipment.

NEC 374-6 Ampacity of bare conductors not to exceed 1000 amps per square inch of cross-sectional area for copper or 700 amps per square inch for aluminum. When 30 or less current-carrying conductors, no derating necessary.

NEC 374-5 Maximum cross-sectional area of all conductors is not to exceed 20% of area of gutter. If over 30 current-carrying conductors (excluding signal and motor-control circuits), the derating factors of Note 8(a) to Ampacity Tables of 0 to 2000 Volts shall apply.

NEC 374-7 Minimum distance between bare conductors mounted on same surface 2 inches. If parts held free in air, 1 inch. Clearance to other metal surfaces, 1 inch.

NEC 374-2 Auxiliary gutter must not extend more than 30 feet beyond the equipment which it supplements, except in elevator work.

NEC 374-8(b) Taps from bare conductors shall leave the gutter opposite their terminal connections and insulated conductors shall not come in contact with bare conductors of different potential.

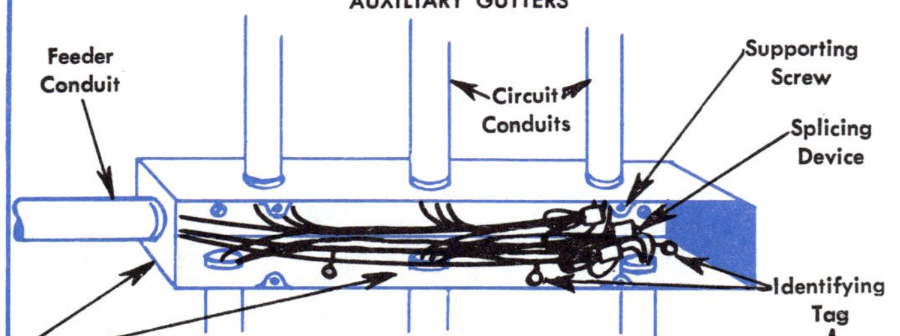

374-3(a)(b) Sheet metal auxiliary gutters shall be supported at intervals not exceeding 5′. Nonmetallic auxiliary gutters shall be supported at intervals not exceeding 3′, and at each end or joint.

374-5(a)(b) Sheet metal gutters shall not contain more than 30 current-carrying conductors at any cross section. Nonmetallic gutters shall not exceed 20% of interior cross-section area.

NEC 374-8 (a) Area of conductors and splices not to exceed 75% of cross section.

NEC 374-8(c) Taps to be identified as to circuits or equipment which they supply.

NEC 374-9 (a, b, c) Gutters shall have electrical and mechanical continuity and be of substantial construction. The interior shall have smooth rounded edges.

NEC 374-9 (d) Where conductors or troughing deflected greater than 30°, 373-6 applies.

NEC 374-1 Switches, overcurrent devices, appliances, etc. not permitted in gutter.

AUXILIARY GUTTERS

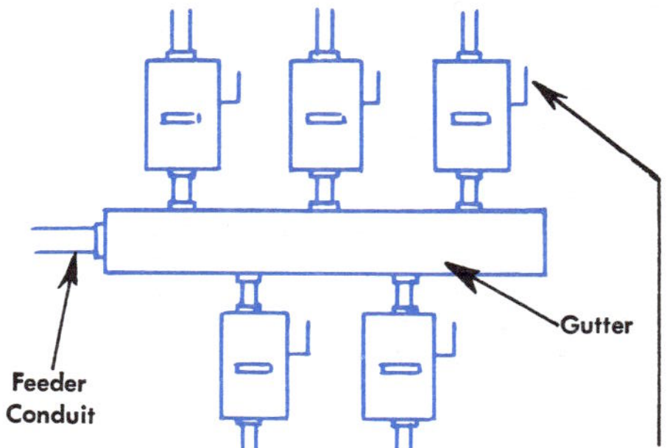

Center of grip of operating handle, when in highest position, shall not be more than 6′-7″ from floor or platform.

NEC 380-8 Switches (or CBs used as switches) to be readily accessible, and center of grip of operating handle when in highest position shall not be more than 6′-7″ from floor or platform.

Note 4 to Tables, Chapter 9, conduit or tubing nipples not over 24″ long may be filled to 60% of internal cross-sectional area. Derating factors do not apply.

SWITCHES -- ACCESSIBILITY

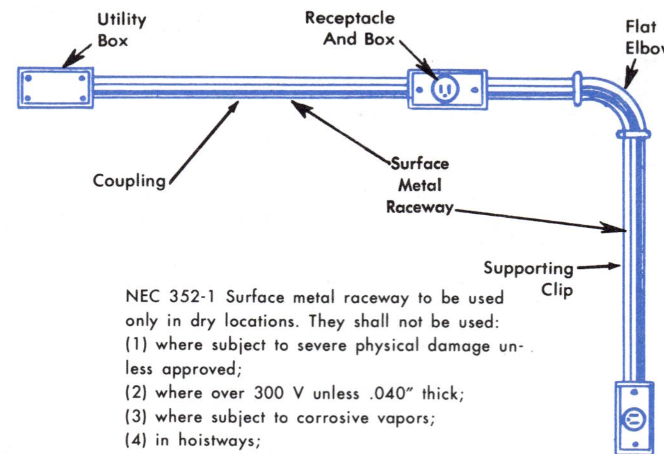

NEC 352-1 Surface metal raceway to be used only in dry locations. They shall not be used:
(1) where subject to severe physical damage unless approved;
(2) where over 300 V unless .040″ thick;
(3) where subject to corrosive vapors;
(4) in hoistways;
(5) in hazardous locations, except as permitted by 501-4(b), Ex.;
(6) concealed, except under raised floors for data processing systems as per 645-5(d)(2).

NEC 352-5 May extend through dry wall, partition, or floor if unbroken length

(Courtesy of Wiremold Co.)

SURFACE RACEWAY

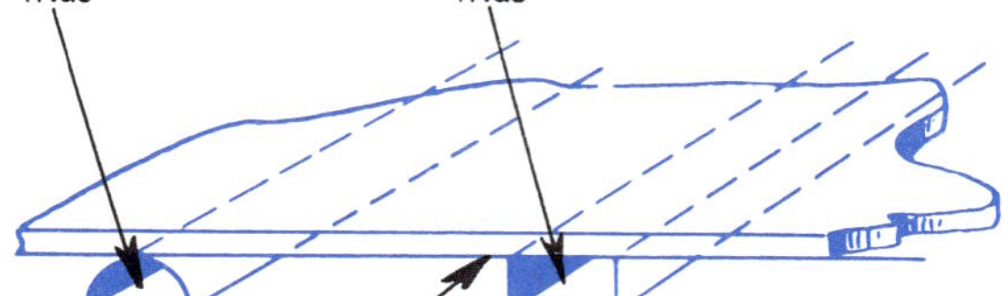

NEC 354-3 (a) ¾-inch concrete or wood required here.

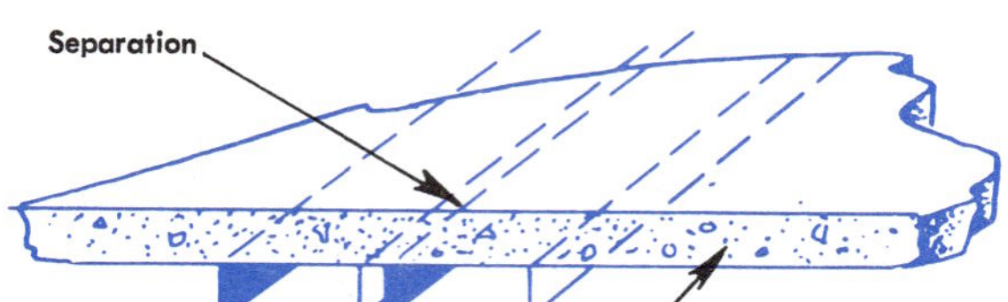

NEC 354-3 (b) With at least 1 inch spacing between 4-inch to 8-inch raceways, 1-inch concrete cover is required. If spacing less than 1 inch, 1½ inches of concrete needed.

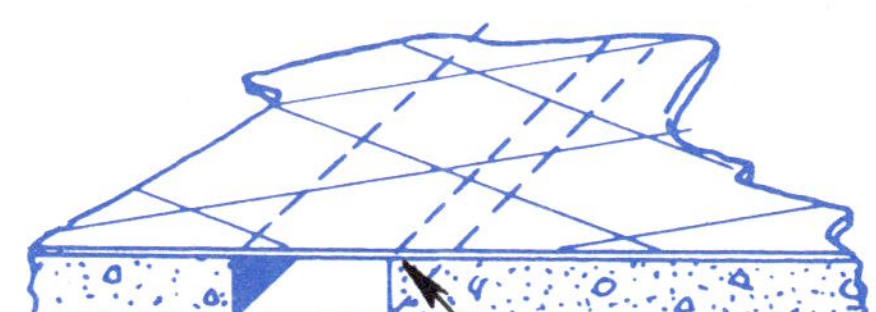

NEC 354-2 This raceway may be installed beneath flooring material, or flush in office occupancies. Not to be used where subject to corrosive vapors or in hazardous locations, except as permitted in 501-4(b), Ex.

OFFICE OCCUPANCY

UNDERFLOOR RACEWAYS

NEC 352-6 Combination raceways permitted for signal, power, and lighting circuits if in different compartments and identified throughout system by sharply contrasting colors of the interior finish.

SURFACE RACEWAY

NEC 354-1 TO 354-15 INCLUSIVE UNDERFLOOR RACEWAYS

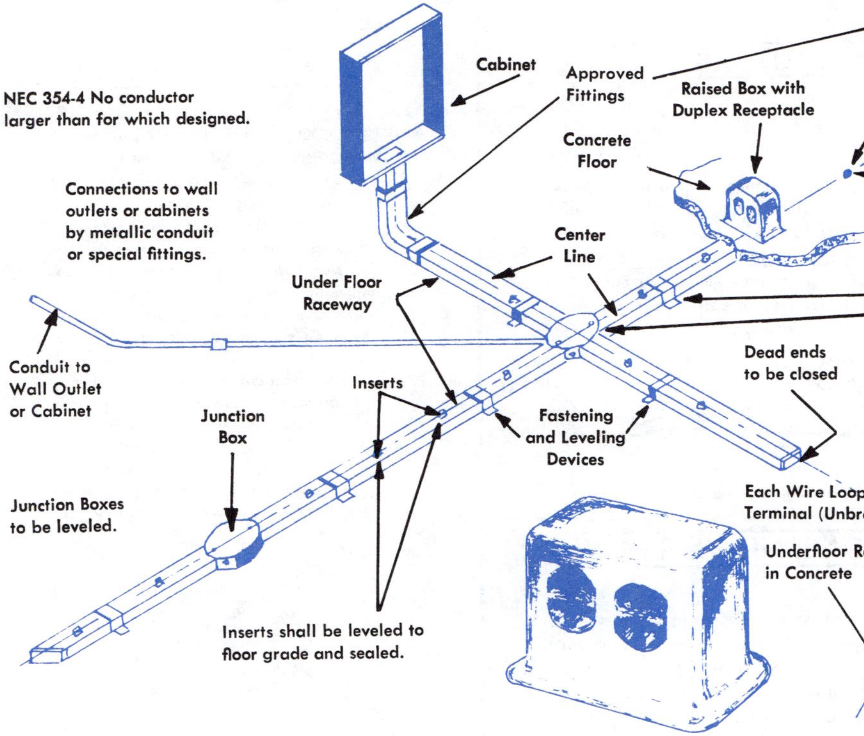

DUPLEX CONVENIENCE OUTLET

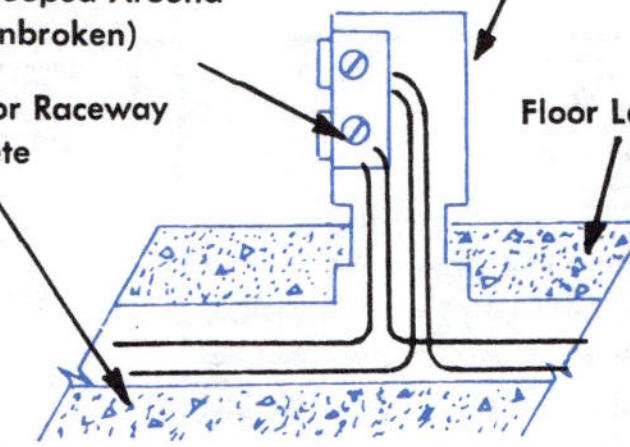

SO-CALLED LOOP WIRING

Loop wiring not considered a splice or tap.

NEC 356-1 TO 356-12 INCLUSIVE CELLULAR METAL FLOOR RACEWAYS

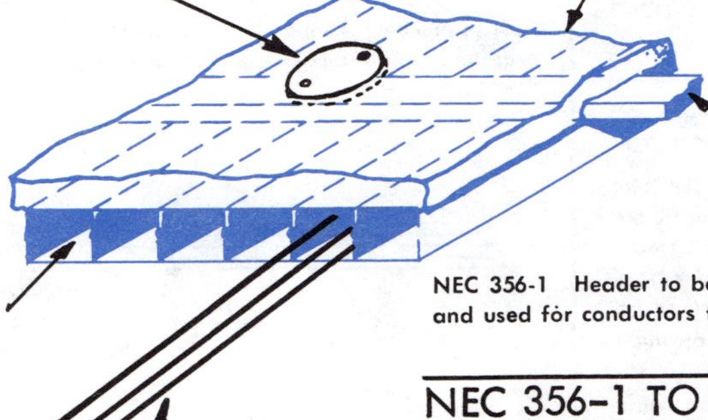

NEC 356-4 No conductor larger than No. 1/0 except by special permission.

NEC 356-2, FPN No electric conductors in same cell as other utilities (water, gas, air, etc.) per 300-8.

NEC 356-2(3) May be used to supply ceiling outlets of floor below in commercial garages.

NEC 358-11 TO 358-13 INCLUSIVE-- CELLULAR CONCRETE FLOOR RACEWAYS

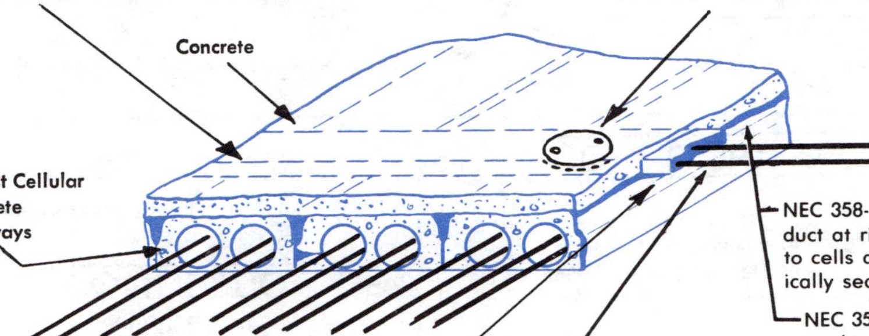

FIXTURES -- SHOW WINDOWS

NEC 210-62 At least one overhead receptacle for each 12 ft of show window.

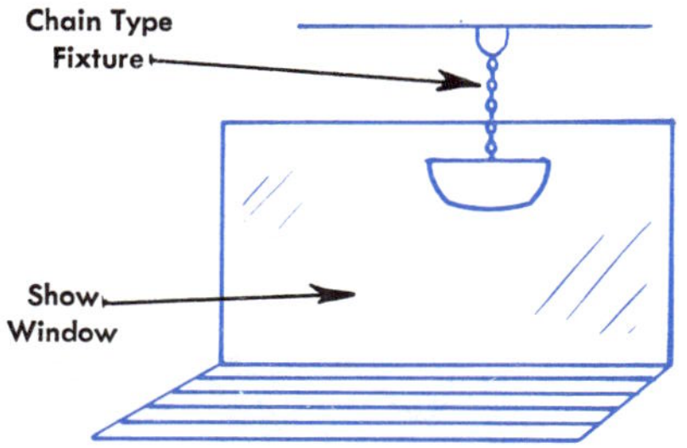

NEC 410-7 No externally-wired fixtures other than the chain type permitted in show windows. (See 400-11 for flexible cords.)

FIXTURE SUPPORTS

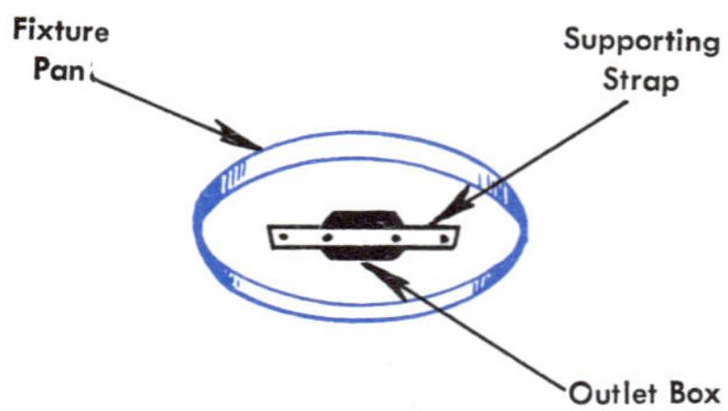

NEC 410-16 (a) (General Rule) Fixture shall be attached to and supported by outlet box.

NEC 410-16 (a) Fixtures weighing more than 50 pounds shall be supported independently of the outlet box.

NEC 410-16 (b) Connections between fixture conductors and circuit conductors shall be accessible for inspection without requiring disconnection of any part of wiring except when connected by attachment plugs and receptacles.

NEC 410-14(b) When electric-discharge lighting fixture is mounted over box, an opening is required in back of fixture so box is accessible.

FLUORESCENT FIXTURES -- CONNECTION OF

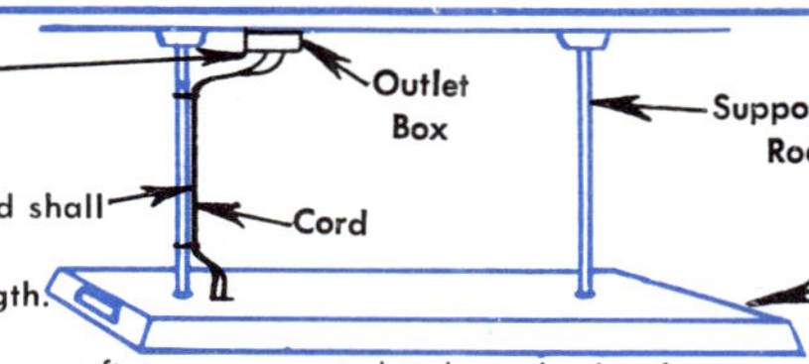

NEC 410-30 (c) Cord shall be continuously visible for entire length.

NEC 410-14 Fluorescent fixtures supported independently of outlet box shall be connected by metal raceway, nonmetallic raceways, type AC cable, type MI cable, metal-clad cable, or nonmetallic cable except cord connections permitted per 410-30(b)(c).

NEC 410-30(c) Cord-equipped fixtures may be suspended directly below outlet box if cord terminates in grounding type plug.

NEC 410-74 Fixtures designed for direct current shall be so marked.

NEC 410-78 Autotransformer ballasts which raise voltage to more than 300 volts shall be supplied only by a grounded system.

FIXTURE MOUNTING

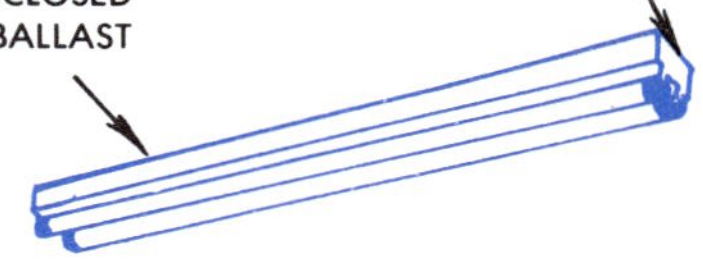

NEC 410-76(b) Surface-mounted fixtures containing ballasts must be listed for mounting on low-density cellulose fiberboard or spaced not less than $1\frac{1}{2}$ inches therefrom. See FPN for definition of combustible low-density cellulose fiberboard sheets, panels, and tiles.

SCREW-SHELL LAMPHOLDERS

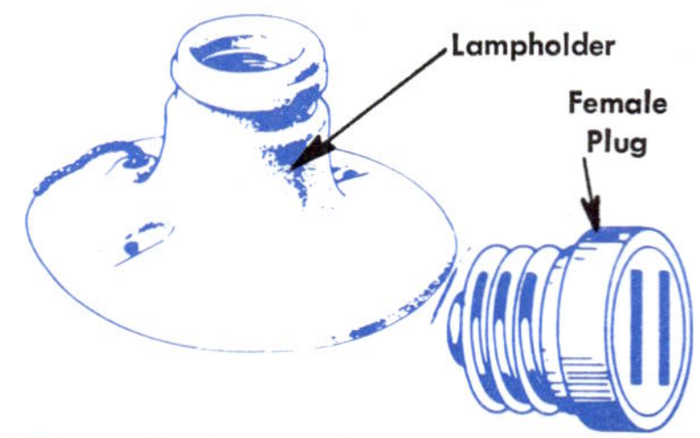

NEC 410-47 Screw-shell lampholders are NOT to be installed for use as plug receptacles. Where supplied by circuit having grounded conductor, then this conductor must connect to screw-shell.

FIXTURES -- USED AS RACEWAY

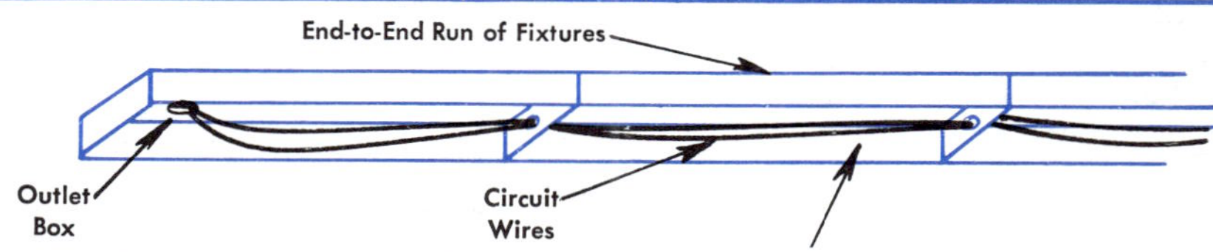

NEC 410-31 (General Rule) Fixtures not to be used as raceways for circuit conductors except: (1) where listed as raceway; (2) conductors of a single or multiwire branch circuit supplying fixtures may be run through an end-to-end assembly of units connected together by recognized wiring methods; (3) in addition, one two-wire circuit that supplies at least one fixture in assembly may be run through fixtures. Conductors within 3″ of ballast shall have 90°C rating. XHHW, THHW, THHN, RHH, FEP, FEPB, Type THW also permitted within 3″ of ballast. See Table 310-13 for special THW rating.

NEC 410-73(e) Fluorescent fixtures installed indoors shall have thermal protection integral within the ballast (P ballast). Replacement ballasts shall also have integral protection. See Exceptions.

COVE LIGHTING

NEC 410-9 Coves to have ample wiring space and to be fully accessible for maintenance.

SHOWCASES -- CORD CONNECTED

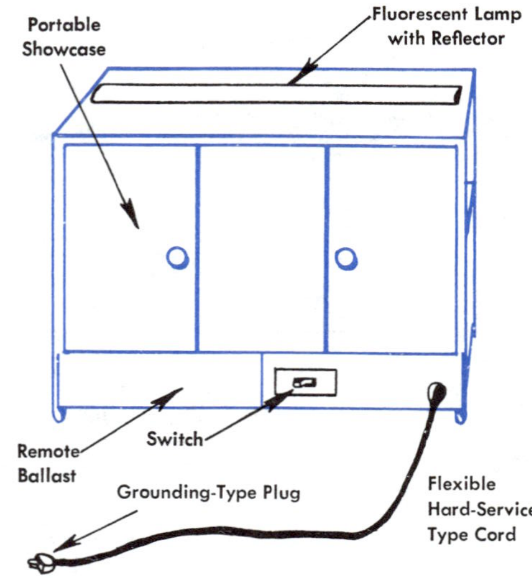

NEC 410-29 Single showcase, not permanently installed, permitted to be cord-and-plug connected to permanent receptacle. Up to six such showcases permitted to be coupled together by flexible cord and locking-type connector, but one showcase of group must connect to permanent receptacle. (See 410-29 (a, b, c, d, e).)

NEC 410-77 (a) Auxiliary equipment not part of light fixture shall be enclosed in accessible permanently installed metal cabinets.

NEC 410-29 (e) In cord-connected showcases the secondary circuit of electric discharge lighting restricted to single case.

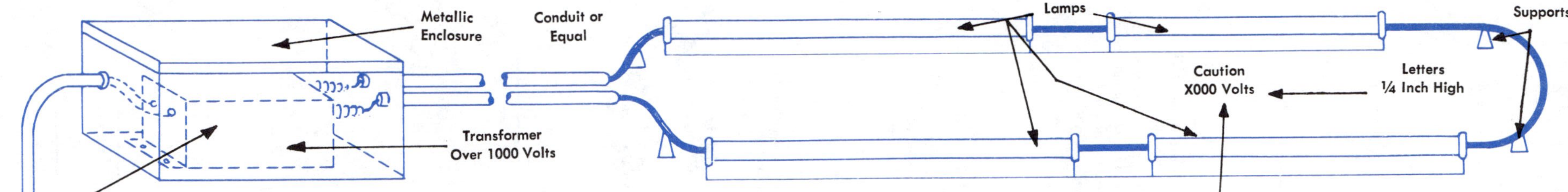

NEC 410-83 Secondary open circuit not over 15,000 volts (plus 1000); secondary current not over 120 milliamperes if voltage over 7500; secondary current not over 240 milliamperes if voltage 7500 or less.

NEC 410-81 (b) Switch or circuit breaker must be within sight of fixture or lamps or else capable of being locked in open position.

NEC 410-84 Transformers to be enclosed and listed.

NEC 410-86 (a, b, c) Transformers must be accessible and must not subject combustibles to temperature over 90°C. Secondary conductors shall be as short as possible.

NEC 410-19 (a) (b) Ex Where circuit operates at over 150 volts to ground, all noncurrent-carrying metal parts shall be grounded except tie wires, screws, clips, etc., at least 1½ inches from lamp terminals.

NEC 410-88 Conductors shall be installed per 600-32.

NEC 410-91 Each fixture or circuit of tubing over 1000 volts shall have a precautionary label. The voltage indicated is open-circuit voltage.

LIGHTING -- ELECTRIC DISCHARGE OVER 1000 VOLTS

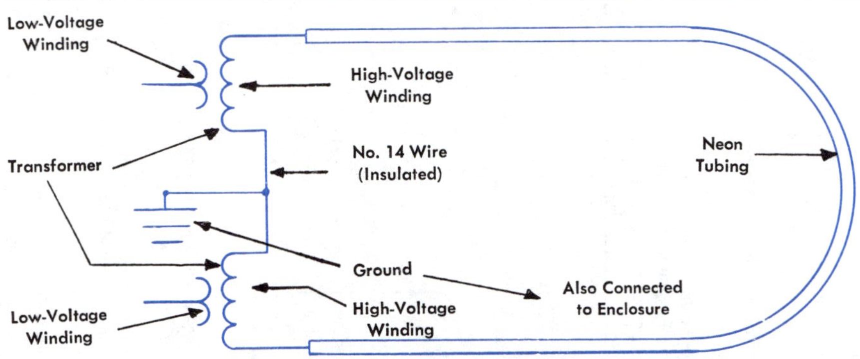

NEC 410-85 High-voltage transformer windings not to be connected in series or in parallel except series connection with mid-point grounded, as shown. Here, grounded ends shall be connected by an insulated conductor not smaller than #14 AWG. See Exception.

NEC 410-88 Conductors shall be installed per 600-32.

SIGN -- ELECTRIC DISCHARGE OVER 1000 VOLTS (NEON)

NEC 422-8(a) Approved heater cord listed in Table 400-4 required for smoothing irons and for similar appliances over 50 VA and produce temperatures over 121°C (250°F) where cord can contact.

NEC 422-12 In other than residential, irons to be installed in connection with a signal unless equipped with integral temperature-limiting device.

NEC 422-2 Current-carrying parts to be protected from contact, except where necessarily exposed as in toasters and grills.

NEC 422-11 Irons and similar appliances shall have an approved stand.

NEC 520-73 In theater dressing rooms, all lights and receptacles shall be controlled by wall switches. Switch controlled receptacles shall be provided with a pilot light to indicate when the receptacles are energized.

SMOOTHING IRONS

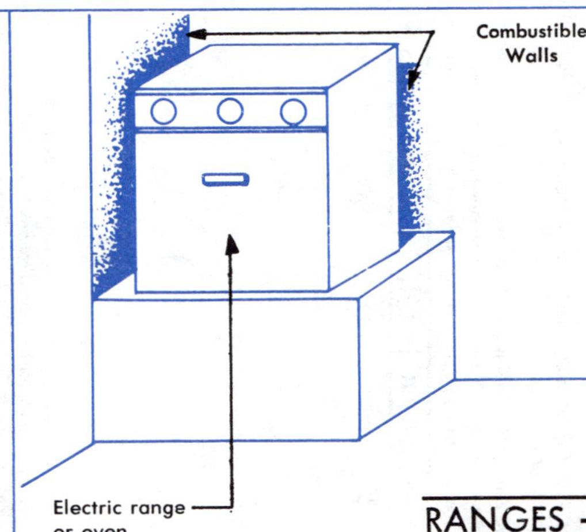

NEC 422-10 Electrically heated appliances shall be placed to provide ample protection between appliance and adjacent combustible material.

NEC 422-6 Appliances to be installed in an approved manner.

NEC 422-16 Exposed metal parts to be grounded per 250. For freezers and refrigerators, see 250-42, 43, and 45. For ranges and clothes dryers, see 250-57 and 250-60.

RANGES -- INSTALLATION OF

NEC 210-4(c) Multiwire branch circuits are permitted by Article 210 and shall supply only line to neutral load, except: (1) supply only one utilization equipment; (2) where branch-circuit overcurrent device simultaneously opens all ungrounded conductors of multiwire circuit.

NEC 210-4(d) If more than one nominal voltage system exists in a building, each ungrounded system conductor shall be identified by phase and system. See FPN.

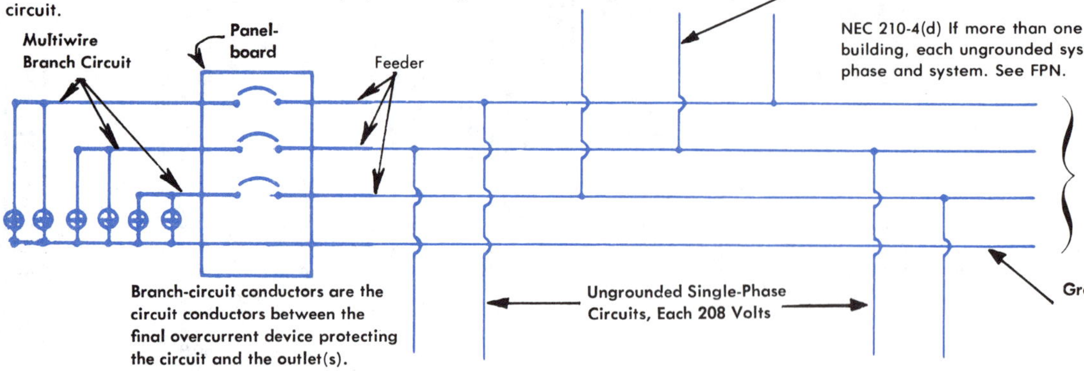

NEC 210-4(a) All multiwire branch circuit conductors shall originate in same panelboard.

NEC 210-4(b) In dwelling units where more than one receptacle on the same yoke (split receptacle) is fed by a multiwire branch circuit, means must be provided at the panelboard to simultaneously disconnect all ungrounded conductors.

NEC 210-10, 215-7 Ungrounded conductors for feeders and branch circuits may be tapped from ungrounded wires of circuits which have identified neutrals. Switching devices in these circuits shall have a pole in each conductor. For branch circuits, all poles must open simultaneously where they serve as disconnecting means as required for appliances (see 422-21 (b)); fixed electric space heating (see 424-20); electric de-icing and snow-melting (see 426-51); motor controller (see 430-85); for motor see 430-103.

BRANCH CIRCUIT -- TAPS FROM

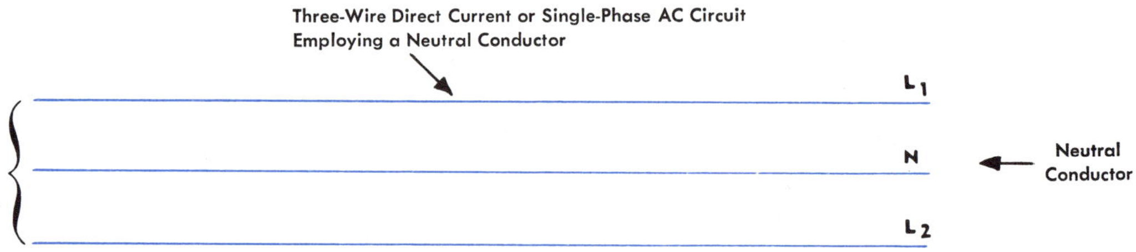

NEC 220-22 Feeder neutral sized for maximum unbalance between neutral and phase conductors, except neutral for household ranges, ovens, and dryers permitted to be reduced to 70% of load as calculated per Table 220-18 and Table 220-19. A further demand of 70% permitted for portion of maximum unbalanced load in excess of 200 amperes. Also applies to 4-wire 3-phase and 5-wire 2-phase systems.

NOTE: Exception. This 70% demand factor does not apply for that portion of the feeder load which consists of nonlinear loads.

NOTE: Where household electric ranges and dryers are supplied by the feeder, this demand factor is additional to that allowed in Table 220-18 and Table 220-19.

Table 220-11 (Footnote) Demand factors in this table do not apply to computed loads on feeders to areas in hospitals, hotels, and motels, where the entire lighting load is used generally at one time.

NEC 220-3(a) Continuous load on branch circuit shall be 125% times the load plus noncontinuous load.

Exception: Does not apply where assembly including overcurrent protection device listed for 100% operation.

NEC 215-2(b), FPN 2. Voltage drop for power, heating, or lighting feeders should not exceed 3%. Voltage drop for feeder plus branch circuit should not exceed 5%. See 210-19(a), FPN 4.

LOAD CALCULATIONS -- FEEDERS AND BRANCH CIRCUITS

2 Sets of Four-Wire Feeders with Common Neutral

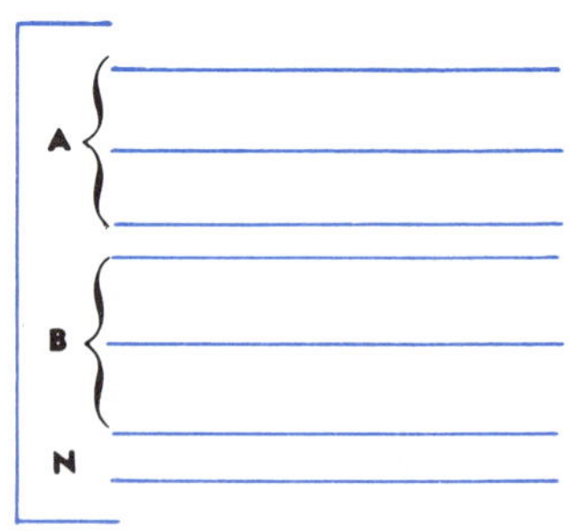

3 Sets of Three-Wire Feeders with Common Neutral

NEC 215-4 (a) A common neutral may be used with two or three sets of 3-wire feeders or two sets of 4-wire or 5-wire feeders.

NEC 225-7(b) The neutral conductor shall have an ampacity not less than the maximum net calculated load current between the neutral and all ungrounded conductors connected to any one phase of the circuit.

NEC 220-22 Neutral sized to carry maximum unbalance as determined by Article 220.

COMMON NEUTRAL FOR FEEDERS

Trade Competency Test 9

Name ______________________________ Date ______________________

Section	Answer	
______	T F	**1.** In a fuseholder, the screw shell shall be connected to the line side of the circuit.
______	______	**2.** Header ducts used with cellular concrete raceways shall be constructed of ________.
______	T F	**3.** Taps not over 18″ long for individual lampholders may be smaller than the branch circuit conductors.
______	T F	**4.** The largest conductor connected to a device on a strap counts toward the fill.
______	______	**5.** Dead ends of underfloor raceways shall be ________.
______	______	**6.** One of the approved methods of identifying the high leg of a 4-wire, delta-connected secondary system is by using a ________ wire. A. white C. red B. green D. none of the above
______	T F	**7.** Openings around electrical penetrations through fire-resistant-rated walls shall be firestopped using approved methods.
______	______	**8.** The minimum feeder allowance for show-window lighting shall be ________ VA per linear foot. A. 100 C. 300 B. 200 D. 500
______	______	**9.** Service conductors of 600 V or less buried under ________″ or more of concrete are considered outside the building.
______	______	**10.** Conductors from discontinued outlets in an underfloor raceway installation shall be ________ from the raceway.
______	______	**11.** Interior raceways installed between areas differing widely in temperature shall be ________.
______	______	**12.** ________ are required at or near the end of each run of underfloor raceway to locate the last insert.
______	______	**13.** The total load on a branch circuit shall not exceed its ________.
______	______	**14.** The unit load for a store is listed as ________ VA per sq ft. A. 2 C. 3 B. 2.5 D. 3.75
______	______	**15.** If the allowable current-carrying capacity of a conductor does not correspond to the rating of a standard size overcurrent protection device, the next larger size may be used provided the current does not exceed ________ A.
______	______	**16.** Branch circuit conductors shall have an ampacity not less than the ________ load served.

__________ __________ **17.** Threaded boxes not over 100 cu in. in size that contain devices need no additional support if two or more conduits, which are properly supported within ________″ of the box, are threaded into the box from the same side.

__________ __________ **18.** Where permissible, the demand factor applied to that portion of an unbalanced neutral feeder load in excess of 200 A is ________%.

A. 40 C. 70
B. 60 D. 80

__________ __________ **19.** In other than dwellings, each 5′ or fraction of multioutlet assembly shall normally be considered as a load of ________ VA.

__________ T F **20.** The neutral feeder conductor shall be the same size as the ungrounded conductors for a resistance load.

__________ __________ **21.** Neon installed on dwelling units shall not exceed ________ V.

__________ __________ **22.** Metal auxiliary gutters shall be supported throughout their entire length at intervals not exceeding ________′.

__________ __________ **23.** The unit load for a hall in other than a one-family dwelling or an apartment is listed as ________ VA per square foot.

__________ T F **24.** Lighting outlets installed in other than dwelling units shall not be connected to 50 A branch circuits.

__________ __________ **25.** A circuit breaker having an interrupting rating other than ________ A shall be marked with the rating of a standard size overcurrent device.

__________ __________ **26.** A(n) ________ is required where underground cable leaves conduit providing physical protection.

__________ __________ **27.** Exposed, non-current-carrying metal parts of fixed equipment shall be grounded if located in a wet location and not ________.

__________ __________ **28.** Where branch-circuit loads are expected to continue for 3 hours or more, the minimum unit loads given in Table 220-3(b) shall be increased by ________%.

A. 10 C. 35
B. 25 D. 80

__________ __________ **29.** Circuit breakers used to switch 120/277 V fluorescent fixtures shall be approved for the purpose and marked ________.

__________ T F **30.** In interior wiring, a common neutral feeder wire shall not be run for two or three sets of 3-wire feeder circuits.

__________ __________ **31.** Circuit breakers shall clearly ________ whether in open or closed position.

__________ __________ **32.** A 15 A receptacle on a standard 15 A or 20 A branch circuit shall not supply a portable appliance load in excess of ________ A.

A. 9 C. 12
B. 10 D. 15

__________ __________ **33.** A No. ________ or larger grounding electrode conductor shall be protected if exposed to severe physical damage.

__________ __________ **34.** The ________ of an underfloor raceway shall determine the largest conductor that can be installed.

__________ __________ **35.** A unit load for each outlet supplying heavy-duty lampholders that are not used for general illumination in industrial occupancies shall be ________.

A. 1½ A C. 180 VA
B. 5 A D. 600 VA

Trade Competency Test 10

Name ______________________ Date ______________

Section	Answer	
______	T F	**1.** Splices are sometimes permitted in service conductors.
______	______	**2.** Auxiliary gutters shall not extend more than ________′ beyond the equipment they supplement.
______	______	**3.** Equipment bonding jumpers are permitted to be routed outside a conduit system if not over ________′ in length.
______	______	**4.** The cross-sectional fill of a wireway shall not exceed ________%.
______	______	**5.** Every commercial building must have a(n) ________ A circuit for a sign outlet.
______	T F	**6.** The maximum rating of a 250 V cartridge fuse is 600 A.
______	______	**7.** Underground service conductors carried up a pole must be protected from mechanical injury to a height of at least ________′. A. 8 B. 9 C. 12 D. 15
______	______	**8.** Metal enclosures for a grounding electrode conductor using a water pipe electrode for a 120/208 V service with three 250 kcmil Cu conductors and a No. 3/0 neutral shall be ________. A. rigid conduit only B. not less than ¾″ in dia. C. bonded D. electrically continuous
______	T F	**9.** Bonding jumpers required for grounding interior equipment on the load side of the service shall be selected from Table 250-94.
______	______	**10.** The area of conductors and splices in a gutter shall not exceed ________% of the cross-sectional area.
______	T F	**11.** Fuses shall never be field connected in parallel.
______	______	**12.** The size of a copper grounding electrode conductor using a water pipe electrode for a 120/208 V service with three 250 kcmil Cu conductors and a No. 3/0 neutral shall be No. ________. A. 2 B. 4 C. 6 D. 8
______	______	**13.** Panelboards equipped with snap switches rated at 30 A, or less, shall have overcurrent protection not in excess of ________ A. A. 100 B. 150 C. 200 D. 300
______	T F	**14.** The service overcurrent protection device shall be an integral part of or shall be located adjacent to the service disconnecting means.
______	______	**15.** The current carried continuously in bare copper bars in metal auxiliary gutters shall not exceed ________ A per sq in.

_________ _________ **16.** No. 1/0 Cu conductors in vertical raceways shall be supported at intervals not exceeding _________′.

A. 50
B. 75
C. 100
D. 125

_________ _________ **17.** Three-way and four-way switches shall be wired so that all switching is done in the _________ conductors.

_________ T F **18.** In general, the disconnecting means shall be within sight of the controller.

_________ _________ **19.** The wiring between 3-way switches and outlets in metal enclosures shall be run with all _________ in the same enclosure.

_________ _________ **20.** The largest standard cartridge fuse rating is _________ A.

A. 100
B. 600
C. 1200
D. 6000

_________ _________ **21.** Screw-shell lampholders shall be installed for use as _________ only.

_________ T F **22.** No. 18 copper wire may be used for grounding a portable device used on a 20 A circuit.

_________ _________ **23.** A nipple having a maximum length of 24″, can be filled to _________% of its internal cross-sectional area.

_________ _________ **24.** Transformers for electric-discharge lighting systems of more than 1000 V shall be _________ after installation.

_________ _________ **25.** The size of the conductors for a(n) _________′ tap shall be equivalent to the terminal to which they are connected.

_________ _________ **26.** A 25′ tap taken from a 100 A feeder shall have an ampacity of not less than _________ A.

A. 15
B. 20
C. 25
D. 33⅓

_________ _________ **27.** When applying the 100′ tap rule, the horizontal run is limited to _________′ or less.

_________ _________ **28.** Tap conductors shall have at least one third ampacity of supply conductors when applying the _________′ tap rule.

_________ _________ **29.** In straight pulls, the length of a box shall be not less than _________ times the diameter of the largest conduit.

_________ _________ **30.** When applying the 25′ tap rule from secondary side of the transformer for industrial installations, overcurrent protection devices shall not _________ output.

_________ _________ **31.** Branch circuit conductors routed within 3″ of a ballast in a fixture shall be rated at least _________°C.

_________ _________ **32.** In U-pulls, the size of the box shall be _________ times the diameter of the largest conduit plus all remaining conduits on the same wall.

_________ _________ **33.** No. 12 conductors enclosed in vertical raceways shall be supported every _________′.

_________ _________ **34.** A length of conduit is considered a nipple if it does not exceed _________″ in length.

_________ _________ **35.** In a building where water pipe is available, the minimum size of the grounding electrode conductor for an ungrounded system with No. 3/0 Cu service conductors shall be No. _________.

A. 2 Cu
B. 4 Cu
C. 6 Cu
D. 4 Al

Trade Competency Test 11

Name ______________________ Date ______________________

Section	Answer	
________	T F	**1.** Conduit nipples not over 24″ long may be filled to 60% of their cross-sectional area.
________	________	**2.** Equipment installed in plenums requiring adjustments can be wired with a(n) ________′ length of flexible metal conduit or liquidtight flexible metal conduit.
________	________	**3.** Coves that enclose lighting units shall have space fully accessible for ________.
________	________	**4.** The secondary OCPD of a transformer may consist of ________ or less CB's or fuses. A. two C. four B. three D. six
________	T F	**5.** Temporary wiring is permitted for Christmas decorative lighting for a period not to exceed 60 days.
________	________	**6.** An autotransformer ballast that raises the voltage above ________ V shall be supplied by a grounded system.
________	________	**7.** The largest conductor permitted in ⅜″ flexible metal conduit is No. ________. A. 10 C. 14 B. 12 D. 16
________	________	**8.** No wiring method of any kind can be installed in ________ used with commercial cooking.
________	T F	**9.** Conductors in underfloor raceway systems may be spliced or tapped at individual outlets.
________	________	**10.** When applying the 100′ tap rule, the junction box containing the tap shall be at least ________′ from the floor.
________	________	**11.** Underfloor raceways over 4″ but less than 8″ wide, and spaced less than 1″ apart, shall be covered with concrete to a depth of not less than ________″. A. ¾ C. 1½ B. 1 D. 2
________	T F	**12.** Unbroken lengths of surface metal raceway may be extended through dry floors.
________	________	**13.** A lighting and appliance panelboard shall not have more than ________ overcurrent protection devices. A. 42 C. 50 B. 46 D. none of the above
________	________	**14.** If a change occurs in the size of the ungrounded conductor, a similar change is permitted in the ________ conductor.

_________ _________ **15.** The combined cross-sectional area of all conductors or cables in an underfloor raceway shall not exceed _________%.

A. 20 C. 40
B. 30 D. 50

_________ _________ **16.** Where more than one voltage system exists in a building, each ungrounded conductor shall be identified by _________ and system.

_________ _________ **17.** No conductor larger than No. _________ can be installed in cellular metal floor raceways, except by special permission.

_________ T F **18.** Header ducts used with cellular concrete floor raceways may be run diagonally across the cells, provided they run in straight lines.

_________ _________ **19.** Cord-equipped fixtures can be suspended directly below an outlet box if the cord _________ in a grounding receptacle.

_________ _________ **20.** Underfloor raceways not over 4″ wide shall be covered by concrete or wood to a thickness of not less than _________″.

A. ¾ C. 1½
B. 1 D. 2

_________ _________ **21.** Transformers shall have a secondary circuit current rating of not more than _________ mA.

_________ _________ **22.** A feeder neutral conductor on a pure resistive load can have a demand factor of _________% applied for all loads above 200 A.

_________ T F **23.** The full cross-sectional area of cellular metal floor raceway may be used if Type AC armored cable is used.

_________ _________ **24.** Fixtures that are designed to be supplied by direct current shall be _________.

_________ _________ **25.** Expansion joints in raceway systems shall be installed where necessary to compensate for _________ movement.

_________ _________ **26.** The largest size conductor allowed in a cellular metal floor raceway without special permission is _________.

A. No. 1/0 C. 250 kcmil
B. No. 3/0 D. 500 kcmil

_________ _________ **27.** Raceway systems that are exposed to widely different temperatures shall be _________.

_________ _________ **28.** More than one grounding conductor entering a box shall be connected so that the removal of a receptacle will not _________ the grounding path.

_________ _________ **29.** The minimum headroom above motor control center is _________′.

A. 5½ C. 6½
B. 6 D. 7

_________ _________ **30.** A fuse or overcurrent protection device shall be connected in _________ with each ungrounded conductor.

_________ _________ **31.** The center of the grip of an operating handle in its highest position shall not exceed _________′ from the floor.

_________ _________ **32.** Each fixture of each secondary circuit of tubing rated over _________ V shall have a warning label.

_________ _________ **33.** An outlet tap for a fixture shall not extend over _________″.

_________ T F **34.** No conduit shall be run through a duct used for dust removal.

Trade Competency Test 12

Name ______________________ Date ______________

Section	Answer	
______	______	**1.** Panelboards with overcurrent protection not above 200 A may contain snap switches of ______ A or less.
______	T F	**2.** Plastic inserts can be used with metal underfloor raceways.
______	______	**3.** Any 15 A or 20 A receptacle shall not supply a cord-and-plug connected load which exceeds ______% of the rating of the receptacle.
______	______	**4.** If the ampacity of a conductor does not correspond with a standard overcurrent protection device, the next larger size up to ______ A can be used. A. 800 C. 1200 B. 1000 D. 1500
______	T F	**5.** In a commercial occupancy, electric discharge lighting transformers of more than 1000 V shall not be connected in series.
______	______	**6.** The OCPD of a phase converter is computed at ______% of the 1ϕ input FLA.
______	T F	**7.** Service-entrance conductors can be spliced where the wiring method is changed.
______	______	**8.** Only ______ mains are permitted in a lighting and appliance panelboard.
______	T F	**9.** Conductors are considered outside the building when encased with at least 2″ of concrete.
______	______	**10.** Splices and wires within any portion of an auxiliary gutter shall not fill the cross-sectional area of the gutter by more than ______%. A. 20 C. 60 B. 50 D. 75
______	T F	**11.** The cross-sectional area of conductors in any section of metal auxiliary gutter shall not exceed 20% of its area.
______	______	**12.** The general rule requires conductors to be protected in accordance to their ______.
______	T F	**13.** AC-DC general-use snap switches controlling ordinary inductive loads shall have an ampere rating twice that of the load.
______	______	**14.** The maximum number of overcurrent protection devices in a lighting and appliance branch-circuit panelboard shall not exceed ______. A. 30 C. 42 B. 36 D. 48
______	T F	**15.** Service-entrance conductors shall not feed one building or structure through another building or structure.

_________ _________ **16.** The general rule requires grounding receptacles to be installed on _________ A and _________ A circuits.

A. 15; 20
B. 20; 25
C. 25; 30
D. none of the above

_________ _________ **17.** Listed extension cord sets with No. _________ or larger conductors can be connected to a 20 A branch circuit.

_________ T F **18.** Handles of circuit breakers shall be clearly marked ON or OFF.

_________ _________ **19.** Fixture wire used on a parking lot lighting pole can extend up to _________′ for No. 18 conductors.

_________ _________ **20.** The load for show windows is computed at a unit load of _________ VA per linear foot.

_________ _________ **21.** A fixture whip can be used if its length does not exceed _________′.

_________ _________ **22.** All 15 A and 20 A, 120 V receptacles used on construction sites shall be _________-protected.

A. GFCI
B. GPF
C. GP
D. none of the above

_________ _________ **23.** Flexible metal conduit does not require support for lengths of _________′ or less at terminals where flexibility is required.

_________ _________ **24.** Receptacles located in floors shall be installed in floor boxes _________ for such applications.

_________ _________ **25.** Angle-pull dimensional requirements apply to junction boxes only when the size of conductor is equal to or larger than No. _________.

A. 1/0
B. 3/0
C. 4
D. 6

_________ _________ **26.** Electric mixers traveling in and out of open-type mixing tanks are considered _________ utilization equipment.

_________ _________ **27.** Flexible cords enclosing No. _________ and larger conductors can be spliced.

_________ _________ **28.** The total load on any overcurrent protection device in a panelboard shall not exceed _________% of its rating if in normal use for over 3 hours.

_________ _________ **29.** For existing installations in residential occupancies, _________ protection is not required in a lighting and appliance panelboard used as service equipment.

_________ _________ **30.** A 120 V circuit breaker used to switch fluorescent fixtures shall be marked _________.

A. SWD
B. SWR
C. SWF
D. FLS

_________ _________ **31.** A subfed panelboard can be protected by the feeder overcurrent protection device if it does not _________ the rating of the panelboard.

_________ _________ **32.** The general rule requires conductors used in flexible cords to be No. _________ or larger.

_________ _________ **33.** The maximum number of quarter bends permitted in a run of flexible metal conduit is _________.

_________ T F **34.** The general rule allows overcurrent protection devices to be inaccessible to occupants of a building.

_________ T F **35.** Where GFCI protection of personnel for temporary electrical power is required, listed cord sets incorporating this protection shall be permitted.

Industrial Locations 4

POWER INSTALLATIONS

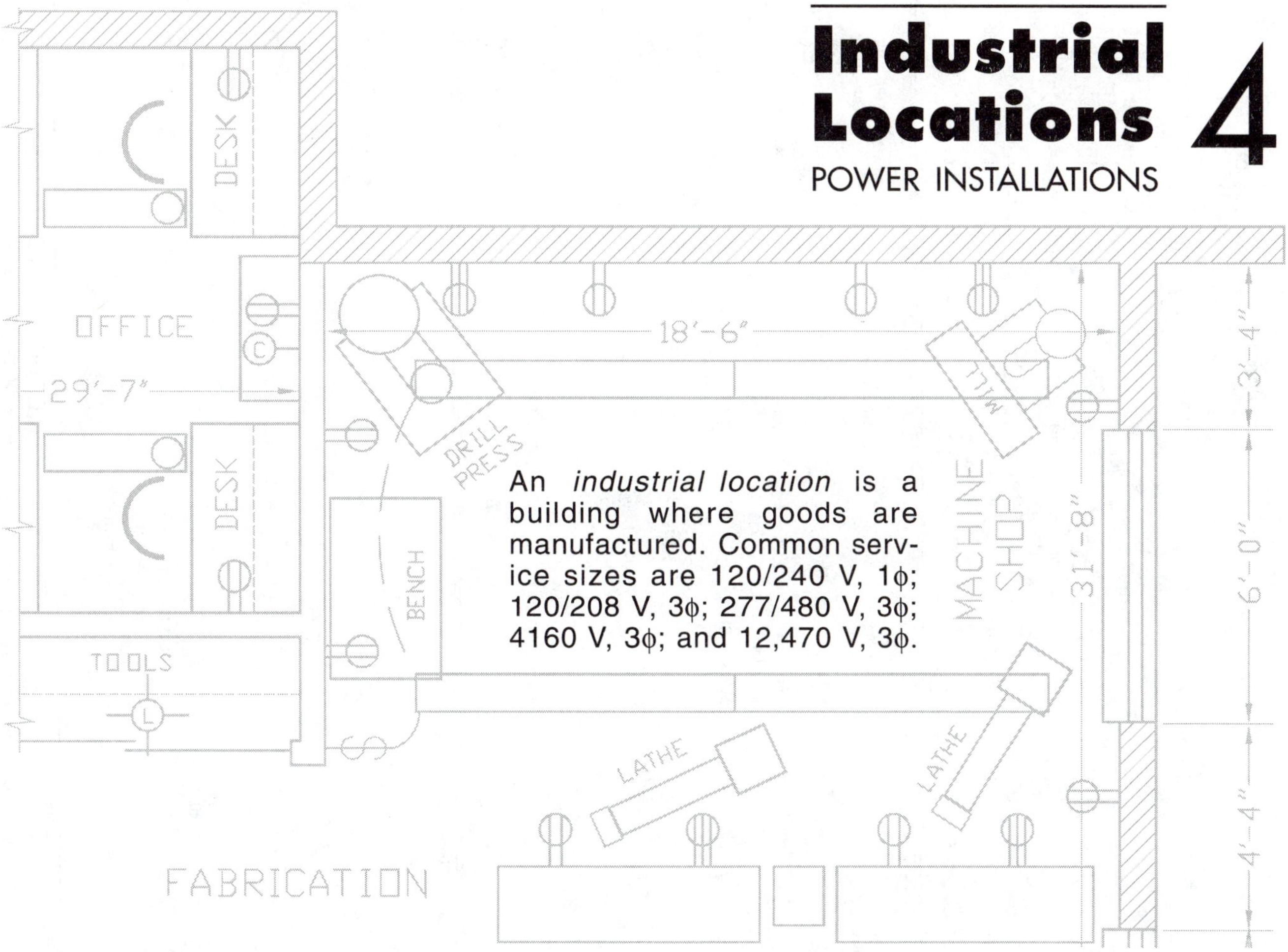

An *industrial location* is a building where goods are manufactured. Common service sizes are 120/240 V, 1ϕ; 120/208 V, 3ϕ; 277/480 V, 3ϕ; 4160 V, 3ϕ; and 12,470 V, 3ϕ.

POWER INSTALLATION FOR MANUFACTURING PLANT

This print shows only power outlets which are supplied by a 4-wire, 3-phase, 265/460 V, WYE distribution system. Lighting and receptacle outlets are supplied by a separate 4-wire, 3-phase, 120/208 V network system and appear on a separate print. Specifications call for Type THW copper conductors throughout the project.

BASIC NEC RULES FOR MOTOR CIRCUITS

Article 430 explains when to use the tables at the end of Article 430 and when to use the motor nameplate rating to determine the ampacities and ratings of the various components of motor circuits. NEC 430-6(a) states that where the current rating of a motor is used to determine the ampacity of conductors or the ampere ratings of switches, controllers, branch circuit overcurrent protection devices, etc., the values given in Tables 430-147, 430-148, 430-149, and 430-150, including notes, shall be used instead of the motor nameplate rating.

If only the full load current is marked on the nameplate, the horsepower rating shall be determined by the corresponding value given in Tables 430-147 through 430-150. In the selection of separate motor-running overcurrent protection which will protect motors, motor-control apparatus, and motor branch circuit conductors against excessive heating due to overloads and failure to start, the full load current rating that appears on the motor nameplate shall be used.

NEC 430-22(a) states that branch circuit conductors supplying a single motor shall have an ampacity not less than 125% of the full-load current of the motor. NEC 430-32(a)(1) specifies that continuous-duty motors rated more than 1 hp shall be protected against overload, in general, by

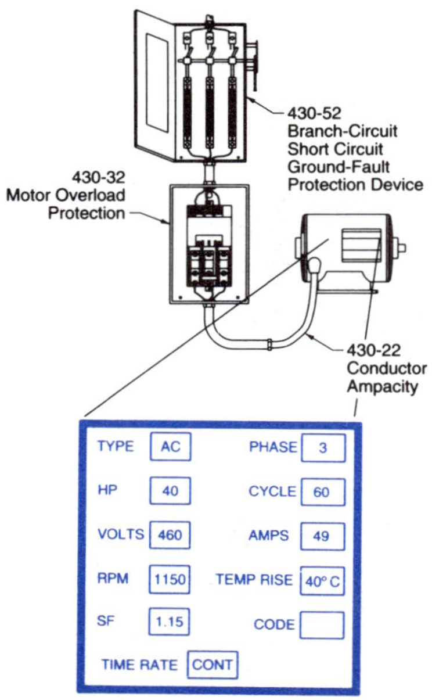

1. Branch Circuit Conductors. 430-22(a) requires that conductors serving a single motor be sized at 125% of motor table current. Table 430-150 lists FLC for a 40 hp squirrel-cage motor at 52 A (52 × 125% = 65 A). Table 310-16 lists nearest size THW conductor as No. 6.

2. Branch-Circuit, Short-Circuit, and Ground-Fault Protective Device. Per 430-52(b), protective device shall be capable of carrying starting current of motor. Protective device shall be selected from Table 430-152. If this value does not correspond to a standard size protective device or protective device setting, next larger size shall be permitted. See Ex. 1. For a polyphase motor using NTDFs, 300% of FLC is permitted (300% x 52 A = 156 A). Next largest standard size overcurrent device is 175 A.

3. Motor Overload Protection. 430-32(a)(1) requires each continuous-duty motor rated more than 1 hp to have overload protection based upon marked service factor or marked temperature rise. For a marked temperature rise of 40°C, use 125%. 430-6(a) requires for separate motor overload protection, current rating of motor, not motor table current shall be used (125% × 49 A = 61.25 A). Maximum setting for a motor overload device is 61.25 A.

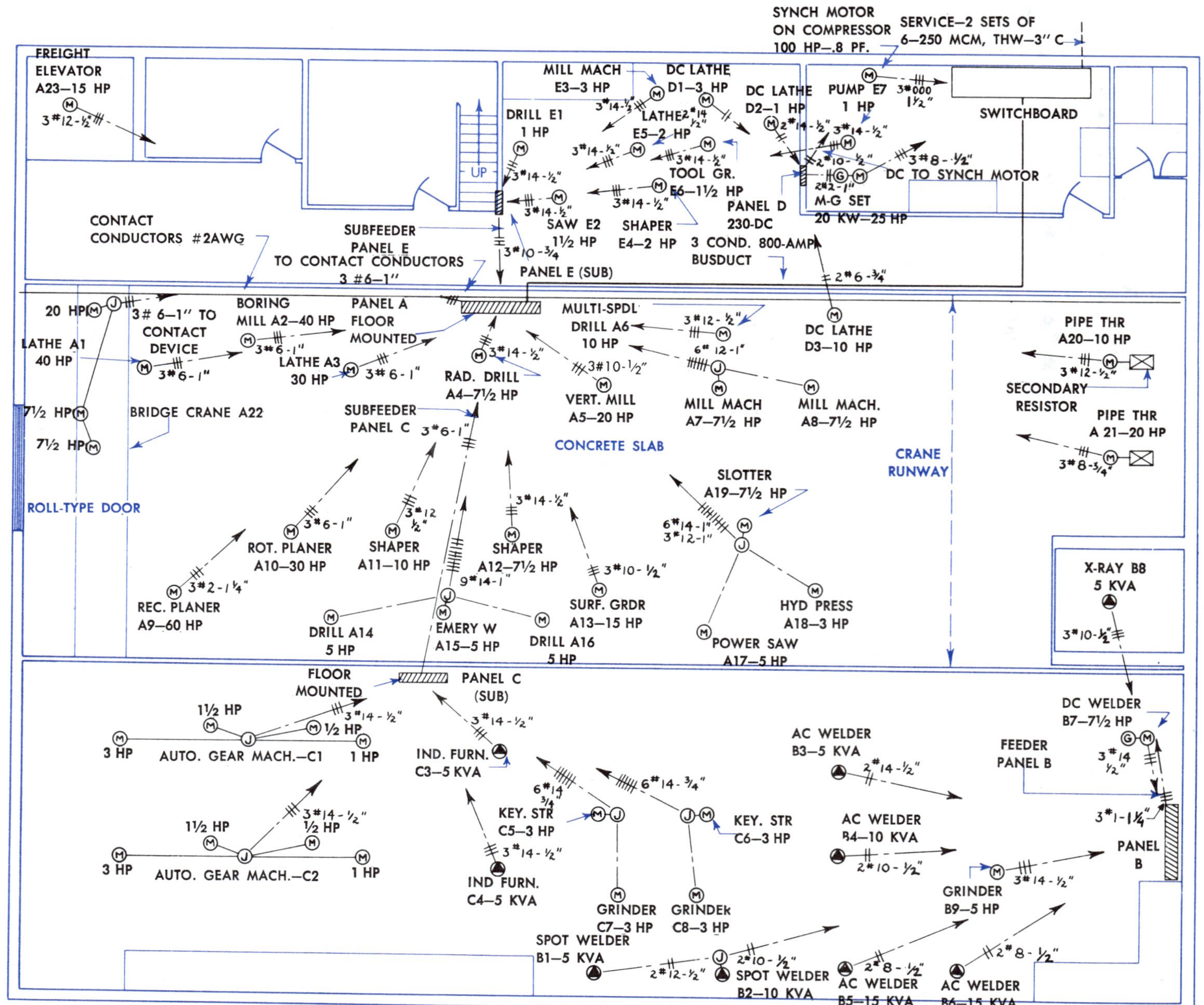

INDUSTRIAL PLANT

PUBLISHER'S NOTE: THESE DRAWINGS WERE ORIGINALLY DRAWN TO THE SCALES SHOWN. THE DRAWINGS WERE REDUCED TO FIT THE PAGE AND CAN NO LONGER BE SCALED.

SCALE: 1/8″ = 1′-0″

a device that trips at not more than 125% of the full-load current of the motor.

NEC 430-52 requires that overcurrent devices protecting branch circuit conductors shall be capable of carrying motor starting current. Motor starting current varies usually between 150% and 300% of full load current, depending on the type of motor. Table 430-152 lists information necessary to calculate the maximum rating or setting of motor branch-circuit, short-circuit, and ground-fault protective devices. In rare cases, starting current runs somewhat higher than values given there. NEC 430-52 provides for this contingency, permitting larger overcurrent protection devices, which permit the motor to start.

Application

Consider unit *A1* on the print, a 40 hp, 3-phase, 460 V, squirrel-cage motor which has no code letter marked on the nameplate a marked service factor of 1.15, a full load current of 49 A, and which is started by means of an autotransformer compensator. Table 430-150 gives the full load current as 52 A. Under 430-22(a), the branch circuit conductors must be able to carry 1.25 × 52 A, or 65 A. Table 310-16 lists the nearest size of Type THW wire as No. 6. Table C8 of App. C permits three No. 6 Type THW conductors in 1″ rigid conduit.

Table 430-152 lists all types of motors including squirrel cage and synchronous types. Column 1 of Table 430-152, "squirrel-cage, other than Design E" applies here. Column 2 shows the branch circuit nontime-delay fuse rating as 300% of full load current. For a motor current of 52 A, circuit protection should be 3 × 52 A, or 156 A. The nearest standard fuse or non-adjustable circuit breaker rating that does not exceed the 300% is 175 A per 240-6.

Motor running overcurrent protection, per 430-32(a)(1) should be not greater than 1.25 × 49 A, or 61.25 A. This protection may be obtained with an adjustable device set at 61.25 A or by the nearest fuse or non-adjustable circuit breaker, which in this case is 60 A.

Vertical milling machine *A5* is driven by a 20 hp motor, Code letter *C*, a marked service factor of 1.15, a full load current of 25 A, and which is protected with an inverse time CB. Table 430-150 gives full load current as 27 A. NEC 430-22(a) calls for a No. 10 conductor (1.25 × 27 A = 33.75 A). Table 3A, Chapter 9 specifies a ½″ conduit. Motor running overcurrent protection as per 430-32(a)(1) should be not greater than 1.25 × 25 A, or 31.25 A. The nearest standard fuse or non-adjustable circuit breaker is rated at 30 A per 240-6. Table 430-152 gives 250% of full load current for branch circuit protection, 2.5 × 27 A = 67.5 A. This allows a 70 A ITCB for branch circuit protection.

The 7½ hp, Code *F*, full-voltage starting motor on radial drill *A4* has a nameplate current of 10 A and a marked service factor of 1.15. Motor running protection is 1.25 × 10 A, or 12.5 A, for an adjustable protector per 430-32(a)(1). Branch circuit protection is 300% of 11 A, or 33 A, which permits a 35 A nontime-delay fuse per Table 430-152.

Pipe-threading machine *A21* uses a 20 hp motor of the wound-rotor type. The size of circuit conductors, conduit, and motor running overcurrent protection are the same as the 20 hp squirrel-cage motor of Circuit *A5*, but the branch circuit protection is different. Referring to Table 430-152, select the wound-rotor motor which allows 150% of full load current, 27 A × 1.50 = 40.5 A. A 45 A nontime-delay, dual element fuse or a 45 A time-limit circuit breaker may be used.

The motor nameplate lists a secondary current of 50 A. NEC 430-23(a) states that for continuous-duty motors, the carrying capacity of wires between the motor secondary (rotor) and the controller shall not be less than 125% of the full load secondary current. This value here is 62.3 A, the nearest conductor being No. 6. Table 430-23(c) specifies ampacities of conductors between controller and resistors. In most cases, high-temperature wire is necessary because of heat that is generated in resistor grids.

Lathe *D3* is driven by a 10 hp, 230 V, direct current motor with a full load current of 37 A and a marked service factor of 1.15. Table 430-147 gives full load current as 38 A, requiring two No. 8 conductors in ¾″ conduit. Motor running overcurrent protection is 1.25 × 37 A, or 46.25 A and branch circuit protection is 1.5 × 38 A, or

57 A per Table 430-152, which permits a 60 A dual element (time-delay) fuse.

Elevator Motor

A23 is a 15 hp, high-reactance, 30-60-minute type motor with full-voltage starting. Table 430-22(a), Ex. states that duty on elevator motors shall be classed as intermittent. NEC 430-33 provides that the ampacity of the branch circuit conductors for intermittent duty motors shall be considered as protected against running overcurrent by a branch circuit device which does not exceed values specified in Table 430-152.

NEC 430-22(a), Ex. 1 states that the ampacity of the branch circuit conductors for intermittent duty motors shall be not less than the percentage of the nameplate current rating of the motor as shown in Table 430-22 (a), Ex. unless the authority having jurisdiction grants special permission for conductors of a smaller size. In this case, the ampacity of the branch circuit conductors for elevator motor *A23* need be only 90% of the nameplate full load current of the motor.

The full load current rating of elevator motor *A23* is listed as 21 A. The ampacity of conductors may be rated on the basis of .9 × 21 A, or 18.9 A. A No. 12 conductor is needed. Table 430-152 shows that the branch circuit protection may be 150% of the full load current or 31.5 A (1.5 × 21 A) which permits a 35 A non-time-delay fuse or ITCB.

Feeder Supplying Two or More Motors

NEC 430-24 states that conductors supplying two or more motors shall have an ampacity of not less than 125% of the full load current rating of the highest rated motor in the group, plus the sum of the full load current of remaining motors. Thus for two motors, one having a full load current rating of 20 A and the other 14 A, the feeder ampacity shall be not less than (1.25 × 20 A) + 14 A, or 39 A. A No. 8 Type THW wire is indicated.

NEC 430-62(a) provides that overcurrent protection for conductors supplying two or more motors shall be not greater than the branch circuit overcurrent for the largest rating or setting plus full load currents of other motors. Assume a 250% protective requirement for the larger motor in the preceding example. The feeder protective device should be set at not more than (2.5 × 20 A) + 14 A, or 64 A. The nearest standard fuse or circuit breaker that does not exceed this value is 60 A per 240-6.

Crane Motors

Bridge crane *A22* has three motors, a 20 hp and two $7\frac{1}{2}$ hp units. The nameplates specify 15 minute ratings. Noncontinuous-duty motors are commonly used on this type of service, the ratings being stated according to the nature of the application as 15 minutes, 30 minutes, or 60 minutes.

Table 610-14(a) has ampacities for conductors supplying this type of motor, permissible currents being larger than those allowed by Table 310-16. Columns of Table 610-14(a) list values according to types of insulation and ambient operating temperatures, the first of each set of two columns dealing with 60 minute ratings, the second with 30 minute ratings. A footnote states that allowable ampacities of conductors for 15 minute motors shall be equal to the 30 minute ratings increased by 12%.

In the present instance, the 75°C columns are applicable. Since the motors are 15 minute units, sizes of conductors are determined on the basis of the 30 minute rating increased by 12%. Full load current of the largest motor is 27 A, and the others are 11 A each. According to the table, a No. 14 Type THW conductor has a 30 minute current-carrying capacity of 26 A. A 12% increase gives this conductor a capacity of 29 A, making it suitable for use with the large motor.

NEC 610-14(c) states that a No. 16 conductor may be used with crane motors, if otherwise suitable. The 30 minute rating of No. 16 is given as 12 A, and the allowable 12% increase raises the

15 minute rating to 13.4 A, a value satisfactory for the two small motors.

NEC 610-14(e) states that the ampacity of the power supply conductors on the crane shall be not less than the combined short-time full-load ampere rating of the largest motor or group for any single crane motion, plus 50% of the short-time full-load ampere rating of the next largest motor or group. Adding the 27 A of the largest motor to 50% of 22 A (full load current of other motors), a No.8 THW conductor may be used. However,the specifications call for a No. 6 THW conductor here.

NEC 610-14(d) lists the sizes of bare contact conductors which, in the present case where the runway is longer than 60′, must be No. 2.

Metal-working Machine Tools

NEC 670-3(a) states that the nameplate shall give necessary data, including the full load current and the ampere rating of the largest motor. NEC 670-4(a) states that the ampacity of supply conductors shall be not less than the full load current marked on the nameplate, plus 25% of the full load current rating of the highest rated motor as indicated on the nameplate. This requirement is equivalent to the general rule stated in 430-24.

Automatic gear machines *C1* and *C2* fall within the metalworking-machine tool classification. The total nameplate rating is 10.2 A and that of the highest rated motor is 4.8 A. Per 670-4(a), the supply conductors shall have a current-carrying capacity of not less than 10.2 A + (.25 × 4.8 A), or approximately 11.4 A. The feeder to Panel *C* consists of three No.14 conductors in ½″ conduit. Short-circuit protection for each feeder is 14.4 A (3 × 4.8 A) + 5.4A (2.6 A + 1.8 A + 1 A), or 19.8 A per 430-62(a). A 15 A nontime-delay fuse is acceptable.

Synchronous Motor

An air compressor in the switchboard room is driven by a 100 hp synchronous motor operating at a leading power factor of 80%. Table 430-150 gives the full load current at unity power factor as 101 A. A footnote states that this figure is to be multiplied by 1.25 when the motor operates at 80% power factor, the current here becoming 126 A. Circuit wires shall have an ampacity of at least 1.25 × 126 A, or 158 A per 430-22(a). Table 310-16 lists No. 2/0 Type THW copper wire at 175 A. Table C8, App. C shows that 1½″ rigid metal conduit is required.

A footnote to Table 430-152 states that for synchronous motors used in applications of this nature, branch circuit protection need not be greater than 200% of the full load current. For a current of 126 A, the overcurrent protection device could be set as low as 252 A. By using an inverse-time breaker, a 300 A breaker is needed. Two No. 10 conductors in ½″ conduit, originating at Panel *D*, supply direct current excitation for the revolving field of the synchronous motor.

X-Ray Unit

The No. 10 conductors specified for this device are larger than might seem to be required. Their size is dictated by a desire to hold voltage drop to a minimum.

Resistance Welders

NEC 630-31(a)(1) states that the ampacity of conductors supplying a resistance welder shall be not less than 70% of the rated primary current for automatically fed welders and 50% for manually operated ones. NEC 630-31(b) requires that rated ampacity of conductors which supply two or more such welders shall be not less than the sum of the above value for the largest welder plus 60% of such values for the smaller welders.

NEC 630-32(a)(b) demand that each welder be protected by an overcurrent protection device set at not more than 300% of the rated primary current, and that a feeder supplying one or more units shall be protected by a device set at not more than 300% of the conductor rating.

Unit *B1* is a 5 kVA spot welder rated at 11 A. Unit *B2* is a 10 kVA spot welder rated at 22 A. Both are manually operated. Conductors supplying *B1* may be No. 14. Those for *B2* may be also No. 14 (.5 × 22 A). *B1* may be protected by a 30 A fuse or non-adjustable circuit breaker and *B2* by a 60 A fuse or non-adjustable circuit breaker. The feeder should not be smaller than (.5 × 22 A) + (.6 × .5 × 11 A), or 14.3 A. The question of voltage drop must be considered if the units are to operate efficiently. Two No. 10 conductors are run in $\frac{1}{2}''$ conduit from the junction box to Panel *B*. Two No. 12 conductors extend from the small welder to the junction box.

Transformer Arc Welders

NEC 630-11(b) provides that conductor rating for a group of welders shall be based on the nature of the operation. The maximum demand for a similar group of welders used on an identical application is 50% of total capacity of the units. The total capacity is equal to 5 kVA + 10 kVA + 15 kVA + 15 kVA, or 45 kVA. One half of this amount is 22.5 kVA, and the current taken when they are balanced across a 460 V, 3-phase supply line is equal to about 30 A per leg. By using the 3-phase formula (kVA = 1.73 × E × I), we find that 1.73 × 460 V divided into 22,500 VA is equal to 28 A.

NEC 630-12(a) requires that each welder have overcurrent protection not greater than 200% of the rated primary current. NEC 630-12(b) requires that feeder protection be not greater than 200% of the conductor rating.

Motor-generator Arc Welder

Unit *B7*, a single-operator, motor-generator, arc welder with a duty cycle of 80% per 630-21(a), which states that conductors supplying such a unit shall have an ampacity equal to 91% of the nameplate rating, which is 11 A. The smallest permissible conductor is No. 14 (91% of 11 A is 10 A.)

Feeder Calculations

The load on the sub-feeder from Panel *C* to Panel *A* totals about 53 A. Three No. 6 THW conductors are used in 1″ conduit. The load on the sub-feeder from Panel *E* to Panel *A* is approximately 21.6 A. Specifications call for a No. 10 feeder. Panel *A* by its own units, including 25% of the largest motor running current, equals approximately 544 A. The total load on Panel *A* equals approximately 544 A plus the full load current of the motors connected to Panels *C* and *E*, or 544 A + 52 A + 22 A for a total feeder load to Panel *A* of 618 A. The print shows an 800 A busway from the main switchboard to Panel *A*.

Feeder conductors for Panel *B* must supply the following loads: 14.3 A, single-phase to resistance welders *B1* and *B2;* 28 A, 3-phase to transformer arc welders *B3, B4, B5,* and *B6;* 10 A 3-phase to motor-generator arc welder *B7;* 11 A, 1-phase to the X-ray unit *B8;* and 7.6 A, 3-phase to the grinder *B9.*

When this load is balanced across the three phases, as equally as possible, the most heavily loaded conductor carries about 70 A. No. 4 Type THW copper conductors would normally be used, but because of the nature of the equipment, especially the X-ray unit, three No. 1 conductors are specified. Table C8, App. C, shows that $1\frac{1}{4}''$ rigid metal conduit is needed.

Service Calculations

The service load is equal to the full load motor currents of all of the motors plus all other equipment on Panels *A* and *B* plus the full load current of the 25 hp motor-generator, plus that of the 100 hp synchronous motor, plus 25% of the full load current of the synchronous motor. The total is: Panel *A*, 525 A + Panel *B*, 70 A + 34 A for 25 hp MG set + 126 A for the 100 hp synchronous motor + 25% of 126 A, or 32 A, or 787 A. This value is larger than that for conductors listed in Table 310-16, so two or more conductors in parallel must be used.

One half of 787 A equals 393.5 A. The nearest listed conductor is 600 kcmil Type THW. The service may consist of two sets of three 600 kcmil Type THW conductors in two $3\frac{1}{2}''$ rigid metal conduits per Table C8, App. C. For ease of handling, smaller conductors can be used. Table 310-16 gives the ampacity of 250 kcmil Type THW as 255 A. With six conductors in one conduit, the ampacity of each is reduced to 80% of 255 A, or 204 A. Two such conductors on each leg furnish capacity equal to 408 A. Table C8, App. C, permits six 250 kcmil Type THW conductors in 3″ rigid metal conduit. The service then consists of two sets of six 250 kcmil Type THW conductors in 3″ conduits.

Note: The above service configuration is one of many possible alternatives. Supply conductors could be run in three or more sets of conduit, to avoid penalization with respect to ampacity, or even in the form of busduct if conditions warranted its use.

POWER INSTALLATION FOR RESTAURANT

The power installation is to be confined to the kitchen area, with the exception of a 5 hp exhaust fan on the roof. The service is 4-wire delta with 3-phase power at 240 V between phase wires and single-phase power at 120/240 V between two phase wires and a neutral conductor, as indicated on the small diagram to the right of the print.

The panelboard is divided into two sections, *A* and *B*. Section *A* supplies 3-phase loads and large single-phase loads which do not require a neutral conductor. Section *B* supplies any 1-phase loads which require a neutral conductor. Type THW copper conductors are used.

The print covers only the power installation. Lighting outlets appear on a separate drawing. This procedure is often used where either the power or the lighting is rather involved, to avoid confusion from crowding both power and lighting onto the same sheet.

Circuit Analysis

Circuits *A-1* and *A-2* are connected to 21 kVA, 3-phase ranges, the nameplate current being 53 A. These loads in normal operation are likely to continue for long periods of time. NEC 384-16(c) requires that conductors serving such loads shall carry not more than 80% of the listed current. The conductors supplying each range are rated at a minimum ampacity of 53 A divided by .8, or 67 A. Table 310-16 rates No. 4 Type THW at 85 A. Table C8, App. C, indicates that 1″ rigid metal conduit is needed for three No. 4 conductors.

Deep fat fryers *A-3* and *A-4* are rated at 10 kVA single-phase, the current being 44 A. These loads are similar to those of the ranges in that they continue for long periods of time so that the 80% factor is used. The conductors have a normal ampacity of 44 A divided by .8 or 55 A. Two No. 6 conductors in 3/4″ conduit will meet the requirements.

Circuit *A-5* supplies a 5 hp, 3-phase blower motor with a nameplate current of 14.8 A and a service factor of 1.15 on the roof. A riser leading to this unit may be noted at the left side of the panelboard. NEC 430-6(a) states that, except for hermetic motors, ampacity of conductors, branch-circuit overcurrent devices, etc., shall be determined from the motor current tables at the end of Article 430 instead of current ratings on motor nameplates. Table 430-150 lists the current of the 5 hp, 3-phase, 240 V motor as 15.2 A.

NEC 430-22(a) requires that branch circuit conductors supplying a single motor have an ampacity of 125% of 15.2 A, or 19 A. No.12 Type THW is rated at 20 A. Table C8, App. C shows that three No. 12 conductors may be run in 1/2″ rigid metal conduit.

NEC 430-52 and Table 430-152 require that the initial rating of a dual-element fuse, used for branch circuit protection, should not exceed 175% of the full-load current of the motor listed in Table 430-150. The rating of the dual-element fuse for the 5 hp motor should not exceed 1.75×15.2 A, or 26.6 A.

NEC 430-52 requires the use of a 30 A fuse. NEC 430-83 states that a controller shall have a horsepower rating not less than that of the motor. NEC 430-109 states a similar requirement for the disconnecting means. A 30 A circuit breaker satisfies the needs of a controller per 430-83, Ex. 2 and of a disconnect per 430-110(a). The rating of a manually operable switch on the roof of the building is 5 hp.

Per 430-32(a), each continuous-duty motor rated at more than 1 hp shall be protected against running overcurrent by a separate overcurrent

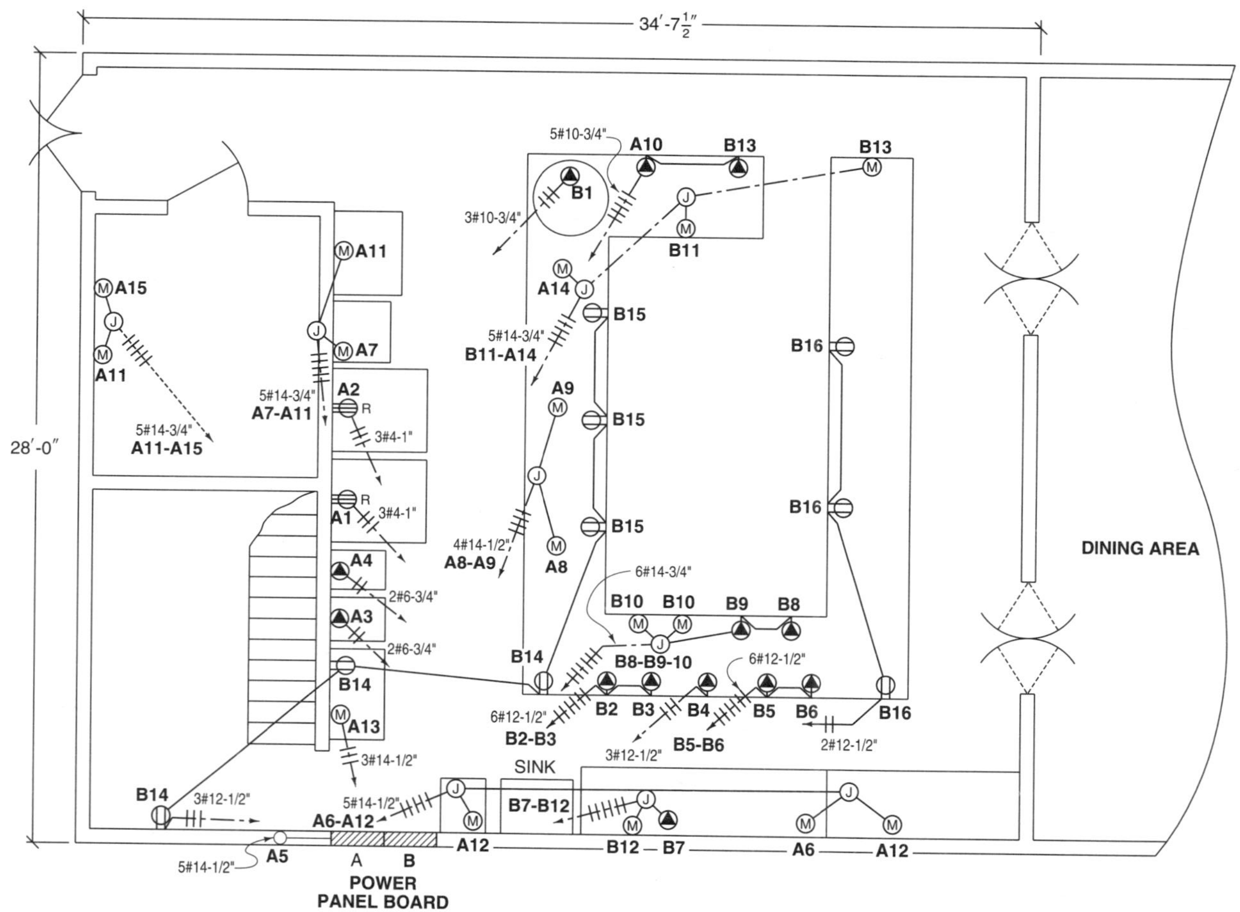

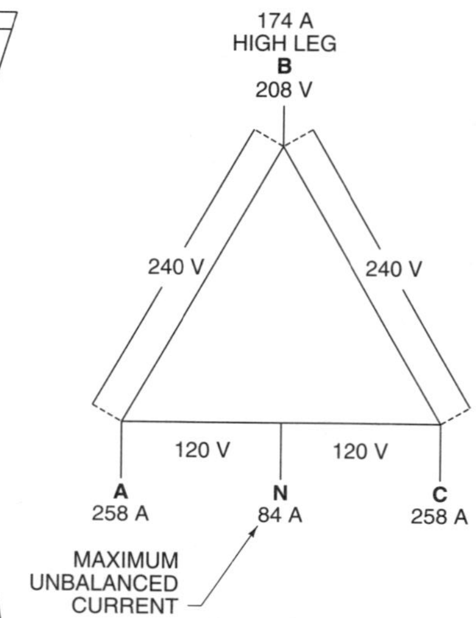

4-WIRE–3ϕ DELTA SERVICE

RESTAURANT KITCHEN

SCALE: $\frac{1}{4}$″ = 1′-0″

MOTOR LOAD ON PANEL A

- **A5** Roof Fan–5 hp, 3ϕ, 240 V, 15.2 A
- **A6** Dishwasher–$1\frac{1}{2}$ hp, 3ϕ, 240 V, 5.2 A
- **A7** Food Mixer–1 hp, 3ϕ, 240 V, 3.6 A
- **A8** Compressor–$\frac{3}{4}$ hp, 3ϕ, 240 V, 2.8 A
- **A9** Compressor–$\frac{3}{4}$ hp, 3ϕ, 240V, 2.8 A
- **A11** Compressor–$\frac{1}{2}$ hp, 1ϕ, 240 V, 4.9 A
- **A11** Chef's Ref–$\frac{1}{2}$ hp, 1ϕ, 240 V, 4.9 A
- **A12** Disposal–$\frac{1}{2}$ hp, 1ϕ, 240 V, 4.9 A
- **A12** Potato Peeler–$\frac{1}{2}$ hp, 1ϕ, 240 V, 4.9 A
- **A13** Food Chopper–$\frac{3}{4}$ hp, 3ϕ, 240 V, 4.9 A
- **A14** Ice Machine–$1\frac{1}{2}$ hp, 3ϕ, 240 V, 5.2 A
- **A15** Hermetic Comp–3ϕ, 240 V, 11 A

COMMERCIAL ELECTRIC COOKING AND KITCHEN EQUIPMENT ON PANEL A

- **A1** Range–21.00 kVA, 3ϕ, 240 V
- **A2** Range–21.00 kVA, 3ϕ, 240 V
- **A3** Deep Fat Fryer–10.00 kVA, 3ϕ, 240 V
- **A4** Deep Fat Fryer–10.00 kVA, 3ϕ, 240 V
- **A10** Hot Table–3.25 kVA, 3ϕ, 240 V

MOTOR LOAD ON PANEL B

- **B10** Compressor–$\frac{1}{4}$ hp, 1ϕ, 120 V, 5.8 A
- **B10** Compressor–$\frac{1}{4}$ hp, 1ϕ, 120 V, 5.8 A
- **B11** Carbonator–$\frac{1}{4}$ hp, 1ϕ, 120 V, 5.8 A
- **B12** Silver Burn.–$\frac{1}{3}$ hp, 1ϕ, 120 V, 7.2 A

COMMERCIAL ELECTRIC COOKING AND KITCHEN EQUIPMENT ON PANEL B

- **B1** Coffee Maker–5 kVA, 1ϕ, 120/240 V
- **B2** Toaster–2.5 kVA, 1ϕ, 120/240 V
- **B3** Toaster–2.5 kVA, 1ϕ, 120/240 V
- **B4** Toaster–2.5 kVA, 1ϕ, 120/240 V
- **B5** Waffle Iron–2.5 kVA, 1ϕ, 120/240 V
- **B6** Waffle Iron–2.5 kVA, 1ϕ, 120/240 V
- **B7** Sterilizer–1 kVA, 1ϕ, 240 V
- **B8** Plate Warmer–1 kVA, 1ϕ, 240 V
- **B9** Bun Warmer–.75 kVA, 1ϕ, 120 V
- **B13** Infrared Cooker–3.5 kVA, 1ϕ, 120/240 V

protection device rated at not more than 125% of the full load current of the motor, or else contain an integral thermal protector. If the blower motor does not have built-in thermal protection, an overcurrent protection device must be supplied (perhaps in a magnetic switch used to start and stop the motor.) The rating of the device should be 1.25 × 14.8 A, or 18.5 A.

NEC 430-52 states that the maximum value of branch circuit protection shall be limited to the value given in Table 430-152 (300% for nontime-delay fuses). NEC 430-58 states that a circuit breaker may be used for motor branch circuit protection if it conforms to 430-110(a) which in turn requires it to have an ampacity of 115% of the full load current of the motor. The 5 hp motor has a full load current of 15.2 A × 115% or 17.5 A. If a standard 20 A circuit breaker at the panel carries starting current of this motor without tripping, it satisfies the need for motor overcurrent protection per 430-52 and 430-58

Circuit *A-6* supplies a 1½ hp, 3-phase, dishwasher motor with a built-in thermal device. Table 430-150 gives the current as 5.2 A, so that No. 14 conductors are large enough. According to 430-83, Ex. 1, a controller located at the operating location may be a general use switch, which has an ampere rating at least twice that of the full load current of the motor.

Three No. 14 conductors in ½″ conduit extend from the motor location to a junction box where they are joined by two No. 14 conductors from a garbage disposal connected to Circuit *A-12*. A ½″ conduit carries the five wires to a second junction box where two conductors from a potato peeler motor are tapped to Circuit *A-12* before the five wires continue to the panelboard.

Two motors are connected to Circuit *A-12* per 430-53(a), which states that motors not exceeding 1 hp each and having a full load current not in excess of 6 A may be connected to a 20 A, 120 V, or less, circuit. Individual running overcurrent protection is required as these motors are permanently installed. See 430-32(b)(2). Usually, motors of this type are provided with a thermal protector integral with the motor which gives required protection.

Other motors are combined in this manner on Circuits *B-10*, *B-11*, and *A-11*. Those on *A-11* are connected at the panelboard because they are run in separate conduits. The ¾ hp compressor motors on Circuits *A-8* and *A-9* may be combined on a single circuit but would require running protection, as they are automatically started per 430-32(c).

Circuit *A-10*, for the hot table, and *B-13*, for the infrared cooker, are taken to the panelboard as five No. 10 conductors in ¾″ conduit. *A-10* is rated at 3250 VA or 13.5 A and *B-13* at 3500 VA or 14.6 A. Since these are continuous loads, 384-16(c) requires that the total load on any overcurrent device not exceed 80% of its rating.

When there are from four to six current-carrying conductors in the same conduit supplying a constant load, they are derated once as per Note 8(a) of Notes to Ampacity Tables of 0 to 2000 Volts. No. 12 conductors are permissible for *A-10* and *B-13*, but in this case a No. 10 Type THW was selected. Table C8, App. C, permits six No. 10 Type THW conductors in ¾″ rigid metal conduit. Derating considerations also enter into calculations for toasters and waffle irons on Circuits *B-2*, *B-3*, *B-5*, and *B-6*.

Circuit *A-15*, whose conduit is run exposed, supplies an 11 A sealed hermetic compressor motor. NEC 440-6(a) states that the full load current marked on the compressor nameplate shall be used to determine ampacity of the branch circuit conductors. NEC 440-4(a) requires that the locked rotor current for such motors shall be marked on the nameplate. Locked rotor current is marked as 53 A. NEC 440-41(a) provides that controllers for sealed hermetic motors shall be selected on the basis of full load current and locked rotor current.

Per Table 430-150, Column 3, the nearest current greater than 11 A, (the full-load current on the motor nameplate) is 15.2 A, which applies to a 5 hp unit. In Table 430-151, the nearest 3-phase locked rotor current to the 53 A value on the nameplate of the motor is 58 A, applying to a 3 hp motor. NEC 440-41(a) states that where two different horsepower ratings are obtained from different tables, the higher of the the values shall be used. Here, the larger of them is used, necessitating a 5 hp controller. NEC 440-12(a) imposes a similar demand with respect to the disconnecting means.

Service Calculation

The load on Panel *A* is obtained by converting motor loads to VA and adding them to the VA of the cooking and heating appliances. A value equal to 25% of the largest motor rating is included per 430-24 as well as a demand factor permitted by Table 220-20 for commercial electric cooking and other kitchen equipment. The total load on Panel *A* is 72,137 VA or 174 A.

MOTOR LOAD ON PANEL *A*

A5 Roof Fan — 5 hp, 3ϕ, 240 V, 15.2 A
VA = 1.73 × 240 × 15.2 = 6311 VA

A6 Dishwasher — 1½ hp, 3ϕ, 240 V, 5.2 A
VA = 1.73 × 240 × 5.2 = 2159 VA

A7 Food Mixer — 1 hp, 3ϕ, 240 V, 3.6 A
VA = 1.73 × 240 × 3.6 = 1495 VA

A8 Compressor — ¾ hp, 3ϕ, 240 V, 2.8 A
VA = 1.73 × 240 × 2.8 = 1163 VA

A9 Compressor — ¾ hp, 3ϕ, 240 V, 2.8 A
VA = 1.73 × 240 × 2.8 = 1163 VA

A11 Compressor — ½ hp, 1ϕ, 240 V, 4.9 A
VA = 240 × 4.9 = 1176 VA

A11 Chef's Ref — ½ hp, 1ϕ, 240 V, 4.9 A
VA = 240 × 4.9 = 1176 VA

A12 Disposal — ½ hp, 1ϕ, 240 V, 4.9 A
VA = 240 × 4.9 = 1176 VA

A12 Potato Peeler — ½ hp, 1ϕ, 240 V, 4.9 A
VA 240 × 4.9 = 1176 VA

A13 Food Chopper — ¾ hp, 3ϕ, 240 V, 2.8 A
VA = 1.73 × 240 × 2.8 = 1163 VA

A14 Ice Machine — 1 ½ hp, 3ϕ, 240 V, 5.2 A
VA = 1.73 × 240 × 5.2 = 2159 VA

A15 Hermetic Comp — 3ϕ, 240 V, 11 A
VA = 1.73 × 240 × 11 = 4567 VA

Total Motor VA = 24,884 VA

VA × 25% of largest motor:
VA = .25 × 6311 = +1578 VA

26,462 VA

COMMERCIAL ELECTRIC COOKING AND KITCHEN EQUIPMENT ON PANEL *A*

A1 Range	21.00 kVA, 3ϕ, 240 V
A2 Range	21.00 kVA, 3ϕ, 240 V
A3 Deep Fat Fryer	10.00 kVA, 3ϕ, 240 V
A4 Deep Fat Fryer	10.00 kVA, 3ϕ, 240 V
A10 Hot Table	3.25 kVA, 3ϕ, 240 V
Total	65.25 kVA

Table 220-20 permits a demand factor of 70% for 5 units of electric cooking and other kitchen equipment in a commercial kitchen, therefore

65,250 VA × .70 = **45,675 VA**

Total VA on Panel *A* = **72,137 VA**

$$I = \frac{VA}{V \times \sqrt{3}}$$

$$I = \frac{72,137}{240 \times 1.73}$$

$$I = 173.7\ A$$

Panel *A* = **174 A**

The load on Panel *B*, found in the same manner, is 20,226 VA or 84 A.

MOTOR LOAD ON PANEL *B*

B10 Compressor — ¼ hp, 1ϕ 120V, 5.8 A
VA = 120 × 5.8 = 696 VA

B10 Compressor — ¼ hp, 1ϕ 120V, 5.8 A
VA = 120 × 5.8 = 696 VA

B11 Carbonator — ¼ hp, 1ϕ 120 V, 5.8 A
VA = 120 × 5.8 = 696 VA

B12 Silver Burn. — ⅓ hp, 1ϕ 120 V, 7.2 A
VA = 120 × 7.2 = 864 VA

Total Motor VA = 2952 VA

+ 25% of largest motor = .25 × 864 = +216 VA

Total Motor VA on Panel *B* = **3168 VA**

COMMERCIAL ELECTRIC COOKING AND KITCHEN EQUIPMENT ON PANEL *B*

B1 Coffee Maker — 5 kVA, 1ϕ, 120/240 V	5000 VA
B2 Toaster — 2.5 kVA, 1ϕ, 120/240 V	2500 VA
B3 Toaster — 2.5 kVA, 1ϕ, 120/240 V	2500 VA

B4 Toaster — 2.5 kVA, 1ϕ, 120/240 V	2500 VA
B5 Waffle Iron — 2.5 kVA, 1ϕ, 120/240 V	2500 VA
B6 Waffle Iron — 2.5 kVA, 1ϕ, 120/240 V	2500 VA
B7 Sterilizer — 1 kVA, 1ϕ, 240 V,	1000 VA
B8 Plate Warmer — 1 kVA, 1ϕ, 240 V	1000 VA
B9 Bun Warmer — .75 kVA, 1ϕ, 120 V	750 VA
B14 Infrared Cooker — 3.5 kVA, 1ϕ, 120/240 V	3500 VA
Total connected load =	23,750 VA

Table 220-20 permits a demand factor of 65% for 6 or more units of electrical cooking and other kitchen equipment in a commercial kitchen, therefore

23,750 VA × .65 =	**15,438 VA**
Nine 120 V receptacles at 180 VA each. 9 × 180 =	**1620 VA**
Total VA on Panel *B* is	**20,226 VA**

$$I = \frac{VA}{V}$$

$$I = \frac{20{,}226}{240}$$

$$I = 84.275\text{ A}$$

Panel B = **84 A**

These values have been marked on the diagram at the right of the drawing, the load on power leg *B* being 174 A, that on *D* or *E*, 258 A and the current in *N* (neutral) being 84 A.

This method of adding single-phase current directly to 3-phase current in order to determine the total current on a particular supply wire is the one customarily used and gives results sufficiently accurate for practical needs.

Table 310-16 gives the nearest ampacities for Type THW copper conductors as No. 2/0 for leg *B*, 300 kcmil for *A* and *C*, and No. 4 for *N*. Table 5, Chapter 9, lists cross-sectional areas as .2781 sq in. for No. 2/0, .5581 sq in. for 300 kcmil, and .1087 sq in. for No. 4.

Total cross-sectional conductor area equals

No. 2/0 THW =	1 × .2781 sq in. =	.2781 sq in.
300 kcmil THW =	2 × .5581 sq in. =	1.1162 sq in.
No.No.4 THW =	1 × .1087 sq in. =	.1087 sq in.
	Total	1.503 sq in.

For 40% fill, the cross-sectional area of conduit must be at least 1.503 sq in. divided by .4, or 3.7575 sq in. Table 4, Chapter 9, lists the nearest size of conduit as $2\frac{1}{2}$″.

Per 250-23(b) the size of a copper neutral conductor is No. 2. The increased size provides an adequate return patch for fault current. The conduit size remains the same in this case.

As in the case of the store building, parallel service conduits could be made use of in order to reduce size of conductors. The circuit breaker may be a 400 A frame, with current settings suited to protect each of the 3-phase conductors. See the exception to 240-6(b) for the conditions that must be adhered to with adjustable trip circuit breakers.

Table 250-94 states that a No. 2 copper grounding electrode conductor is required for this application.

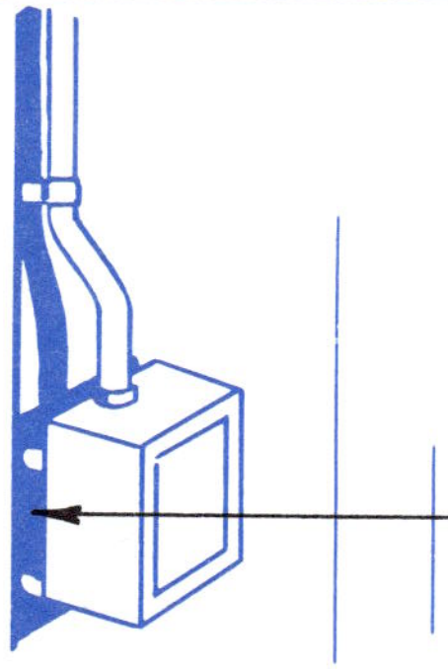

NEC 240-32 Enclosures for overcurrent devices to have at least 1/4-inch air space from surface in damp or wet locations and shall be identified for use in such locations.

NEC 300-6 (a) (b) (c) Ferrous raceways, cabinets, etc., must be suitably protected from corrosion. Enameled conduit not permitted outdoors or in wet locations. Unless suitable for existing conditions, conduit, boxes, etc., are not to be installed in concrete, in direct contact with the earth, or where subject to corrosion. In frequently washed areas such as canneries or dairies, the entire wiring system shall have at least 1/4-inch air space between it and the wall or supporting surface.

CONDUIT & EQUIPMENT IN CONCRETE & WET LOCATIONS

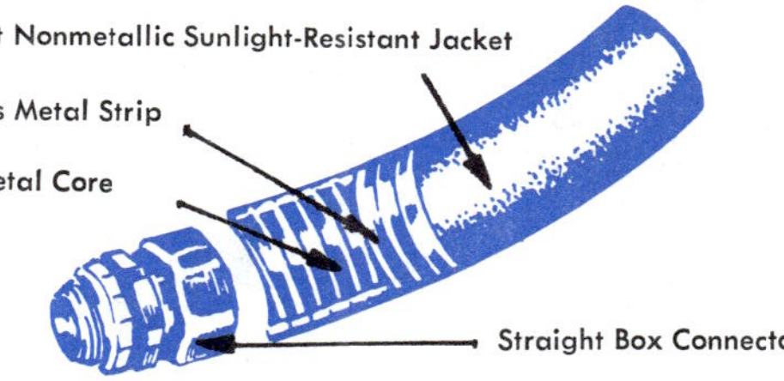

NEC 351-4(a) Permitted for direct burial in earth where listed and marked for purpose and exposed or concealed work. May be used where flexibility required or protection from liquids, vapors, or solids and some hazardous areas. 501-4(b), 502-4, and 503-3.

NEC 351-8 Permitted as fixed raceway if secured every 4½ feet and within 12 inches of box except where – (1) fished, (2) length not over 3 feet at terminals where flexibility is necessary, and (3) not over 6 feet for fixtures.

NEC 351-5 (b) Maximum size shall be 4 inches.

NEC 351-9 Permitted as grounding means where both conduit and fittings are approved for grounding. See Exception.

LIQUIDTIGHT FLEXIBLE METAL CONDUIT

NEC 330-3 Permitted for services, feeders, branch circuits, in wet locations, indoors or outdoors, exposed or concealed. (See 330-3.)

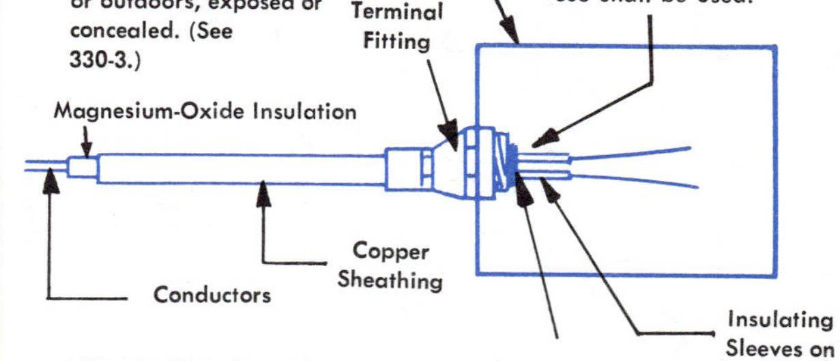

NEC 330-14 Fittings properly identified for use shall be used.

NEC 330-13 Radius of bends at least 5 times cable diameter up to 3/4", 10 times the diameter from 3/4" to 1".

NEC 330-15 Terminating seal required to prevent entrance of moisture.

NEC 300-15(b), Ex. 3 Straight-through splice for MI cable is allowed without box provided the splice is accessible.

MINERAL-INSULATED METAL-SHEATHED CABLE

NEC 300-5 (c) Underground cable run under a building shall be in raceway that extends beyond the outside wall of the building.

NEC 339-1 UF cable in sizes No. 14 copper or No. 12 aluminum thru 4/0 permitted.

NEC 339-5 Ampacity based on 60°C (140°F) basis per 336.

Per Note 8 of Notes to Ampacity Tables of 0 to 2000 Volts, bundled conductors (without spacing) are required to be derated.

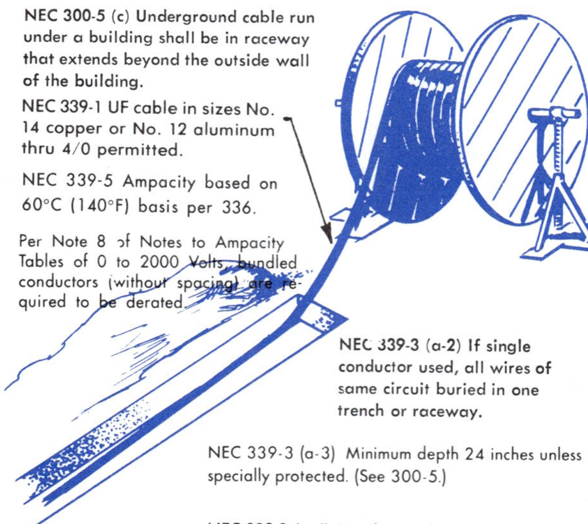

NEC 339-3 (a-2) If single conductor used, all wires of same circuit buried in one trench or raceway.

NEC 339-3 (a-3) Minimum depth 24 inches unless specially protected. (See 300-5.)

NEC 339-3 (a-4) May be used for interior wiring.

NEC 339-3 (b) Not permitted for: service-entrance cables, commercial garages, theaters, motion picture studios, storage battery rooms, hoistways, hazardous locations, concrete or cement except for nonheating leads as per Article 424, where exposed to the sun, unless identified.

UNDERGROUND FEEDER AND BRANCH-CIRCUIT CABLE

Press trigger button to open jaws. Encircle one conductor with transformer jaws. Read amps on amp scale.

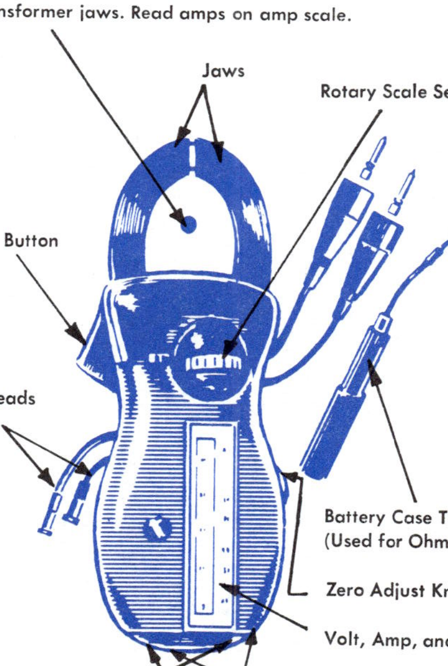

VOLTMETER--AMMETER--OHMMETER

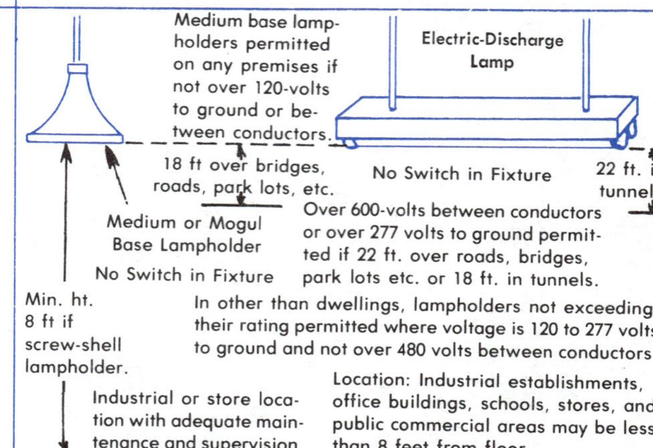

NEC 210-6(a) Voltage to ground or between conductors not to exceed 120 volts to medium base lampholders or 125-volt devices. (b) Circuits from 120 to 277 volts to ground and not over 480 volts between conductors permitted to supply: 1. Cord and plug loads not over 1440 VA. 2. In other than dwellings: (1) lampholders not exceeding their rating or mogul base screw-shell lampholders or; (2) auxiliary equip. of discharge lamps or; (3) receptacles rated for circuit voltage. Except: 1. infrared industrial lamps as per 422-15(c); 2. railroads as per 110-19; 3. circuits as per 210-6(c).

NEC 210-6(d) Circuits not over 600 volts between conductors or over 277 volts to ground permitted to supply: (1) receptacles rated for voltage; (2) permanently wired utilization equip.; (3) permanent electric-discharge fixtures when mounted: 1. on poles etc. to light outdoor areas such as roads, bridges, parking lots etc. if at least 22 ft. high; 2. at least 18 ft. in tunnels.

MAXIMUM VOLTAGE TO LTG. FIXTURE

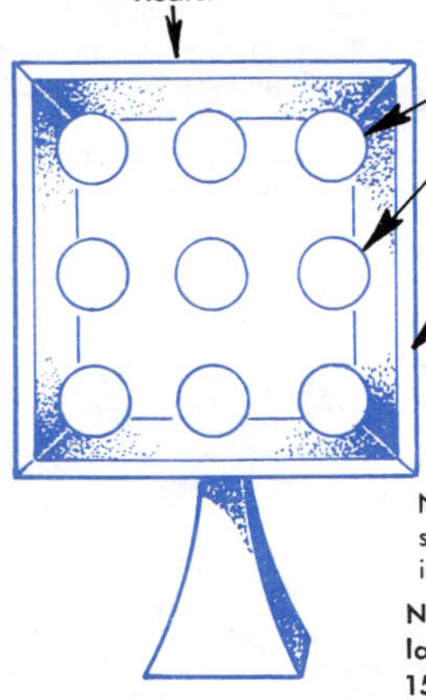

NEC 422-15 (c) Each section considered an appliance and each terminal connection considered individual outlet.

NEC 422-15(a) Lamps 300 VA or less may use medium-base unswitched porcelain lampholders or other types approved for that purpose.

NEC 422-15(b) Over 300 VA require special lampholders (not screw-shell) except if identified as suitable for infrared over 300 VA.

NEC 422-15 (c) In industrial location, lamps may be in series on circuits over 150 volts to ground if lampholder approved for circuit voltage.

NEC 210-6(d), Ex. 1 Voltage supplying lampholders of infrared heating appliances limited to 300 volts to ground in other than dwelling units as per 422-15(c).

NEC 422-28(c) Circuit protection not to exceed 50 A.

INFRARED LAMP HEATING APPLIANCES

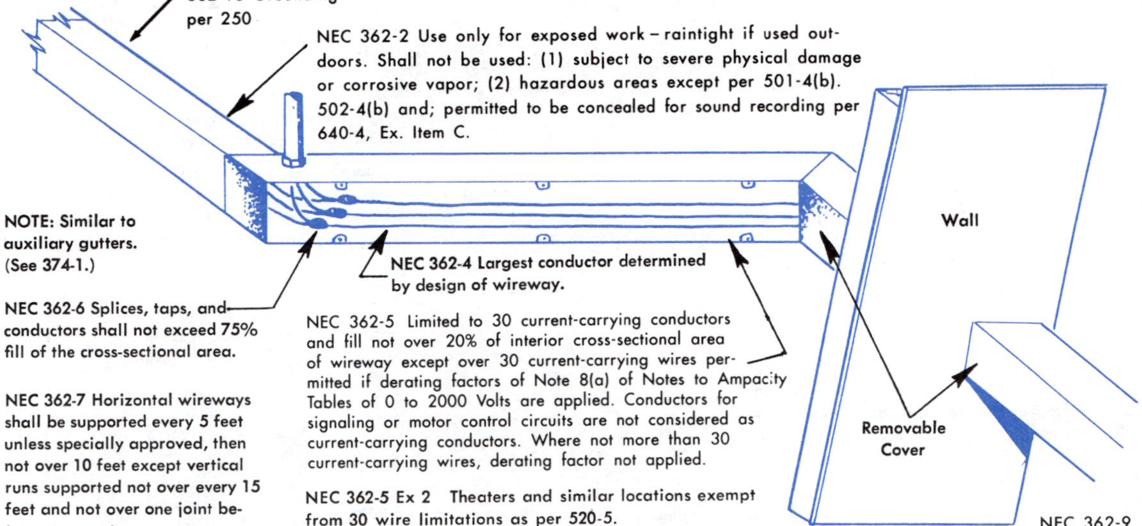

NEC 362-2 Use only for exposed work – raintight if used outdoors. Shall not be used: (1) subject to severe physical damage or corrosive vapor; (2) hazardous areas except per 501-4(b). 502-4(b) and; permitted to be concealed for sound recording per 640-4, Ex. Item C.

NOTE: Similar to auxiliary gutters. (See 374-1.)

NEC 362-6 Splices, taps, and conductors shall not exceed 75% fill of the cross-sectional area.

NEC 362-7 Horizontal wireways shall be supported every 5 feet unless specially approved, then not over 10 feet except vertical runs supported not over every 15 feet and not over one joint between supports.

NEC 362-1 Installation to be complete before wires laid in place. Constructed of sheet metal with hinged or removable covers.

NEC 362-4 Largest conductor determined by design of wireway.

NEC 362-5 Limited to 30 current-carrying conductors and fill not over 20% of interior cross-sectional area of wireway except over 30 current-carrying wires permitted if derating factors of Note 8(a) of Notes to Ampacity Tables of 0 to 2000 Volts are applied. Conductors for signaling or motor control circuits are not considered as current-carrying conductors. Where not more than 30 current-carrying wires, derating factor not applied.

NEC 362-5 Ex 2 Theaters and similar locations exempt from 30 wire limitations as per 520-5.

NEC 362-5 Ex 3 Elevators and dumb waiters exempt from 20% fill limitation as per 620-32.

NEC 362-5, Ex. 1 More than 30 current-carrying conductors are allowed if correction factors of Note 8(a) of Notes to Ampacity Tables of 0 to 2000 V are applied.

NEC 362-9 May extend through dry wall if in unbroken lengths.

METAL WIREWAYS

NEC 364-6 May extend through dry walls unbroken lengths or dry floor if unventilated and if protected to height of 6 feet.

NEC 364-5 Unless otherwise designed and marked, busways shall be supported at least every 5 ft.

NEC 364-4 Must be located where in the open and visible except permitted behind removable panels if: (a) no overcurrent except for one fixture; or other loads (b) not air-handling area; (c) totally enclosed, nonventilating type of busway; (d) joints are accessible for maintenance.

NEC 364-11 For industrial establishments only, overcurrent devices may be omitted where size reduced if smaller busway is not over 50 feet long, and at least ⅓ the rating of setting of next overcurrent device back on line, provided not in contact with combustible material.

NEC 364-13 If used as branch circuit, overcurrent device required here to afford protection to outlet devices as per Article 210.

NEC 364-12 Ex 2 When used to supply lighting units, branch circuit overcurrent device may be part of fixture cord cap or may be made part of the fixture.

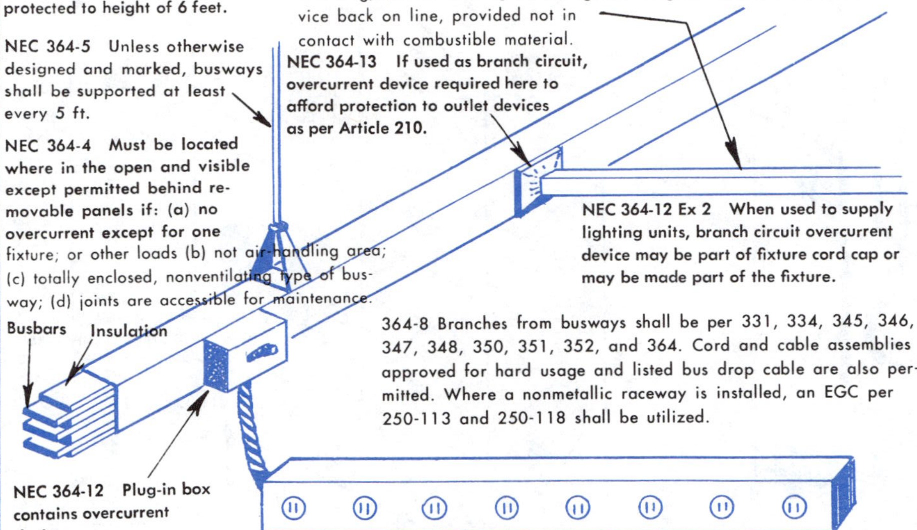

364-8 Branches from busways shall be per 331, 334, 345, 346, 347, 348, 350, 351, 352, and 364. Cord and cable assemblies approved for hard usage and listed bus drop cable are also permitted. Where a nonmetallic raceway is installed, an EGC per 250-113 and 250-118 shall be utilized.

NEC 364-12 Plug-in box contains overcurrent devices.

BUSWAYS

(Photo courtesy of Burndy Husky)

NEC 365-3 (a) Conductors shall have 75 degrees C rating or higher and shall be suitable for application as per 310 and 710.

NEC 365-3 (d) Conductors supported at least every 3 feet horizontal and every 1½ feet vertical. Space between conductors not less than diameter of conductor at points of support.

NEC 365-1 A complete wiring method consisting of insulated conductors and terminations in a completely enclosed, ventilated metal housing. Designed to withstand available fault currents and may be used at any voltage or current for which spaced conductors are rated.

NEC 365-2 Use only for exposed work. Permitted outdoors or in corrosive, wet or damp locations if approved for purpose. May be used for services, branch circuits or feeders.

NEC 365-2 Not permitted in hoistways or hazardous areas unless specifically approved.

NEC 365-3 (b) Ampacity of conductors same as in free air. (Tables 310-17, 310-19)

NEC 365-3 (c) Conductors not smaller than No. 1/0.

365-9 Cablebus shall be grounded and bonded per 250, excluding 250-33, Ex. 2.

NEC 365-10 Each section marked: manufacturer's name or trade designation and maximum diameter, number, voltage rating and ampacity of conductors. Marking to be visible after installation.

CABLEBUS SYSTEM

SWITCHBOARD RULES (ESSENTIAL PROVISIONS)

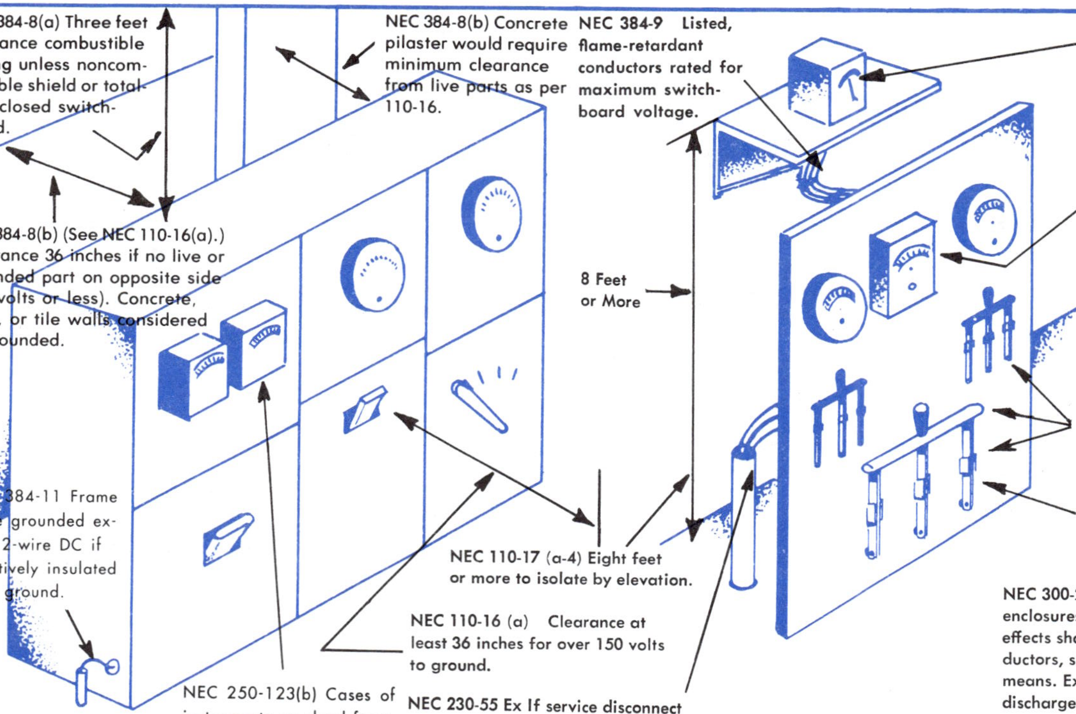

NEC 384-8(a) Three feet clearance combustible ceiling unless noncombustible shield or totally enclosed switchboard.

NEC 384-8(b) Concrete pilaster would require minimum clearance from live parts as per 110-16.

NEC 384-9 Listed, flame-retardant conductors rated for maximum switchboard voltage.

NEC 384-8(b) (See NEC 110-16(a).) Clearance 36 inches if no live or grounded part on opposite side (600 volts or less). Concrete, brick, or tile walls considered as grounded.

NEC 384-11 Frame to be grounded except 2-wire DC if effectively insulated from ground.

NEC 250-123(b) Cases of instruments on dead-front boards to be grounded.

NEC 110-17 (a-4) Eight feet or more to isolate by elevation.

NEC 110-16 (a) Clearance at least 36 inches for over 150 volts to ground.

NEC 230-55 Ex If service disconnect is on switchboard having exposed busbars on the back, raceway may terminate at bushing.

NEC 250-124 Ex Electrostatic ground detector isolated by elevation.

NEC 384-7 Not to be installed close to readily ignitible material.

NEC 250-123 (c) Instrument cases, live front-boards not to be grounded. Suitable insulated floor or rubber mats provided for operator if over 150 volts to ground.

NEC 110-22 Each disconnect means shall be clearly identified.

NEC 110-16(e) Minimum headroom of 6 ½′ required above switchboards, panelboards, control centers, and service equipment, except for residential of 200 A or less.

NEC 110-17, 110-18 Live or arcing parts to be adequately isolated or guarded.

NEC 384-36 Spacing between bare metal parts: 600 V, 2″ mounted on same surface or 1″ in air. See Table 384-36.

NEC 300-20 Where AC passes through metallic enclosures or magnetic partitions, inductive effects shall be minimized by grouping conductors, slotting the metal or similar effective means. Exception: X-ray circuits and electric-discharge lighting conductors.

INSTRUMENT CIRCUITS

384-32 In general, switchboard devices shall be protected by standard overcurrent devices not larger than 15 A. Where necessary to prevent interruption of power because of a hazard, overcurrent protection greater than 15 A shall be permitted. Special fuses may be used for fuses rated 2 A or less.

NEC 230-82 Ex 4 Current and potential instrument transformers permitted on supply side of service disconnect.

NEC 450-3(c) Potential transformer, installed indoors or enclosed requires primary fuses. See 384-32.

INSTRUMENT TRANSFORMERS

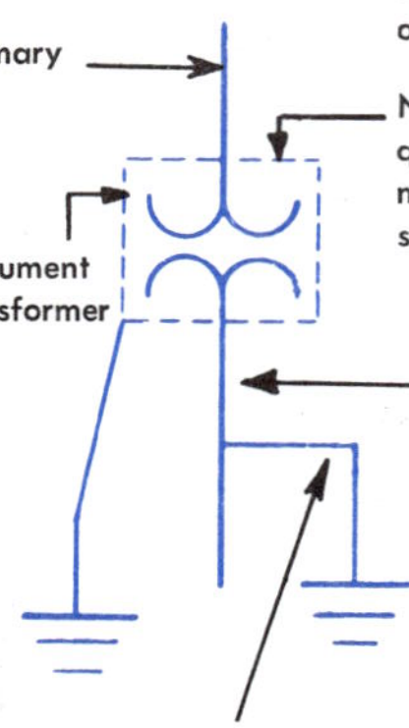

NEC 250-123 (a) Cases of portable instruments to be grounded if voltage to ground 300 or over and operating voltage 1000 or less.

NEC 250-122 Where accessible to other than qualified persons, cases and frames of instrument transformers to be grounded, except to supply meters of not over 150 volts to ground.

NEC 250-121 Secondaries of instrument transformers to be grounded if primary over 300 to ground. On switchboards, only if primary voltage over 1000 and no exposed live parts.

NEC 250-125 Grounding conductor No. 12 copper or No. 10 aluminum.

NEC 250-124 If operating voltage over 1 kV, instruments to be isolated or guarded. See Exception for electrostatic ground detectors.

FARM LOADS – ARTICLE 220, PART D

NEC 230-21 Overhead wires to a building or other structure from another building or structure (such as a pole), where a disconnect or meter is installed, shall be treated and installed as a service drop.

NEC 220-40 (a) Dwelling load computed as normal dwelling, using Part B or C of Article 220.
NEC 220-40 (b) Use Table 220-40 to compute the feeder or service to each farm building other than dwelling.

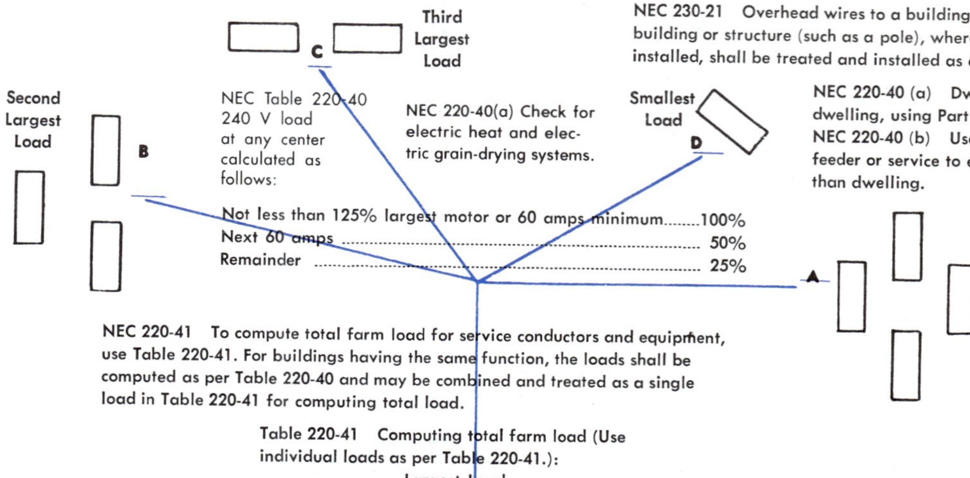

NEC Table 220-40 240 V load at any center calculated as follows:

NEC 220-40(a) Check for electric heat and electric grain-drying systems.

Not less than 125% largest motor or 60 amps minimum	100%
Next 60 amps	50%
Remainder	25%

NEC 220-41 To compute total farm load for service conductors and equipment, use Table 220-41. For buildings having the same function, the loads shall be computed as per Table 220-40 and may be combined and treated as a single load in Table 220-41 for computing total load.

Table 220-41 Computing total farm load (Use individual loads as per Table 220-41.):

Largest Load	______	×	100%	= ______
Second Largest Load	______	×	75%	= ______
Third Largest Load	______	×	65%	= ______
Remaining Loads	______	×	50%	= ______
Plus Dwelling Unit as per Article 220, Part B or C	______	×	100%	= ______
Size of Service Entrance Conductors & Equipment			Total	______

MOTORS -- GENERAL

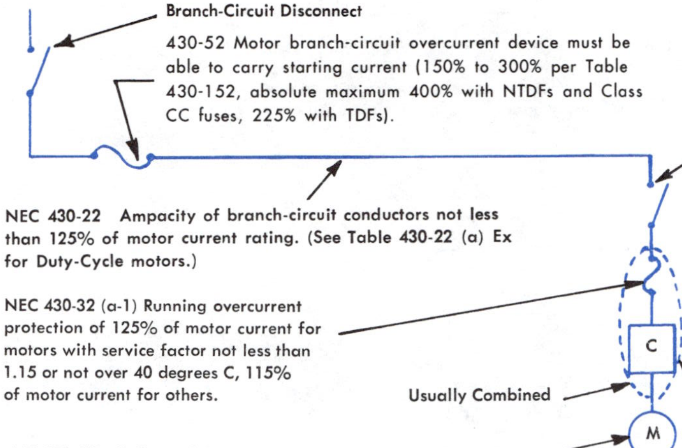

430-52 Motor branch-circuit overcurrent device must be able to carry starting current (150% to 300% per Table 430-152, absolute maximum 400% with NTDFs and Class CC fuses, 225% with TDFs).

NEC 430-22 Ampacity of branch-circuit conductors not less than 125% of motor current rating. (See Table 430-22 (a) Ex for Duty-Cycle motors.)

NEC 430-32 (a-1) Running overcurrent protection of 125% of motor current for motors with service factor not less than 1.15 or not over 40 degrees C, 115% of motor current for others.

440-52(a)(1–4) Hermetic-type motor overload and failure to start protection.

NEC 430-40 Thermal cutouts and overload relays for motor-running protection not capable of opening short circuits shall be protected as per 430-52 unless approved for group installation and marked with maximum size of required protection.

NEC 430-33 Application to be deemed continuous duty unless positive information to the contrary.

430-84 Controllers are not required to open all conductors unless they also serve as disconnecting means per 430-111.

430-82(b) If autotransformer, must be arranged so cannot remain in starting position.

430-102(b) Controller not in sight of motor must be individually lockable, or else disconnect switch within sight of motor is required.

MOTOR AND CONTROLLER DISCONNECT MEANS

NEC 430-102(b) Must be in sight of controller location. For over 600 volts see Exception.

NEC 430-107 One or more to be readily accessible and in sight of controller.

NEC 430-108 Every disconnect means in motor circuit shall comply with 430-109 & 430-110.

NEC 430-109 Hp-rated switch, circuit breaker, or a molded-case switch, and shall be a listed device, except if 2 hp or less and 300 volts or less, may be general-use twice motor current. Over 100 hp, a general-use or isolating switch marked "Do Not Operate Under Load."

430-110(a) Motor disconnects shall have an ampere rating of at least 115% of motor's FLC, taken from FLC tables per 430-6(a).

430-111 Suitable switch or CB may serve as both disconnect and controller.

430-83(a) Controller shall have hp rating at the application voltage not lower than hp rating of motor, with few exceptions. 430-83, Ex. 1, 2, 3, 4, 5.

MOTORS -- RUNNING PROTECTION

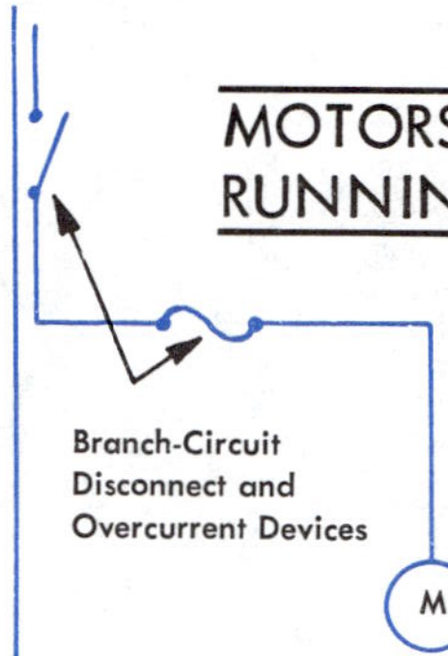

NEC 430-32 (b) (1) Motors 1 hp or less, not permanently installed, manually started within sight of controller, need no running protection if branch circuit is 120 volts or less and 20 amps or less.

NEC 430-81 (c) Attachment plug and receptacle may be controller for portable motor of 1/3 hp or less.

NEC 430-32 (c-3) Automatically-started motors, 1 hp or less, if part of approved non-overloading assemblies, viz., oil burners, need no running overcurrent protection.

NEC 430-32 (c-4) Motors with high impedance windings, viz., clock motors, need no running overcurrent protection. Does not include most split-phase motors.

NEC 430-81 (b) Stationary motors 1/8 hp or less normally left running may be controlled by branch-circuit overcurrent device provided they will not be damaged by overload or failure to start.

MOTOR DISCONNECT -- TYPE

430-109, Ex. 3 A general-use AC only snap switch may be used to disconnect an AC motor 2 hp or less and 300 V or less whose FLC does not exceed 80% of switch amp rating. (See definition of AC general-use snap switch in 100.)

430-109, Ex. 3 For stationary motor 2 hp or less, 300 V or less, disconnecting means may be general-use switch having ampere rating not less than twice full-load rating of motor.

430-109, Ex. 5 For stationary motors over 100 hp, disconnect need not be hp rated. May be general-use or isolating switch. Isolating switches marked "Do not open under load."

430-109, Ex. 6 Cord- and plug-connected motors permitted to use plug and receptacle and need not be hp rated except when required by 422-22 or 440-63.

MOTOR GENERATOR -- DISCONNECT MEANS

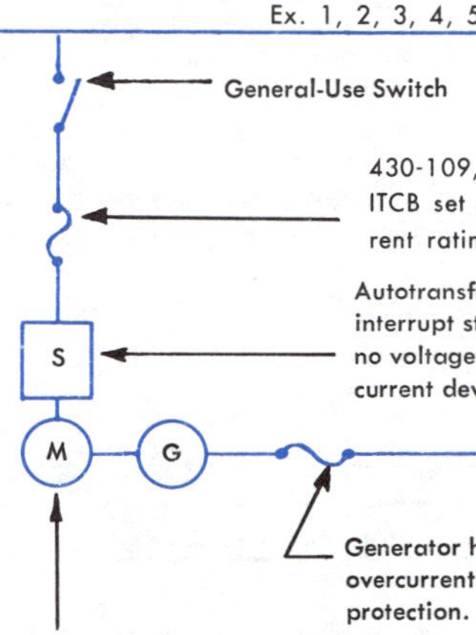

430-109, Ex. 4(c) Separate fuses or ITCB set at not over 150% motor current rating.

Autotransformer starter which can interrupt stalled rotor current has no voltage release and 125% overcurrent device.

430-109, Ex. 4(a, b, c) Special exception allowing general-use switch for motor which drives a generator that has overload protection.

MOTOR -- MAXIMUM RUNNING OVERLOAD PROTECTION

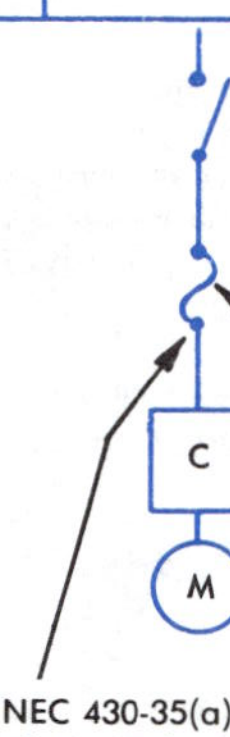

NEC 430-34 Where overcurrent value does not correspond to standard size, next larger size may be used—not to exceed 140% full-load current for motors with service factor not less than 1.15 or motors not over 40°C. All others 130%.

NEC 430-35(a) Running overcurrent device may be shunted at starting of manual-start motor if no hazard is introduced and branch circuit device of not over 400% is operative in the circuit during starting period. (b) Shunting not permitted if automatically started. (See Exception)

NEC 430-113 Where equipment is supplied from more than one source, each disconnect shall be adjacent to the equipment. (See Ex 1, 2.)

MOTOR -- PART WINDING

NEC 430-3 Where separate over-current devices, sizes of conductors and devices based on one-half full-load current rating.

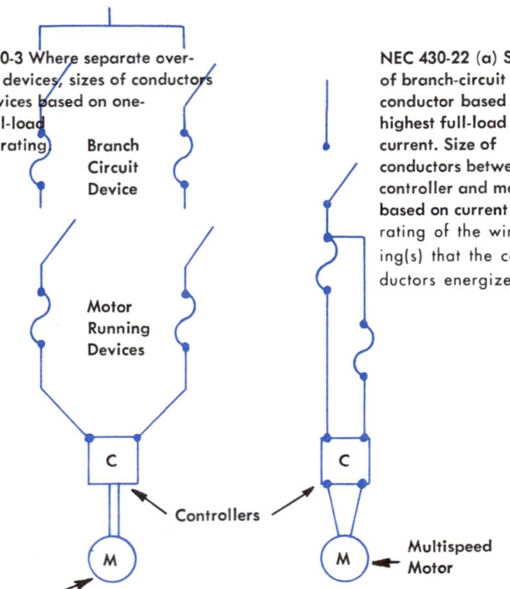

NEC 430-22 (a) Size of branch-circuit conductor based on highest full-load current. Size of conductors between controller and motor based on current rating of the winding(s) that the conductors energize.

Standard part-winding motor (does not include a hermetic-refrigerant type refrigerator, compressor motor).

MOTOR CONTROL CIRCUITS

NEC 430-72 (a) Motor-control circuit tapped from controller to be protected as per 430-72 (b, c).

NEC 430-72 (b) General Rule: Motor-control conductor overcurrent protection shall not exceed values listed in Column A of Table 430-72 (b).

NEC 430-72 (b) Ex 1 Control conductors that do not leave the controller are permitted to be protected by the motor branch-circuit protective device not exceeding the values listed in Column B of Table 430-72 (b).

NEC 430-72 (b) Ex 2 Control conductors that extend beyond the controller are permitted to be protected by the motor branch-circuit protective device not exceeding the values listed in Column C of Table 430-72 (b).

NEC 430-72(b), Ex. 3 Transformer-fed, 2-wire primary, 2-wire secondary, control conductors are considered protected if motor branch-circuit overcurrent device protects secondary wires per Table 430-72(b). Also see 240-3(i).

NEC 430-72 (b) Ex 4 Motor branch-circuit overcurrent devices permitted where opening of control circuit creates a hazard. (Example: The control circuit of a fire pump motor.)

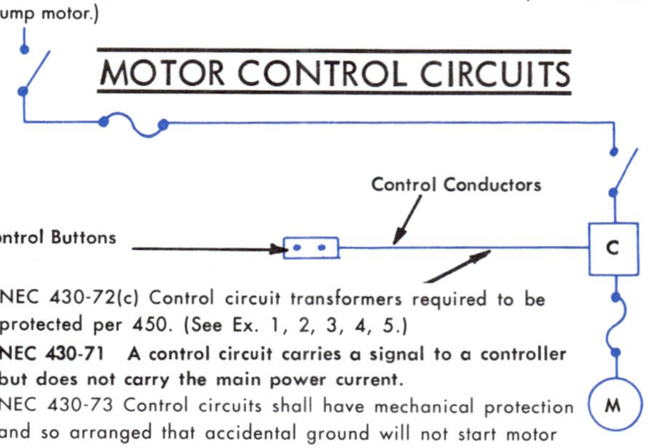

NEC 430-72(c) Control circuit transformers required to be protected per 450. (See Ex. 1, 2, 3, 4, 5.)

NEC 430-71 A control circuit carries a signal to a controller but does not carry the main power current.

NEC 430-73 Control circuits shall have mechanical protection and so arranged that accidental ground will not start motor or bypass shutdown devices.

NEC 430-73 Control circuits shall have mechanical protection and so arranged that accidental ground will not start motor.

NEC 300-3(c)(2), Ex 3 Excitation, control, relay, and ammeter conductors used in connection with motor or starter may occupy the same enclosures as motor circuit conductors.

NEC 430-74 (a) Control circuit shall be disconnected from supply when motor circuit opened. If two separate devices are used, they must be immediately adjacent. (See Ex 1, 2.)

NEC 430-74 (b) Control transformers located inside controller to be on load side of disconnect for control circuit.

HERMETIC MOTORS

NEC 440-6(a) Full-load current marked on equipment nameplate is used to determine size branch circuit conductors.

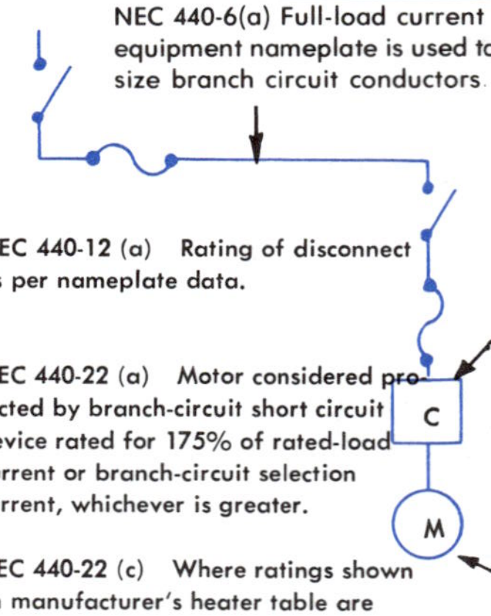

NEC 440-4(a) Locked-rotor current to be marked on compressor nameplate of all polyphase motors and single-phase over 9 amps, 115 volts or 4.5 amps, 230 volts.

NEC 440-12 (a) Rating of disconnect as per nameplate data.

NEC 440-6(a) Data marked on nameplate of equipment containing compressor used to determine rating of controller. If equipment not marked, data from compressor or to be used. (See worked example later.)

NEC 440-22 (a) Motor considered protected by branch-circuit short circuit device rated for 175% of rated-load current or branch-circuit selection current, whichever is greater.

NEC 440-22 (c) Where ratings shown on manufacturer's heater table are less than those permitted by 440-22 (a, b), then the lower values listed on manufacturer's heater table shall be used.

NEC 440-52 (a) (1, 3) Overload setting not over 140% full-load current for overload relays and 125% for other devices.

WOUND-ROTOR MOTOR

NEC 430-22 (a) Ampacity not less 125% of full-load motor current.

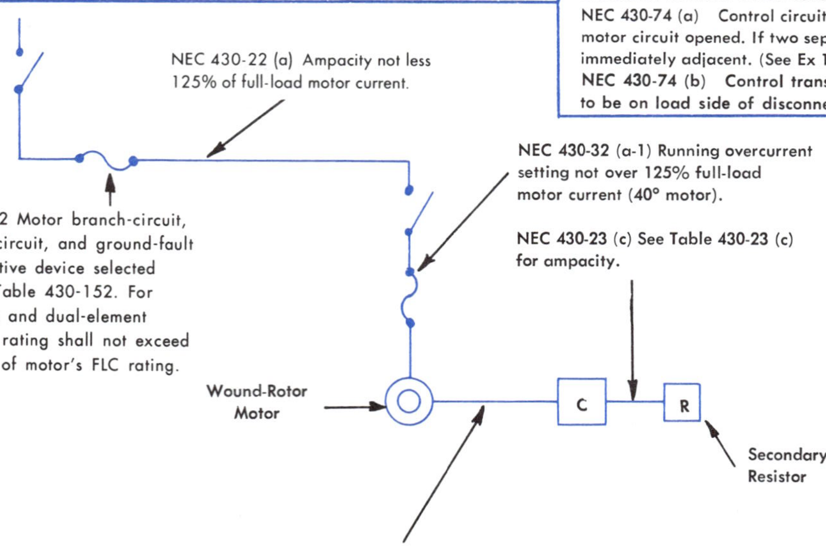

430-52 Motor branch-circuit, short-circuit, and ground-fault protective device selected from Table 430-152. For NTDFs and dual-element fuses, rating shall not exceed 150% of motor's FLC rating.

NEC 430-32 (a-1) Running overcurrent setting not over 125% full-load motor current (40° motor).

NEC 430-23 (c) See Table 430-23 (c) for ampacity.

NEC 430-23 (a) Ampacity 125% for continuous-duty. (For other duty see Table 430-22 (a) Ex.)

NEC 430-32 (d) Secondary circuit requires no overcurrent devices.

DIRECT CURRENT MOTORS

NEC 430-22(a) Ex. 2 DC motors supplied from rectified single-phase supply shall have conductor ampacity of: (a) 190% for single-phase half-wave rectifier; (b) 150% for single-phase full-wave rectifier.

NEC 430-22(a) Ampacity 125% of continuous-duty. (For other duty see Table 430-22(a) Ex.)

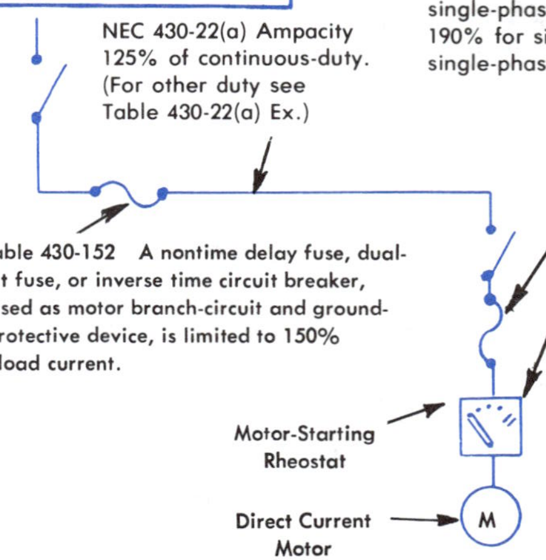

NEC Table 430-152 A nontime delay fuse, dual-element fuse, or inverse time circuit breaker, when used as motor branch-circuit and ground-fault protective device, is limited to 150% of full-load current.

NEC 430-32 (a-1) Overload setting not more than 125% of full-load motor current (40° motor).

NEC 430-82 (c) (1) Rheostat designed so contact arm cannot be left on intermediate segments.

NEC 430-82(c)(2) Under-voltage release required to interrupt supply when speed falls to 1/3 its normal rate.

NEC 430-88 Adjustable-speed control to be arranged so motor cannot start on weakened field except where so designed.

NEC 430-89 Speed limiting device sometimes required. (See a, b, c.)

MOTOR FEEDER CALCULATIONS

NEC 430-24 Minimum feeder ampacity 125 percent full-load current largest motor plus full-load currents of others. Where motors for types of service listed in 430-22(a), Ex. 1, similar procedure followed, taking into consideration values obtained from Table 430-22(a), Ex.
Where circuitry prevents all motors from operating at same time, conductor size is determined on basis of maximum load at any given time.

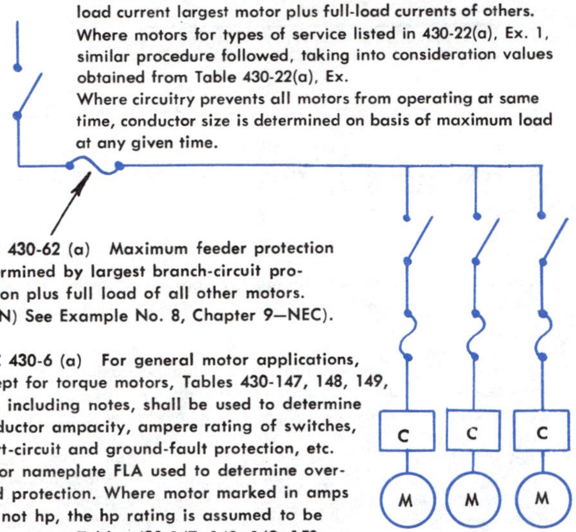

NEC 430-62 (a) Maximum feeder protection determined by largest branch-circuit protection plus full load of all other motors. ((FPN) See Example No. 8, Chapter 9—NEC).

NEC 430-6 (a) For general motor applications, except for torque motors, Tables 430-147, 148, 149, 150, including notes, shall be used to determine conductor ampacity, ampere rating of switches, short-circuit and ground-fault protection, etc. Motor nameplate FLA used to determine overload protection. Where motor marked in amps but not hp, the hp rating is assumed to be value as per Tables 430-147, 148, 149, 150, interpolated if necessary.

TWO OR MORE MOTORS ON ONE BRANCH CIRCUIT

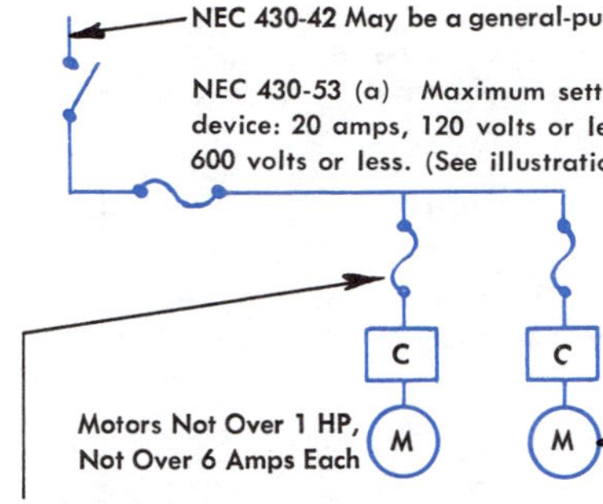

NEC 430-42 May be a general-purpose branch circuit.

NEC 430-53 (a) Maximum setting overcurrent device: 20 amps, 120 volts or less; 15 amps, 600 volts or less. (See illustrations.)

NEC 430-53 (a) Two or more motors not over 1 hp permitted on same 120-volt, 20-amp branch circuit or 600-volt, 15-amp branch circuit if: (1) not over 6 amps, (2) controller max. branch circuit O.C. device is not exceeded; (3) each overload is per 430-32. (b) Where all motors have overload protection and smallest motor protected as per 430-52 and where determined that the branch-circuit device will not open under most severe normal service conditions that might be encountered.

NEC 430-112 Individual disconnects not required where provisions of NEC 430-53(a) are complied with. (See above illustrations.)

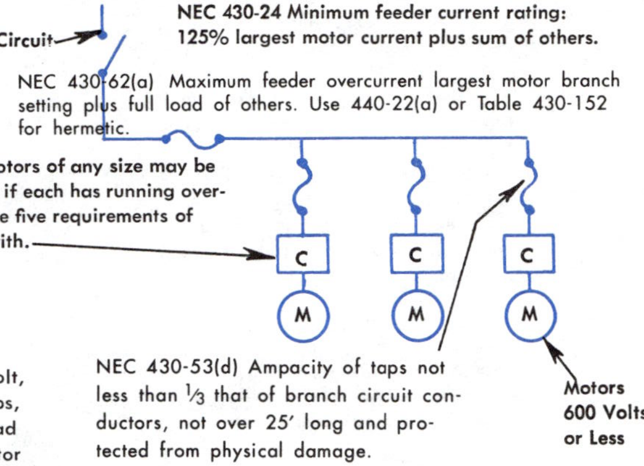

NEC 430-24 Minimum feeder current rating: 125% largest motor current plus sum of others.

NEC 430-62(a) Maximum feeder overcurrent largest motor branch setting plus full load of others. Use 440-22(a) or Table 430-152 for hermetic.

NEC 430-53 (c) Two or more motors of any size may be connected to one branch circuit if each has running overcurrent protection and all of the five requirements of NEC 430-53 (c) are complied with.

NEC 430-53(d) Ampacity of taps not less than ⅓ that of branch circuit conductors, not over 25′ long and protected from physical damage.

NEC 430-112 Ex (a) (c) Individual disconnects required unless group is part of single machine or if group is in a single room in sight of disconnect.

NEC 430-12(a) Terminal housings to be metal except nonmetallic permitted in other than hazardous area. (b) Table 430-12(b) gives size for wire-to-wire connections.

NEC 430-142 Frames of stationary motors to be grounded if: (1) metal enclosed wiring, (2) wet location, (3) hazardous location, or (4) over 150 volts to ground. If frame not grounded it shall be insulated from ground.

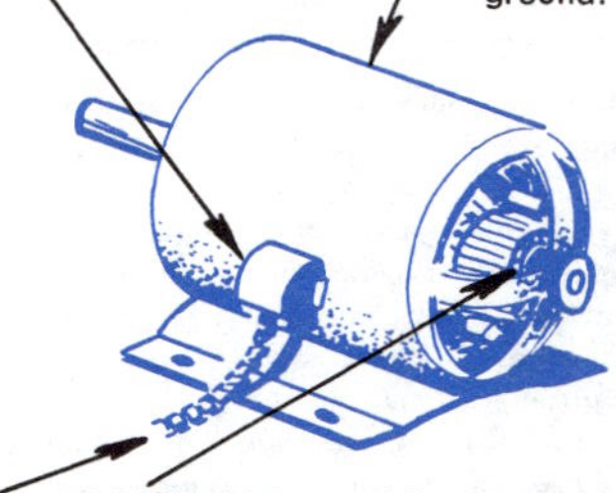

NEC 430-132 Exposed live parts to be guarded if over 50 volts.

NEC 430-14 (a) Motor shall be located where adequate ventilation is provided; also so maintenance can be accomplished.

NEC 430-16 To be protected from accumulation of dust.

NOTE: Motor grounded by separate EGC.

NEC 430-43 An overload which can automatically restart a motor shall be approved for the motor and if motor restarts could not result in injury.

MISCELLANEOUS MOTOR RULES

NEC 430-145 (a) Where motor terminals of fixed motors are in metal raceways or metal enclosed cable, junction boxes for motor terminals required. Metal raceways and Type AC cable connected as per Article 250. (FPN See 430-12(e)) for grounding requirements.)

NEC 430-145(b) Junction box not more than 6′ from motor under conditions specified.

NEC 430-145 (c) Instrument transformer secondaries and exposed metal parts of instruments or relays used in connection with motors shall be grounded as per 250-121 thru 250-125.

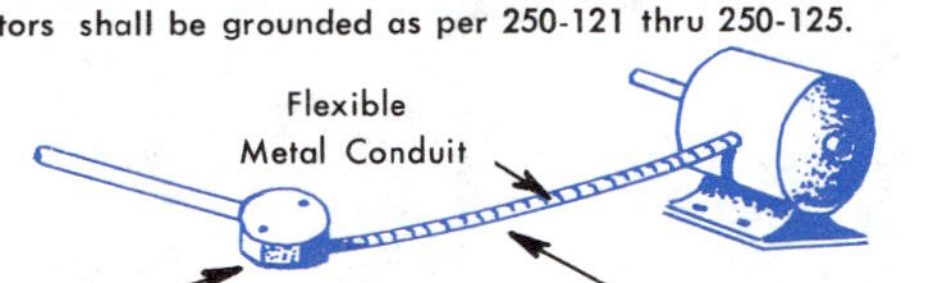

MOTOR CONTROLLERS

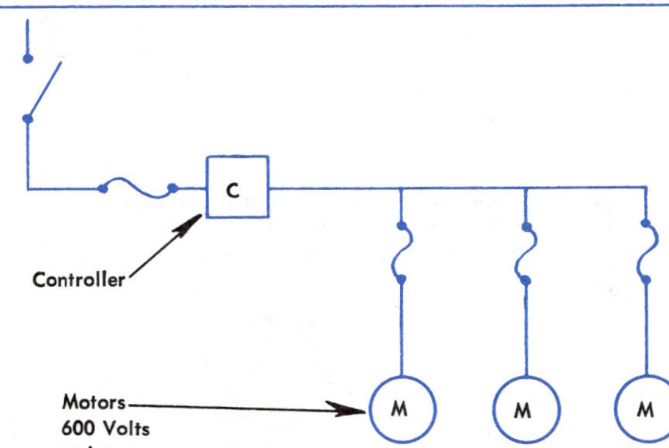

NEC 430-87 (General Rule) Each motor of 600 volts or less to have its own controller.
Exceptions (a) (b) (c): Single controller permissible if motors are part of same machine, or if group within provisions of NEC 430-53 (a) or if group in single room in sight of controller.

GENERATORS

NEC 445-4(a) Constant potential DC generators (other than exciters) shall have overcurrent protection.

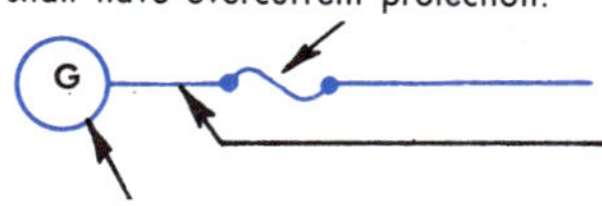

NEC 445-5 Ampacity of conductors from generator to first overcurrent device not less than 115% of nameplate amps, except: (1) Not less than 100% where design and operation prevent overloading; (2) Where factory-wired unit has overcurrent device that is an integral part of generator set.

NEC 445-1 Generators and associated wiring and equipment shall comply with NEC 230, 250, 695, 700, 701, 702, and 705

NEC 445-4 (b) Two-wire generator may have overcurrent device in one lead; (c) If generator 65 volts or less, driven by motor, motor overcurrent device satisfies requirements if setting limits DC output to 150% of normal; (d) Control devices must be arranged to disconnect balancer sets if they are part of system; (e) Three-wire generators must have overcurrent devices in each armature lead, devices so interlocked as to disconnect generator and equalizers in case of trouble.

NEC 445-3 Nameplate shall show kVA or kW, amps, volts, frequency, power factor, maker's name, rpm, ambient temperature and other.

MOTORS OVER 600 VOLTS

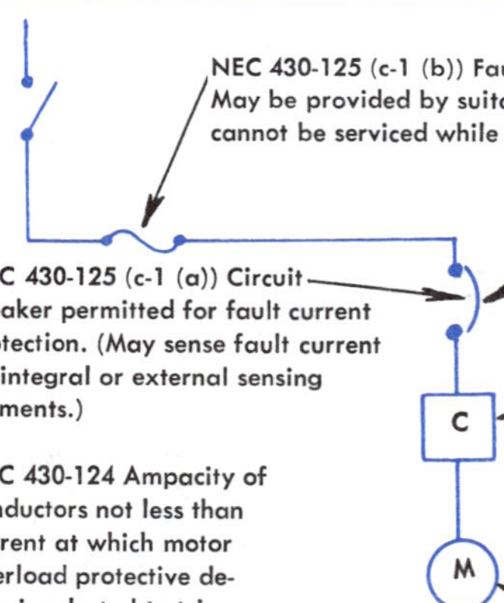

NEC 430-125 (c-1 (b)) Fault current protection required. May be provided by suitable type and rated fuses that cannot be serviced while energized.

NEC 430-125 (c-1 (a)) Circuit breaker permitted for fault current protection. (May sense fault current by integral or external sensing elements.)

NEC 430-127 Disconnect for controller shall be able to be locked in open position.

NEC 430-122 Controllers marked with maker's name, voltage, current, hp rating and other necessary data (see 430-8) and control voltage.

NEC 430-124 Ampacity of conductors not less than current at which motor overload protective device is selected to trip.

Motor Over 600 Volts

NEC 430-125 (a) Motor shall have overload protection and motor, motor circuit conductors and control apparatus shall have fault current protection. This protection shall be coordinated and automatically operated except vital motor may use supervised annunciator or alarm.

NEC 430-125 (b-1) Each motor to be protected against overloads or failure to start by thermal protector integral with motor or external sensing devices or both.

NEC 430-125 (b-2) Secondary circuits of wound-rotor motors are considered to be protected by motor overcurrent protection.

NEC 430-125 (b-3) Operation of overload device shall open all ungrounded conductors at same time.

NEC 430-125 (b-4) Overload sensing devices shall not automatically restart motor unless restarting is no hazard.

MOTOR-RUNNING OVERLOAD PROTECTION

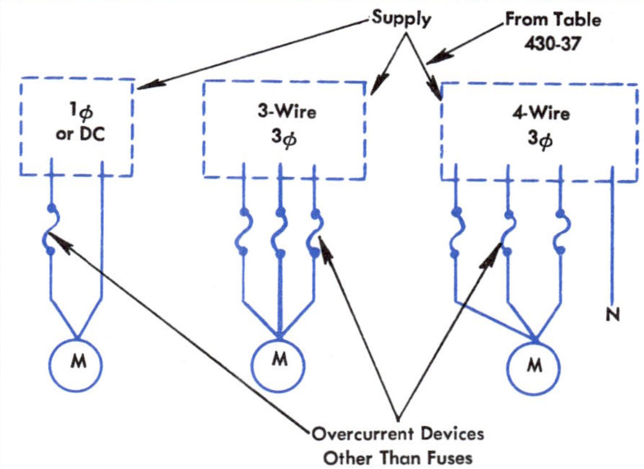

Table 430-37 Three motor-running protective devices shall be used to protect three-phase motors unless protected by other approved means.
NEC 430-36 Where fuses are used for motor-running protection in a 3-wire, 3-phase AC system with one wire grounded, a fuse shall be used in the grounded conductor.
NEC 430-38 The number of overcurrent devices other than fuses is governed by Table 430-37. (See above illustration.)

METAL WORKING MACHINE TOOLS

NEC 670-3(a) Nameplate of a metalworking machine tool must show full-load current ampere rating of largest motor, and short-circuit interrupting capacity of machine overcurrent device if one is furnished.

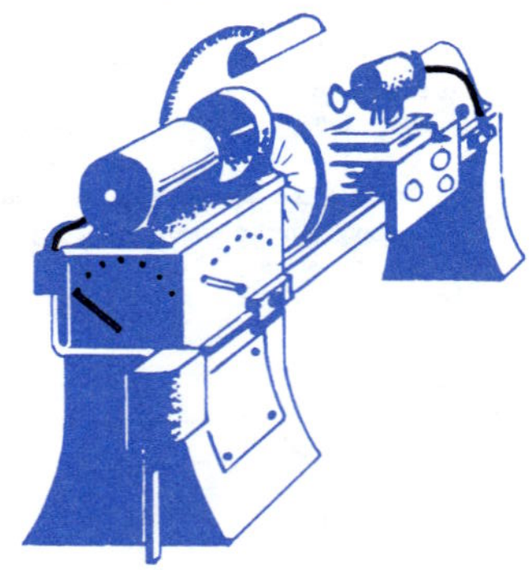

NEC 670-4(b) A machine tool is considered a single unit and shall have a disconnect which may be fed by branch circuits protected by breakers or fuses.

CRANES AND HOISTS

NEC 610-11 Supply conductors shall be in raceways, AC cable with insulated EGC, MI or MC cable, except: (1) Bar contact conductors; (2) Short lengths of open conductors at resistors, collectors, etc.; (3) For flexibility, flexible stranded wire installed in flex metal conduit, liquidtight flex metal conduit, liquidtight flex nonmetallic conduit, multiconductor cable, or approved PVC enclosure; (4) Multiconductor cable to pendant pushbutton station.

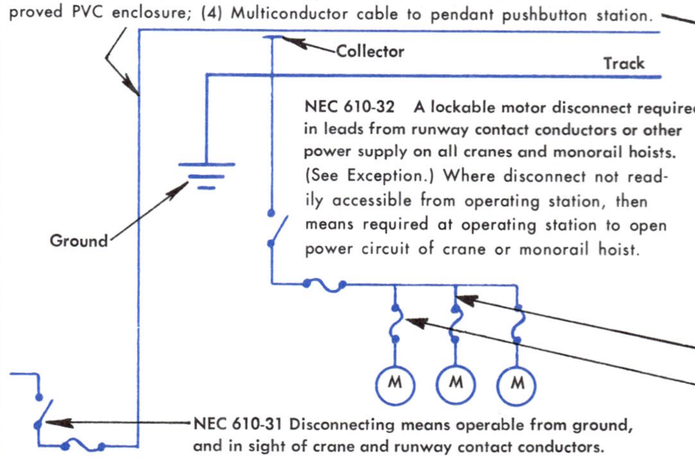

NEC 610-32 A lockable motor disconnect required in leads from runway contact conductors or other power supply on all cranes and monorail hoists. (See Exception.) Where disconnect not readily accessible from operating station, then means required at operating station to open power circuit of crane or monorail hoist.

NEC 610-31 Disconnecting means operable from ground, and in sight of crane and runway contact conductors.

NEC 610-33 Minimum rating of disconnecting means: 50% of combined short-time ampere ratings of all motors and 75% of sum of short-time ampere ratings of motors for single crane operation.

NEC 610-14 (e-2) Ampacity of supply conductors shall be not less than the sum of short-time full-load current for largest motor, or group, for a single crane motion, plus 50% of that for the next larger motor or group.

NEC 610-14 (d) Bare conductors No. 6 span 0-30 feet; No. 4 span 30-60 feet; No. 2 over 60 feet.

NEC 610-21 (f) Grounded track may serve as a conductor if supply by way of insulating transformer not over 300 volts and other conductors insulated.

NEC 610-14 (a) and Table 610-14 (a) Provides increased ampacity for conductors to crane motors.

NEC 610-15 Where more than one motor a common-return conductor is permitted.

NEC 430-112 Ex (a) Individual disconnects not required.

NEC 610-42 (a) Each motor to have branch-circuit protection. Ex 1 Two motors operating as a unit may have single overload protector.

NEC 610-61 Grounding of metal parts required.

NEC 610-55 Limit switch or other device required on hoist unit.

NEC 610-41 Overcurrent device required for main contact conductors.

NEC 620-51 Identified disconnect required for power supply to each unit.

620-52(b) For power from more than one source and multiple disconnects, a clearly legible warning sign shall read, "WARNING- PARTS OF THE CONTROLLER ARE NOT DE-ENERGIZED BY THIS SWITCH." Sign shall be mounted on or adjacent to disconnecting means.

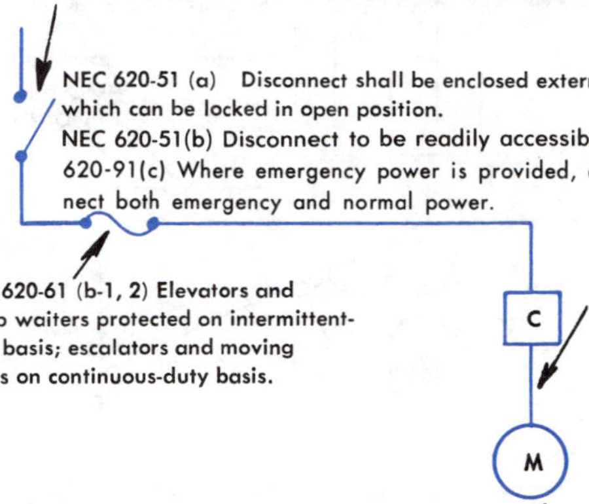

NEC 620-51 (a) Disconnect shall be enclosed externally operable motor switch or circuit breaker which can be locked in open position.

NEC 620-51(b) Disconnect to be readily accessible to qualified persons.

620-91(c) Where emergency power is provided, disconnect required by 620-51 shall disconnect both emergency and normal power.

NEC 620-61 (b-1, 2) Elevators and dumb waiters protected on intermittent-duty basis; escalators and moving walks on continuous-duty basis.

620-21 In general, all elevator wiring to be in rigid conduit, IMC, EMT, rigid nonmetallic conduit, wireways, or in Types MC, MI, or AC cable. See Ex. 1–5.

NEC 620-82 In general, operating equipment to be grounded as per Article 250.

620-61(b-1) Elevator motors classified as intermittent duty and protected against overload per 430-33.

620-13(a) Ampacity of conductors for a single motor shall conform to 430-22(a) and 430-22, Ex. 1.

620-61(b-2) Escalator and moving-walk driving motors shall be classified as continuous and protected against overload per 430-32.

620-13(a)(b) Conductors supplying several motors shall comply with standard rule governing motor feeders unless demand factor is allowed per 620-14.

OPERATING REQUIREMENTS

NEC 620-53 A single disconnecting means shall be provided for disconnecting all ungrounded car light and accessories power supply conductors for each unit.

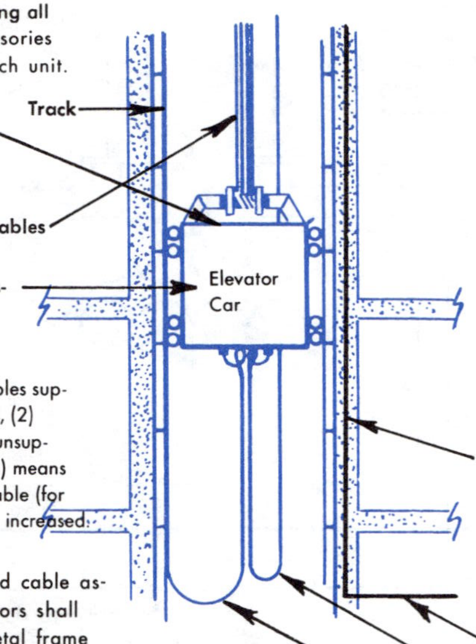

NEC 620-41 Traveling cables suspended at car and hoistway ends to reduce strain.

NEC 620-41 (1, 2, 3) Traveling cables supported by: (1) steel supporting fillers, (2) looping cables around supports for unsupported lengths less than 100 feet, (3) means that automatically tighten around cable (for cables up to 200 ft.) when tension is increased.

620-81 Metal raceways and cable assemblies attached to elevators shall be bonded to grounded metal frame of car they contact.

620-21(a)(1) Flexible metal conduit, liquidtight flexible metal conduit, liquidtight flexible nonmetallic conduit permitted in hoistway or in car for limit switches, interlocks, doors, etc.

620-12(a) Traveling cables (1) For lighting circuits, minimum #14, or equivalent in parallel conductors #20 or larger; (2) For other circuits, minimum size No. #20.

620-3 Control and signal voltage limited to 300 V except for special limited-energy circuits.

620-37(a)(c) Only wiring directly related to elevator or related systems shall be permitted inside hoistway and machine room. Main power feeders supplying elevator or dumbwaiters shall be installed outside hoistway unless necessary for elevators with driving machine motors located in hoistway, on car, or on counterweight, or by special permission.

NEC 620-11(b) Traveling cables used as flexible connections shall be of the types listed in Table 400-4 as elevator cable or be of other approved types.

TRAVELING CABLES AND CONTROL HOISTWAY

NEC 620-32
50% fill permitted in wireway. Vertical runs require support at least every 15 ft.
NEC 362-5 shall not apply.

NEC 620-35 Thirty-foot length restriction of 374-2 and number of wires restriction, 374-5, do not apply.

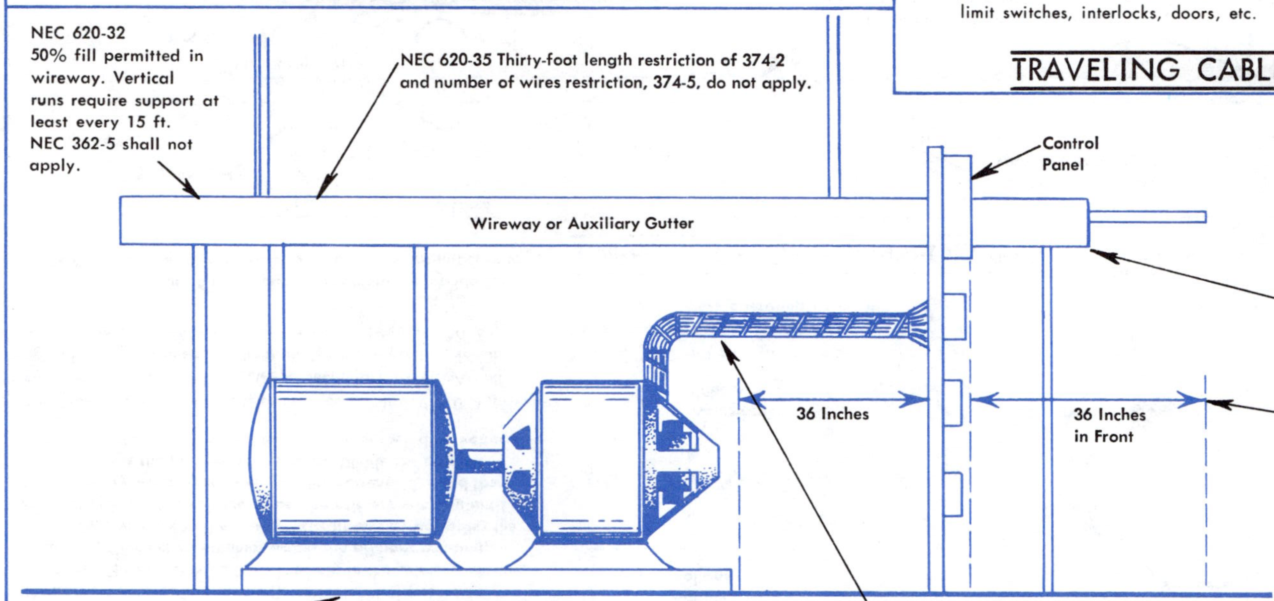

MACHINE ROOM

620-21(a)(3)(d) Conductors may be grouped and bound together if supported every 3′ free from physical damage.

ELEVATORS, DUMBWAITERS, ESCALATORS, AND MOVING WALKS

NEC 620-36 Conductor of power, control, lighting and signal systems not over 600 volts may be run in same raceway or cable if insulated for maximum voltage found in the cable or raceway system.

620-5 Sufficient, clear working space around control panels to permit safe and convenient access to all live parts for maintenance and adjustment. Minimum space as per NEC 110-16. Where minimum working space cannot be provided for escalator and moving-walk panels and panel is in same space as drive machine, the clearances may be waived where the entire panel and disconnecting means are arranged to be readily removed from machine space and has flexible leads to all external connections.

IMPEDANCE. **The impedance that appears on the nameplate of a transformer is determined by applying the following steps: (1) Connect an ammeter across (in series) with the secondary winding. (2) Connect a variable rheostat in the primary winding circuit. (3) Connect a voltmeter across the primary winding. (4) Apply a voltage to the primary circuit and by adjusting the rheostat increase the voltage across the primary winding until the secondary winding reaches its rated current. (5) Read the voltage across the primary winding. (6) Divide the applied primary winding voltage by the rated primary winding voltage.**

For example, a certain transformer has a 2400 V, 10 A primary rating and a 240 V, 100 A secondary rating. A voltage of 120 V is needed across the primary winding for the ammeter in the secondary winding circuit to read 100 A.

By using step number 6 in the above procedure, we divide 120 by 2400 and arrive at a figure of 0.05 or 5 percent impedance.

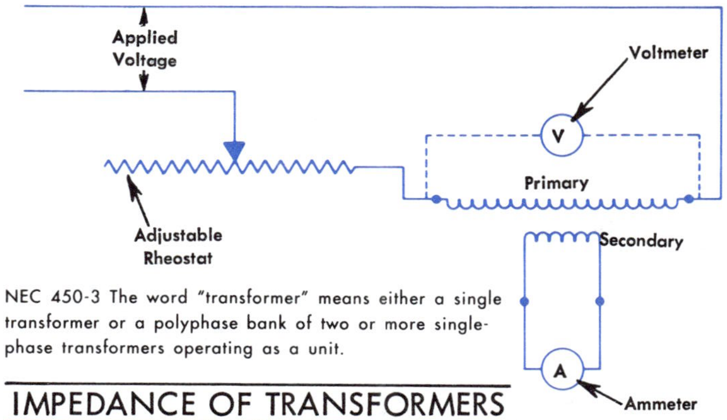

NEC 450-3 The word "transformer" means either a single transformer or a polyphase bank of two or more single-phase transformers operating as a unit.

IMPEDANCE OF TRANSFORMERS

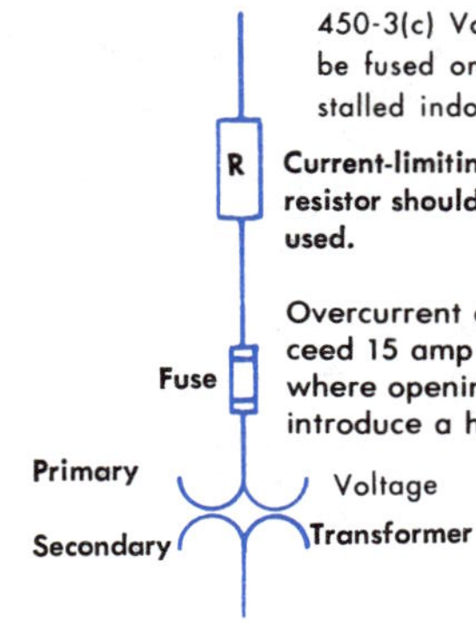

450-3(c) Voltage transformers shall be fused on primary sides when installed indoors. See 384-32.

Current-limiting resistor should be used.

Overcurrent device not to exceed 15 amp rating except where opening of circuit could introduce a hazard.

VOLTAGE TRANSFORMERS

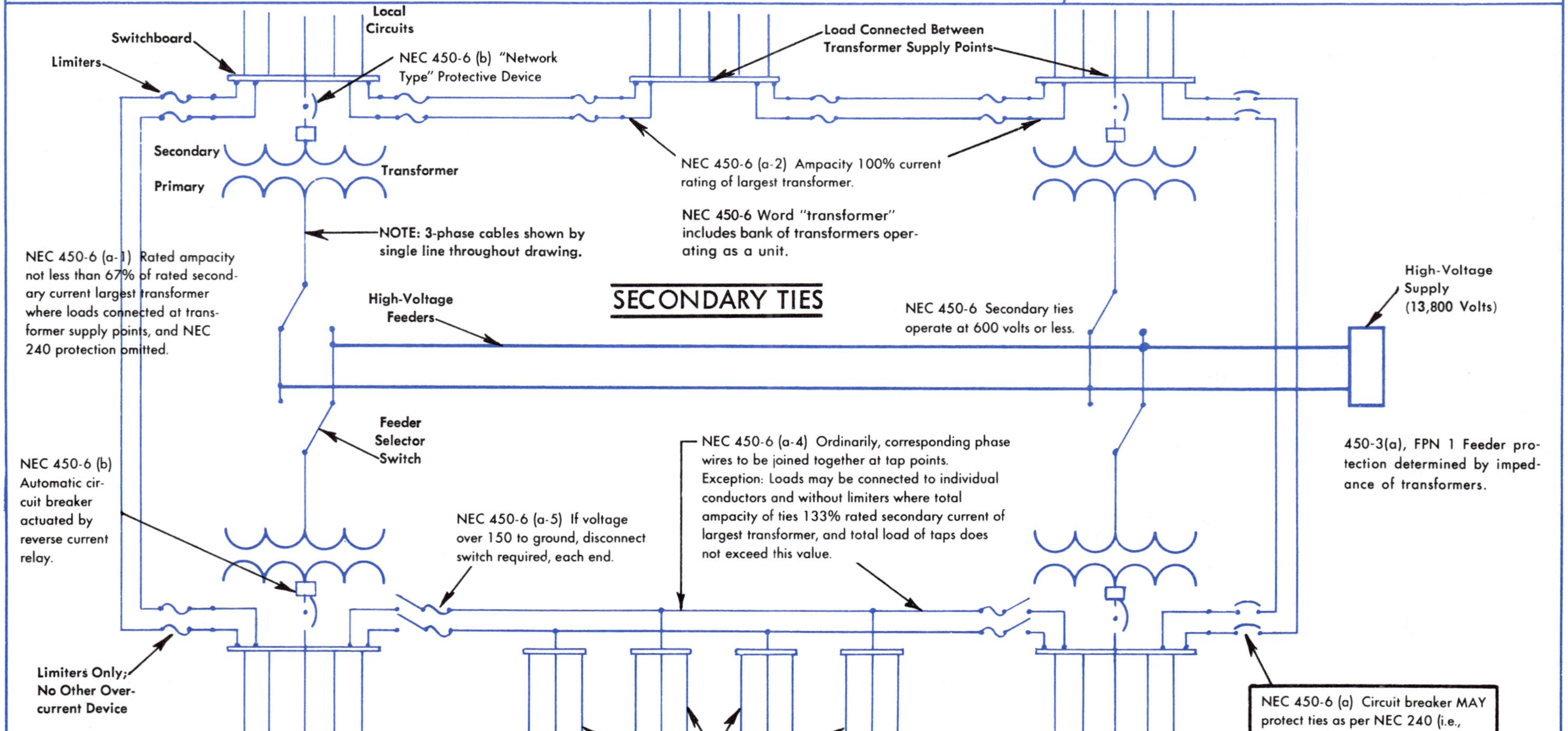

SECONDARY TIES

DEFINITION OF SECONDARY TIES – NEC 450-6 Secondary tie is circuit of not over 600 volts between phases that connects sources of power or power supply points, such as two transformer secondaries, and the tie may consist of one or more conductors per phase.

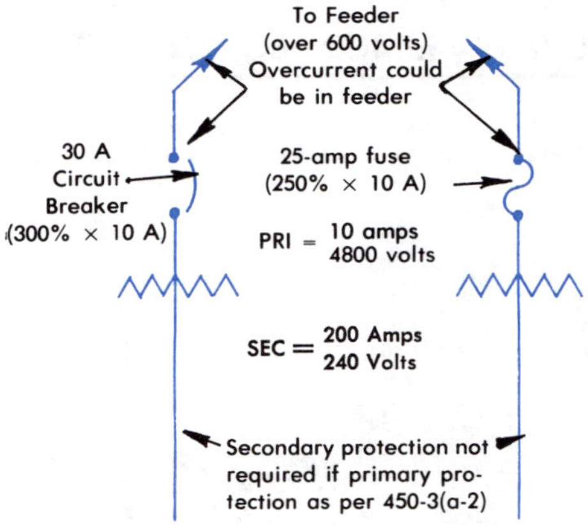

450-3(a)(2) Supervised installations require individual protection on primary side. (a) Primary fuse continuous rating not over 250% of transformer and for circuit breakers rating not over 300%. EX. 1 Next higher fuse or circuit breaker rating permitted if not corresponding to next standard size. EX. 2 Circuit protection permitted to protect transformer.

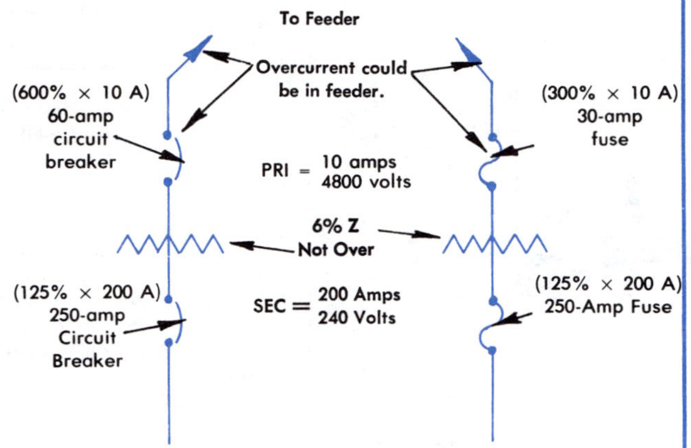

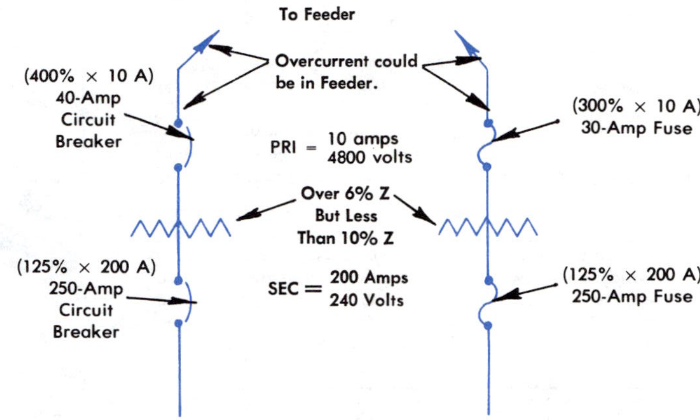

NEC 450-3(a)(1) Transformers over 600 V to have primary and secondary protection per Table 450-3(a)(1). Where fuse or circuit breaker rating exceeds standard rating, the next higher rating is permitted. Table 450-3(a)(1) permits primary protection of 600% for circuit breakers and 300% for fuses if not over 6% Z and 400% for circuit breakers and 300% for fuses if over 6% but not over 10% Z. See Table 450-3(a)(1) for secondary protection. See 450-3(a)(1), Ex. 2.

NEC 450–3 (a) (1) (2) OVERCURRENT PROTECTION OF TRANSFORMERS (OVER 600 VOLTS)

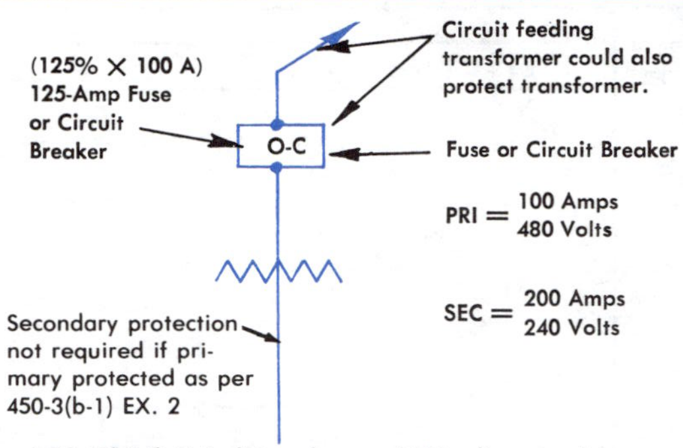

NEC 450-3 (b-1) Each transformer of 600 volts or less is required to have individual primary overcurrent protection. This overcurrent device is to be rated or set at not over 125% of rated primary current except: (1) if primary current is 9 amps or more and 125% is not standard fuse or circuit breaker, then next larger device may be used. Where primary is less than 2 amps, overcurrent device may be rated or set at not more than 300% of rated primary amps, (2) primary circuit protection may also protect transformer if of proper size; (3) if secondary is protected as per 450-3 (b-2).

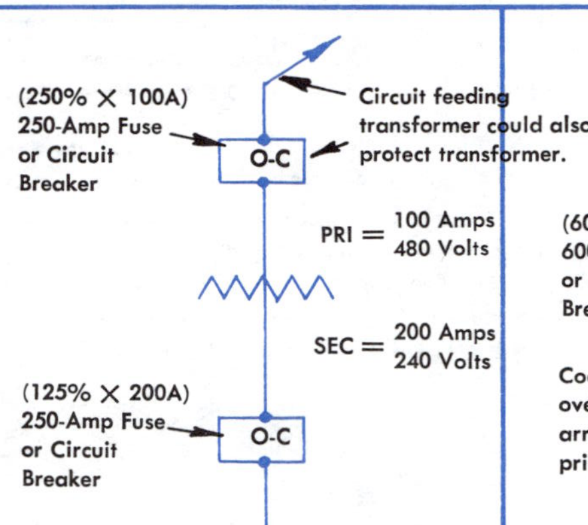

NEC 450-3 (b-2) Individual primary overcurrent not required if primary feeder protected at 250% of primary rated current and secondary protected not over 125% of rated secondary current.

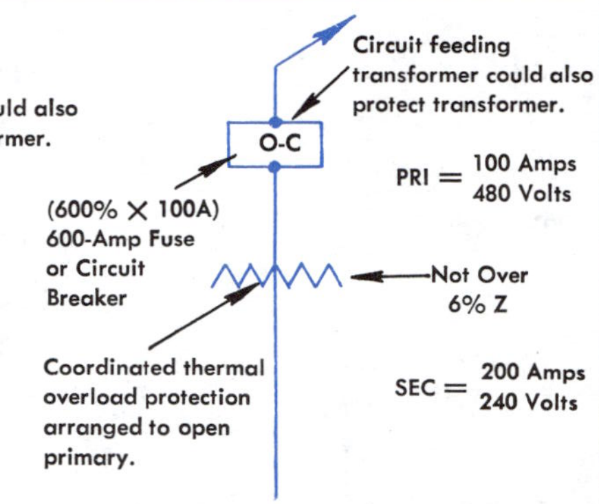

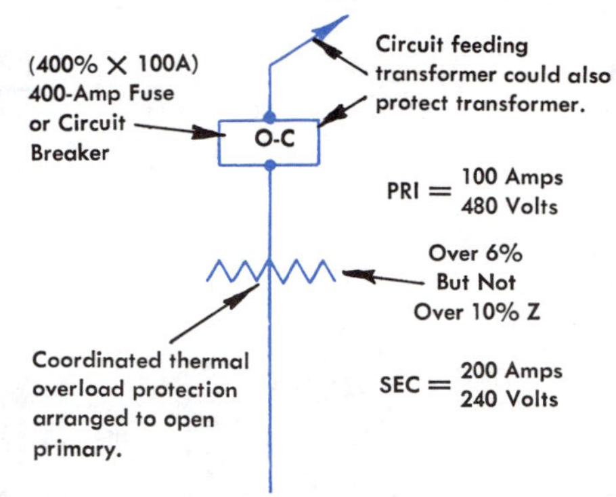

NEC 450-3 (b-2) Individual transformer primary protection not required if transformer has coordinated thermal overload protection by manufacturer, arranged to open primary circuit, and if primary overcurrent device is not over 6 times rated primary current for transformers with not more than 6% impedance and not more than 4 times rated primary current for transformers having more than 6 but not more than 10% impedance except: where secondary current is 9 amps or more and 125% is not standard fuse or circuit breaker, then next larger device may be used. Where secondary is less than 9 amps the overcurrent device may be rated or set at not more than 167% rated primary current.

All conductors feeding in primary or out of secondary shall be protected at their rated ampacities as per 240-3. This protection shall be provided where the conductors receive their supply as per 240-21.

NEC 450-3 and 450-6 indicate that when the word "transformer" is used it shall mean a transformer or polyphase bank of 2 or more single-phase transformers operating as a unit.

Where secondary conductors feed a lighting and branch-circuit panelboard, each panelboard shall be individually protected on supply side at ampacity not exceeding rating of panelboard. (See NEC 384-16.)

NEC 450–3 (b) (1) (2) OVERCURRENT PROTECTION OF TRANSFORMERS (600 VOLTS OR LESS)

TRANSFORMERS IN PARALLEL

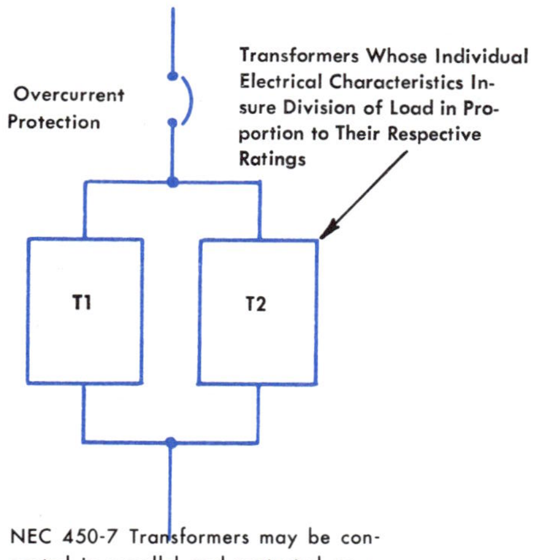

NEC 450-7 Transformers may be connected in parallel and protected as a unit when their electrical characteristics are suitable, provided overcurrent protection for each transformer meets requirements of 450-3(a)(1) or (b)(2).

TRANSFORMERS -- DRY-TYPE INSTALLED INDOORS

NEC 450-21 (a) Dry-type transformers permitted indoors if rated 112½ kVA or less and 12 in. separation from combustible material unless separated by a fire-resistant, heat-insulating barrier, or unless rated 600 volts or less and totally enclosed except for ventilating openings.

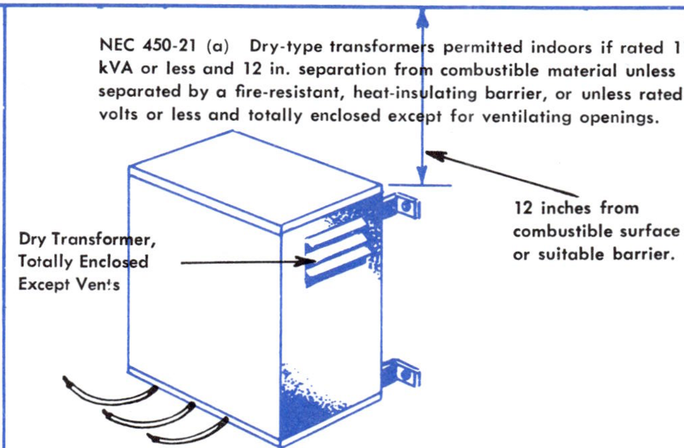

NEC 450-21(b), Ex. 1, 2 If over 112½ kVA then required to be in transformer room of fire-resistant construction. Ex. 1 Transformer room not required if transformer has an 80°C rise or higher rating and separated from combustibles by at least 12′ vertically and 6′ horizontally unless (Ex. 2) totally enclosed and ventilated-type construction.

NEC 450-21 (c) Dry-type transformers over 35,000 volts require vault as per Article 450 Part C.

NEC 110-31 (a) Where over 600 volts, openings in dry transformer cases to be guarded if necessary for protection of persons.

AUTOTRANSFORMER

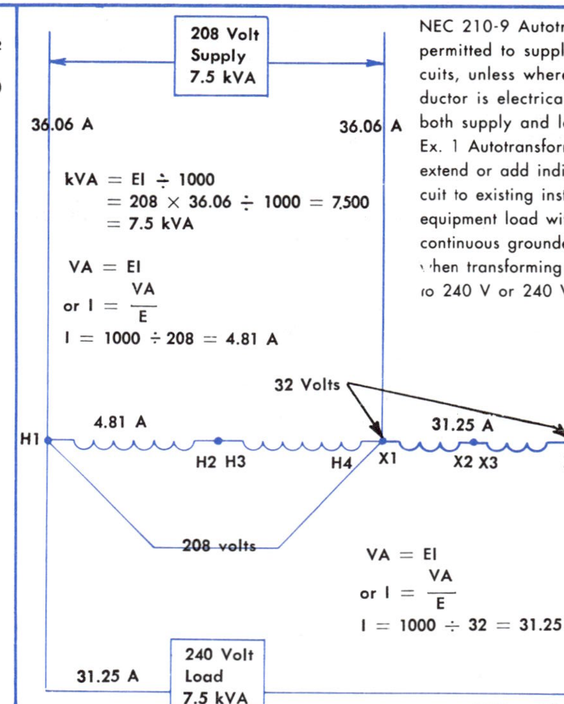

NEC 210-9 Autotransformers not permitted to supply branch circuits, unless where grounded conductor is electrically connected to both supply and load.

Ex. 1 Autotransformer permitted to extend or add individual branch circuit to existing installation for equipment load without electrically continuous grounded conductor when transforming nominal 208 V to 240 V or 240 V to 208 V.

210-9, Ex. 2 Autotransformers, in industrial occupancies, are permitted to supply nominal 600 V loads from nominal 480 V systems and nominal 480 V loads from nominal 600 V systems, without connection to a similar grounded conductor, if installed by a qualified person.

$$kVA = EI \div 1000 = 240 \times 31.25 \div 1000 = 7.5 \text{ kVA}$$

ASKAREL OR NONFLAMMABLE TRANSFORMERS INSTALLED INDOORS

NEC 450-25 Where askarel-insulated transformers are installed in poorly ventilated area, means shall be provided to absorb any gas, or a chimney or flue connected to carry such gases outside buildings.

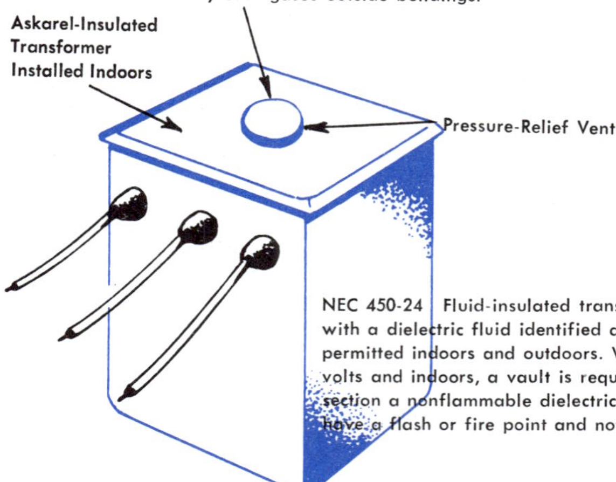

NEC 450-24 Fluid-insulated transformers insulated with a dielectric fluid identified as noflammable permitted indoors and outdoors. When over 35,000 volts and indoors, a vault is required. (In this section a nonflammable dielectric fluid does not have a flash or fire point and not flammable in air.)

NEC 450-25 Askarel-insulated transformers 25 kVA or over must have pressure-relief vent. If more than 35,000 volts, they shall be installed in an approved vault.

OIL-INSULATED TRANSFORMERS--INSTALLED INDOORS AND OUTDOORS

NEC 450-26 (General Rule) Transformer vault required for oil-insulated transformer indoors. Ex. 1 if not over 112 ½ kVA, thickness of vault wall need be only 4″. See Exceptions 2, 3, 4, 5.

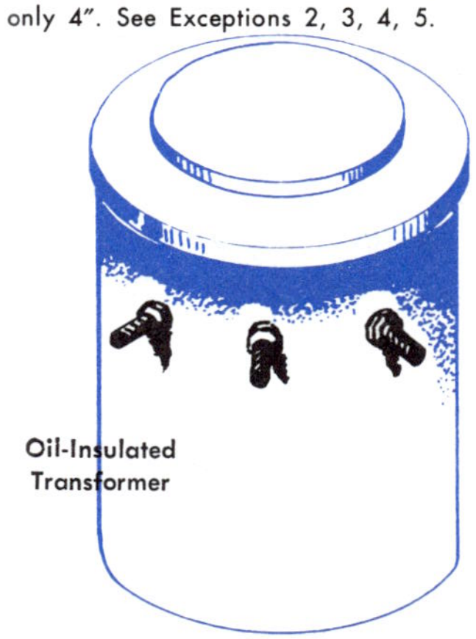

NEC 450-27 Oil-insulated transformers installed outdoors shall be installed so combustible material, combustible buildings, parts of buildings, door and window openings, fire escapes are safeguarded from transformer oil fires when transformer is installed, attached to, or adjacent to a building or combustible material, or on roofs. (See National Electrical Safety Code ANSI C2-1981)

NEC 450-26 Ex 3 Electric furnace transformers totalling not more than 75 kVA may be installed in fire-resistive location if provisions are made to prevent spread of oil fire.

NEC 450-26 Ex 2 Vault not required if not over 600 volts, not over 10 kVA in combustible area, or 75 kVA in fire-resistant area.

NEC 450-26 Ex 4 Vault not required if in suitable detached building accessible only to qualified persons.

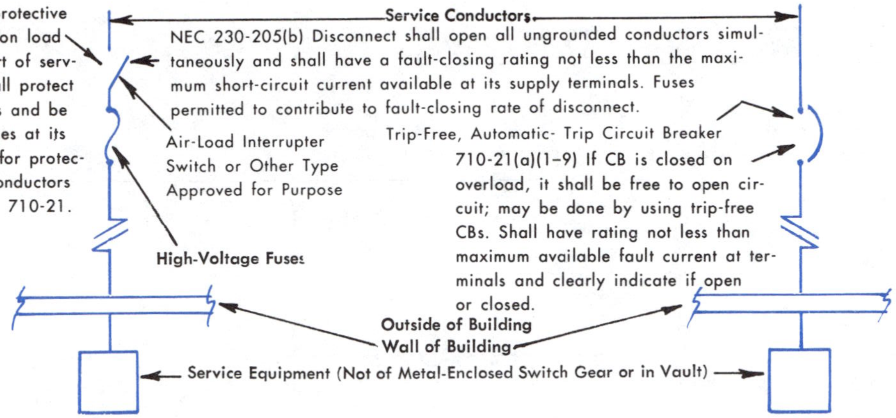

SERVICES -- OVER 600 VOLTS -- OVERCURRENT PROTECTION

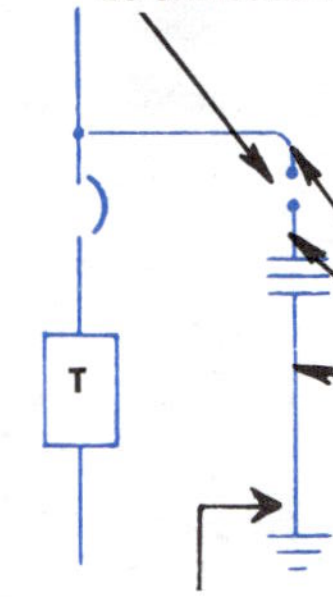

SURGE ARRESTERS

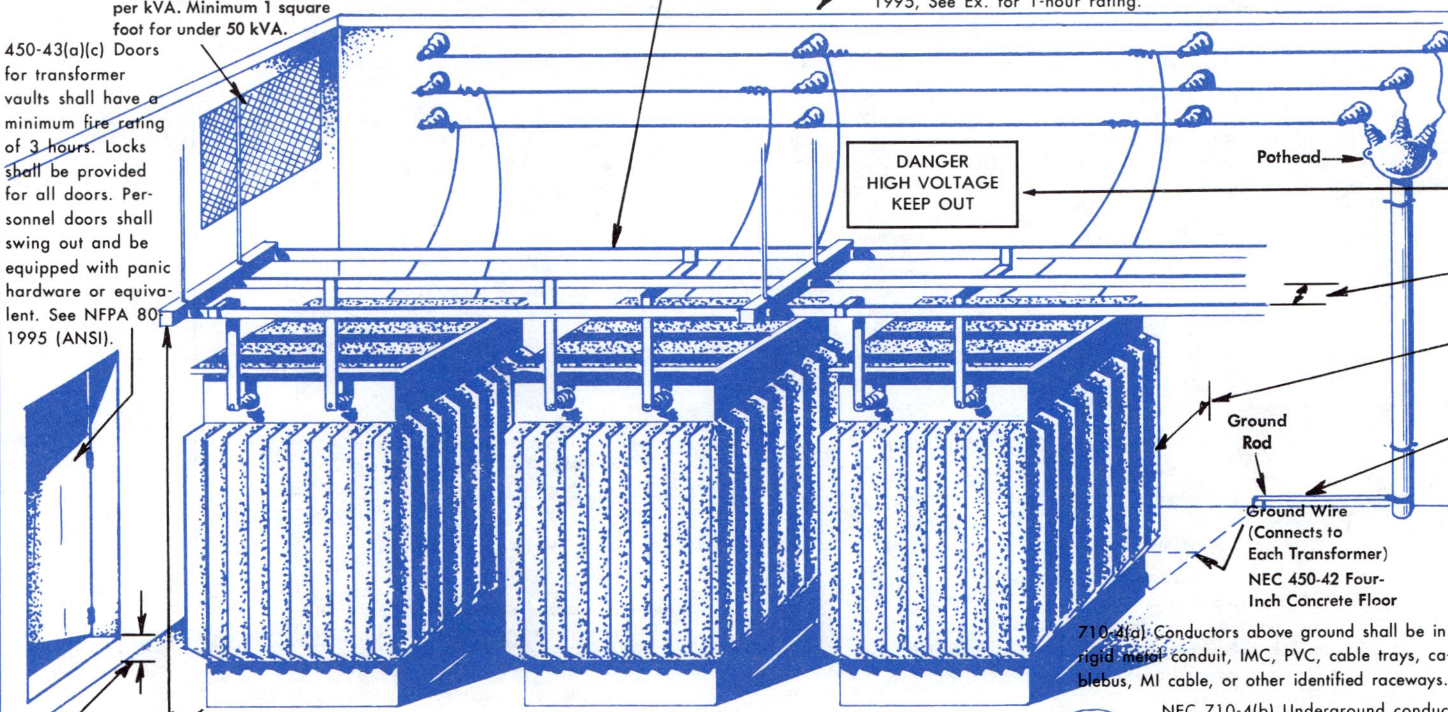

HIGH-VOLTAGE INSTALLATION IN APPROVED TRANSFORMER VAULT

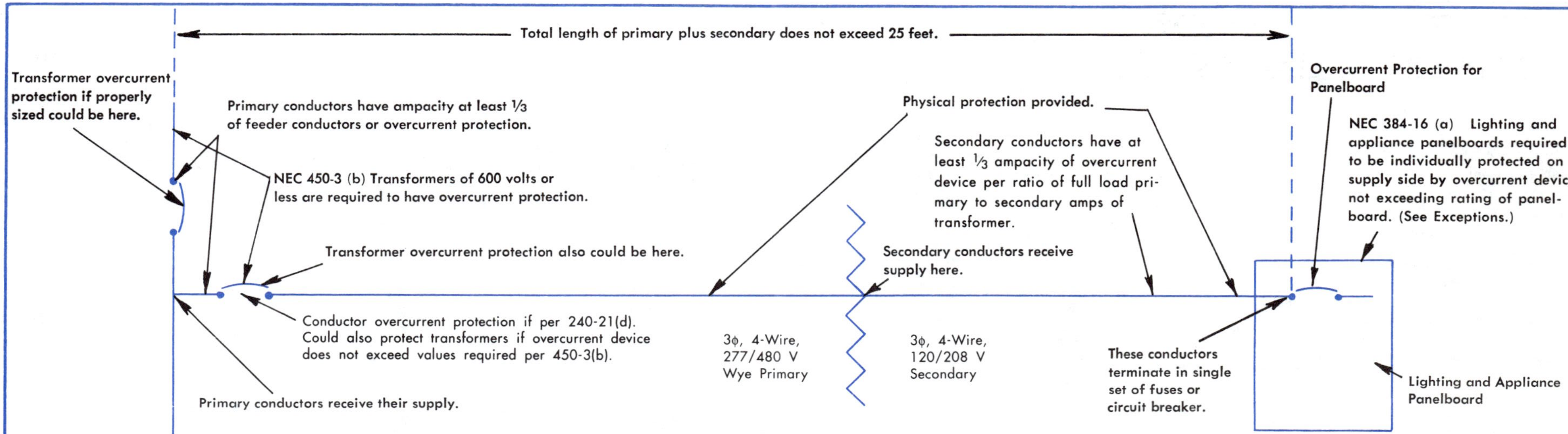

NEC 240-21 All conductors are required to be protected where they receive their supply.

240-21(d) A transformer feeder tap is permitted if the primary plus secondary conductors are not over 25′ long and all of the following conditions are met:

(1) Primary conductors have ampacity of at least one third of the rating of the overcurrent device protecting the feeder conductors.

(2) The transformer secondary conductors have ampacity that when multiplied by the ratio of the secondary to primary voltage is not less than one third the ampacity of the rating of the overcurrent device protecting the feeder conductors.

(3) The total length of one primary plus one secondary conductor that is protected at its ampacity is no over 25′.

(4) Physical protection provided for both primary and secondary conductors.

(5) Secondary conductors terminate in single set of fuses or CBs which limits load to ampacity per 310-15.

SUMMARY: The above drawing and notes are confusing in that we are dealing with two entirely different NEC requirements. One is that we shall have overcurrent protection for all conductors where they receive their supply, (NEC 240-21) and the other is that we shall provide proper transformer overcurrent protection as per NEC 450-3 (b). Because we are dealing with two different and complex requirements it is not unusual to concentrate on providing overcurrent protection for a transformer and neglect the required overcurrent protection of the conductors or the panelboard. The only time that a secondary conductor of a transformer is protected by the primary overcurrent device is where a single-phase, 2-wire (single-voltage) transformer is used and the primary is protected as per 450-3(b) and the voltage ratio of the transformer protects the secondary conductors to their ampacity. See 240-3(i).

Single-phase, 120/240 V, 2-Wire, (Single Voltage) Secondary

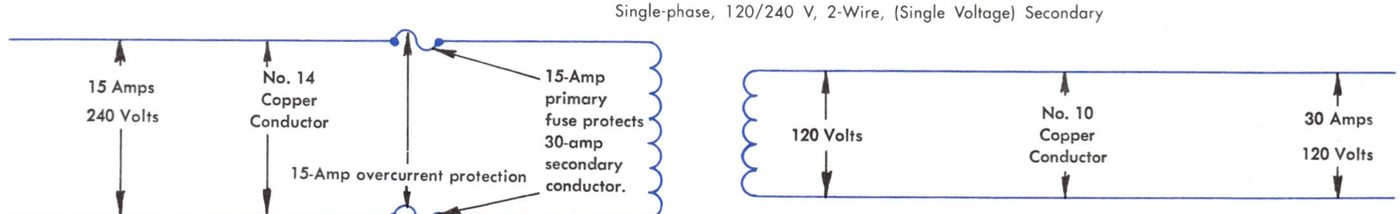

240-3(i) The secondary conductors of a 1ϕ, 2-wire (single-voltage) secondary are considered protected if the primary overcurrent conforms to 450-3 and is not more than the current determined by multiplying the secondary conductor amps by the secondary-to-primary transformer voltage ratio.

TRANSFORMER FEEDER TAP 25 FOOT RULE

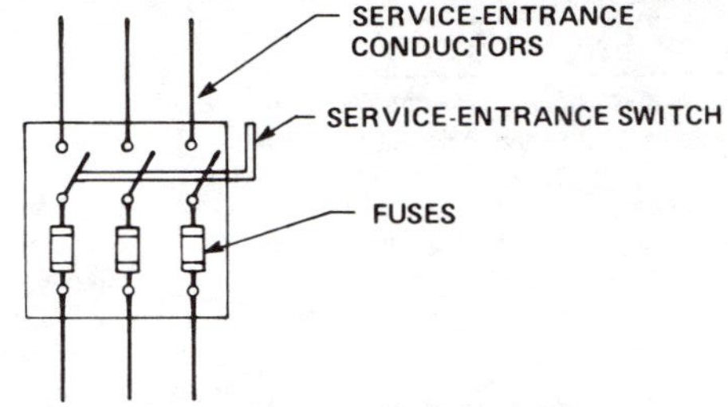

CONTINUOUS RATING OF EACH FUSE MUST NOT EXCEED 300% OF SERVICE CONDUCTORS' AMPACITY RATING

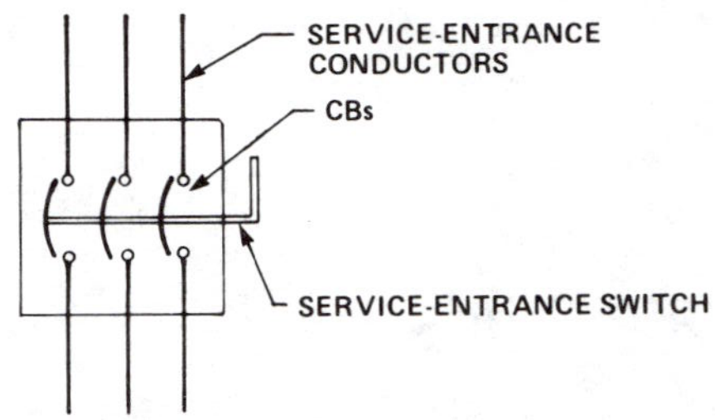

CB's TRIP SETTING MUST NOT EXCEED 600% OF SERVICE CONDUCTOR'S AMPACITY RATING

230-208 Fuses used for short-circuit protection shall not exceed three times ampacity of conductor. For CB, trip setting shall not exceed six times ampacity of conductors.

NEC 230-202 (a) Service wires over 600 volts not smaller than No. 6 except in cable No. 8.

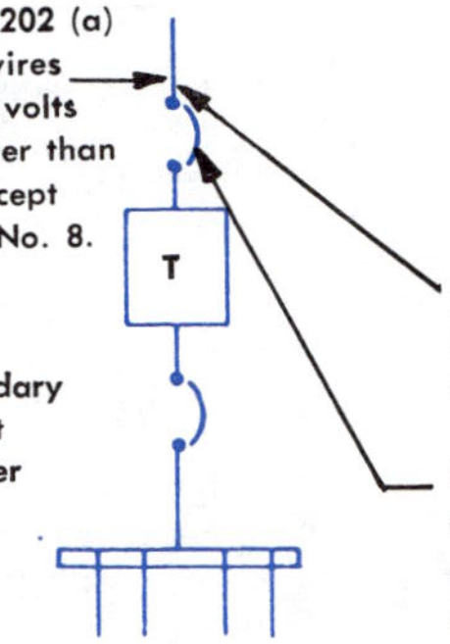

NEC 230-6 Raceways in 2 inches of concrete or brick are considered outside the building.

NEC 230-205(a)(b), 206 Circuit breaker or equivalent disconnecting means shall have a fault-closing rating not less than the maximum short-circuit current available at its supply terminals.

230-202(b) Wiring methods for services over 600 V per 710-4. For aboveground conductors, wiring methods shall be (1) rigid metal conduit, (2) IMC, (3) rigid nonmetallic conduit, (4) cable tray, (5) busbars, (6) cablebus, (7) other identified raceways, (8) open runs of MC cable suitable for use and purpose. Underground conductors shall be identified for voltage and conditions of use.

SERVICES OVER 600 VOLTS -- WIRING METHODS

SERVICES -- OVER 600 VOLTS

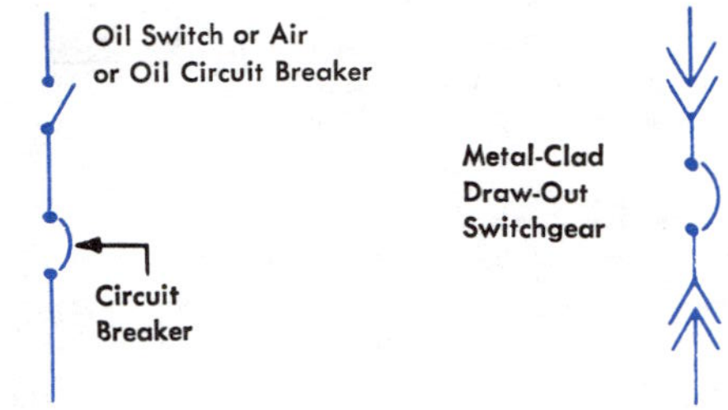

NEC 230-204 (a) Where service disconnect is oil switch or air or oil circuit breaker, then air-break isolating switch required on line side of disconnect and all service equipment.

NEC 230-204 (c) (d) Air-break isolators shall be accessible to qualified persons only and arranged for grounding on load side.

SERVICES OVER 600 VOLTS -- ISOLATING SWITCHES

Disconnect Type of Fuse

230-204(b) Fuses shall be permitted to serve as isolating switch if they can be operated as a disconnecting switch.

SERVICES OVER 600 VOLTS -- FUSED ISOLATING SWITCH

Nonautomatic Oil Switch Oil Fuse Cutout or Air Interrupter Switch

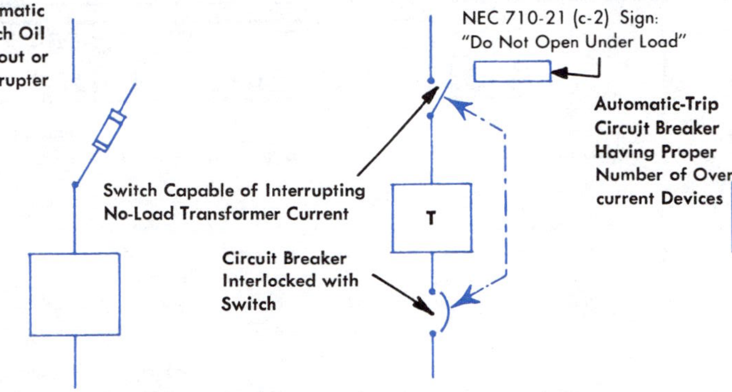

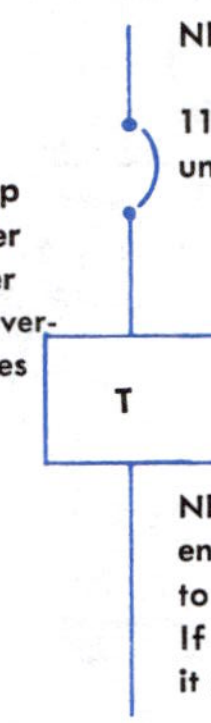

NEC 710-12 refers to 110-31 (b) "Outdoor Installations."

110-31 (b-1) Shall comply with Article 225 if open to unqualified persons.

110-31 (b-2) Outdoor installations with exposed live parts shall be accessible to qualified persons only and shall also comply with 110-34, 710-32 and 710-33.

NEC 710-13 refers to 110-31 (c) which permits metal-enclosed equipment to be installed outdoors and accessible to the public, if bolts and nuts are difficult to remove. If bottom of enclosure is less than 8 feet from floor it shall be locked.

NEC 710-34 refers to 110-34 that gives the working space required for live parts.

230-208(a) Equipment used to provide fault protection for service-entrance conductors over 600 V per 710, Part C. Disconnecting means for over 600 V services shall open all ungrounded conductors of circuit simultaneously and shall have a sufficient fault-closing rating for maximum available short-circuit current. See 230-205 and 710-21. Some acceptable forms of disconnecting means and fault protection are (1) oil fuse cutout, (2) nonautomatic oil switch, (3) air load interrupter switch, (4) automatic trip CB.

NEC 710-22 Isolating switches are not required where equipment may be safely de-energized through metal-clad switchgear units and removable truck panels.

SERVICES OVER 600 VOLTS -- DISCONNECT MEANS

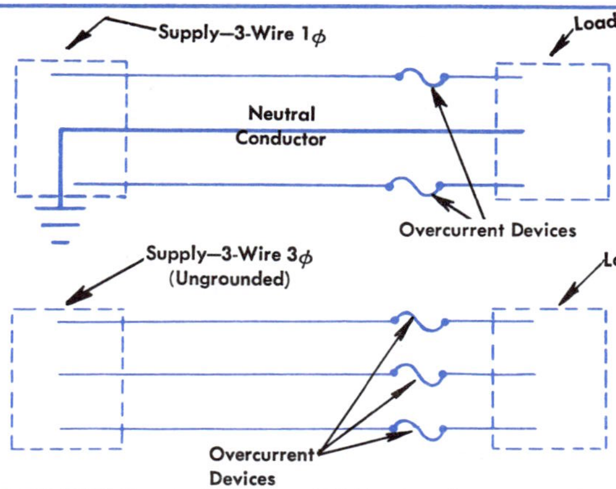

NEC 710-20(b) For circuits of over 600 V, where fuses are used for overcurrent protection, a fuse is required in each ungrounded conductor.

NEC 710-20 (a) (General Rule) On circuits over 600 volts, where circuit breakers are used for AC 3-phase circuits, they shall have at least 3 overcurrent relays operated from 3 current transformers. (See Ex 1, 2.)

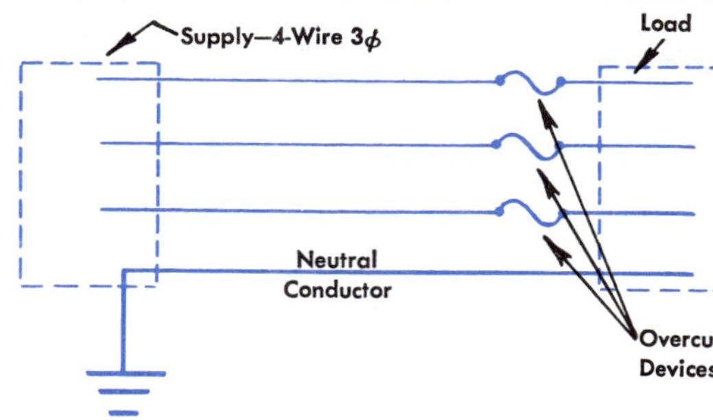

NEC 240-20 (a) **For circuits of 600 volts or less overcurrent protection required in series with each ungrounded conductor.**

NEC 240-20(b) For circuits of 600 V or less, circuit breakers required to open all conductors of circuit, except single-pole breakers with handle ties, permitted for 3-wire DC or single-phase circuits, and lighting and appliance branch circuits connected to 4-wire 3-phase systems or 5-wire 2-phase systems, if circuits supplied from system having a grounded neutral, and voltage does not exceed that permitted by 210-6.

NEC 710-20 **For circuits of over 600 volts, overcurrent protection required for each ungrounded conductor.**

NUMBER OF OVERCURRENT DEVICES TO PROTECT CIRCUIT CONDUCTORS: 600 VOLTS OR LESS -- NEC 240-20; OVER 600 VOLTS -- NEC 710-20

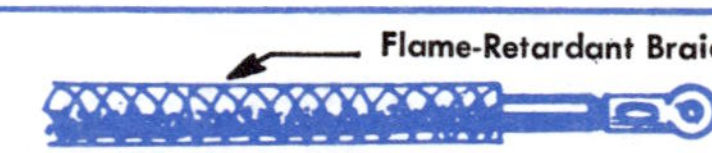

NEC 710-5 Where braid-covered insulated conductors are used for open runs for circuits over 600 V, then braid shall be flame retardant.

NEC 710-5 Braid shall be stripped back a safe distance. Shall be 1″ for each kilovolt to ground in circuit.

OVER 600 VOLTS - BRAID-COVERED INSULATED CONDUCTORS, OPEN RUNS

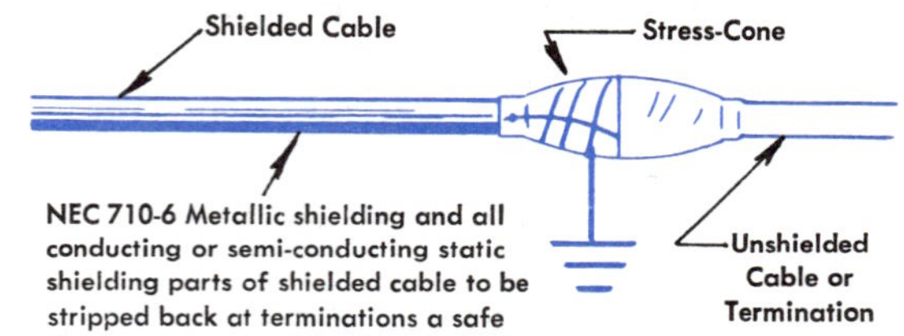

NEC 710-6 Metallic shielding and all conducting or semi-conducting static shielding parts of shielded cable to be stripped back at terminations a safe distance determined by circuit voltage. All metallic shielding components to be grounded.

OVER 600 VOLTS -- GROUNDING OR SHIELDING TAPE

NEC 460-24 (a) Group-operated switch rated to carry continuously at least 135% of rated capacitor bank current, rated to interrupt load of bank, rated for maximum in rush currents and rated to carry capacitor fault currents.

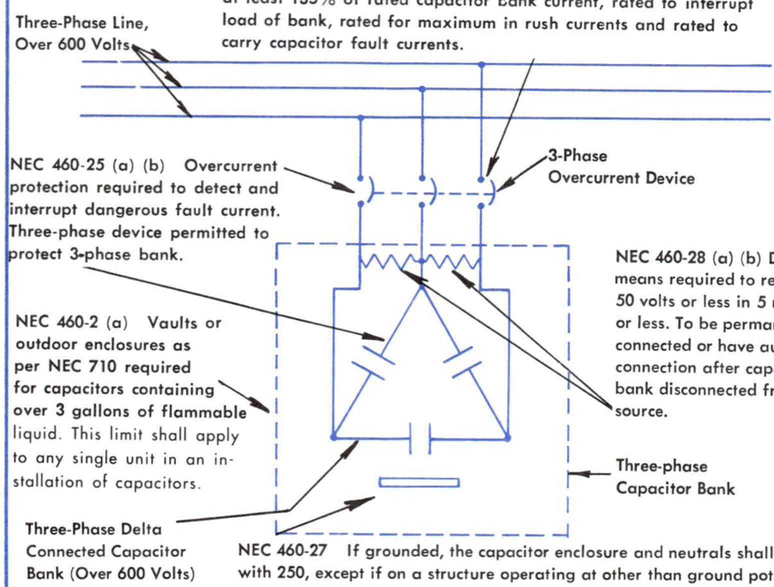

NEC 460-25 (a) (b) **Overcurrent protection required to detect and interrupt dangerous fault current. Three-phase device permitted to protect 3-phase bank.**

NEC 460-2 (a) **Vaults or outdoor enclosures as per NEC 710 required for capacitors containing over 3 gallons of flammable** liquid. This limit shall apply to any single unit in an installation of capacitors.

NEC 460-28 (a) (b) Discharge means required to reduce to 50 volts or less in 5 minutes or less. To be permanently connected or have automatic connection after capacitor bank disconnected from source.

NEC 460-27 **If grounded, the capacitor enclosure and neutrals shall comply with 250, except if on a structure operating at other than ground potential.**

NAMEPLATE—NEC 460-26 **Lists maker's name, rated voltage, kVar or amps, frequency, phases, and type and amount of liquid.**

CAPACITORS -- OVER 600 VOLTS

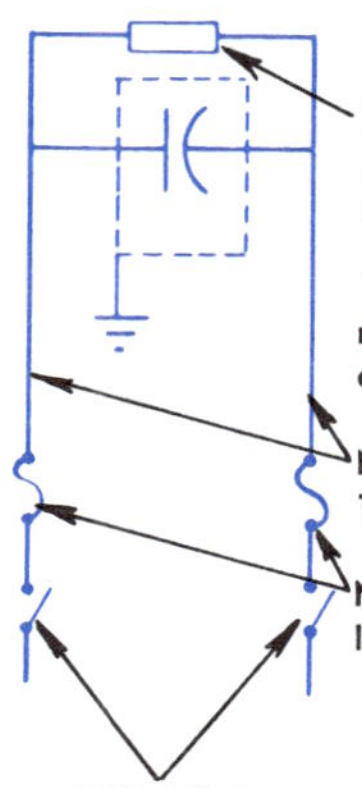

NEC 460-6 (a) (b) Discharge resistor connected permanently or automatic to drain charge to 50 volts in 1 minute.

NEC 460-28 (a) For over 600 volts discharrge means required to reduce to 50 volts or less in 5 minutes or less.

NEC 460-8 (a) Ampacity of conductors at least 135% of capacitor current rating.

NEC 460-8 (b-1, 2) Overcurrent protection as low as practicable.

NEC 460-8 (c) 1, 2, 4) **Disconnecting means shall open all ungrounded conductors simultaneously. Its continuous ampacity must not be less than 135% of rated capacity current.**

NEC 460-10 **Capacitor cases to be grounded as per 250, except if on a structure operating at other than ground potential.**

NEC 460-2 (b) **Live parts to be enclosed or guarded except when enclosed and accessible only to qualified personnel.**

CAPACITORS -- 600 VOLTS OR LESS

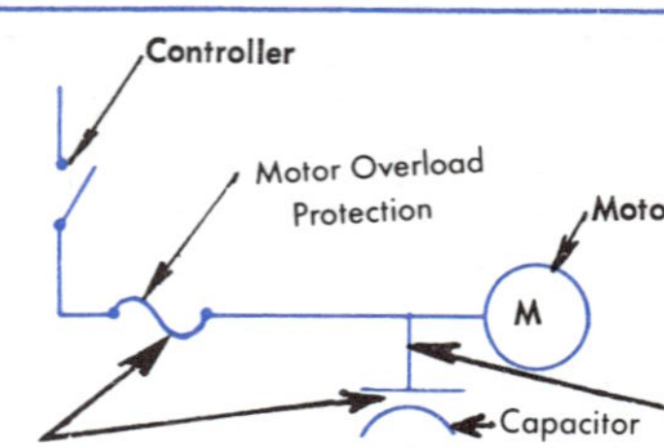

460-9 If a capacitor is connected to the load side of a motor overload device, the rating or setting of the device is computed from the improved power factor circuit. For the purposes of 430-22, the effect of the capacitor shall be disregarded.

NEC 460-8 (c-1) Ex No separate disconnect required for capacitor.

NEC 460-8(b-1), Ex. Overcurrent device not required in each conductor if connected on load side of motor controller.

NEC 460-8 (a) Rating of conductors at least ⅓ that of motor circuit conductors and 135% of capacitor current rating.

CAPACITORS -- MOTOR CIRCUITS

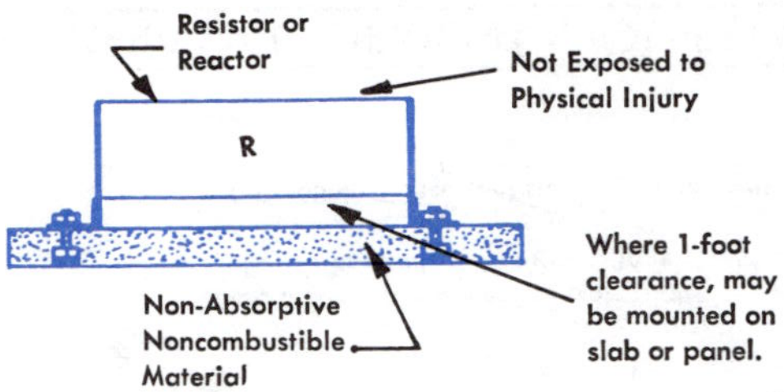

NEC 470-2 Resistors and reactors are not to be installed where exposed to physical damage.

NEC 470-3 A space of 12 inches is required between resistors and reactors and any combustible material. If space less than 12 inches, a thermal barrier is required. (See above for alternate method where 12 inches cannot be obtained.)

NEC 470-4 Insulated conductors shall be rated for 90°C when used for connections between resistance elements and controllers, except, for motor starting service, other insulations permitted.

RESISTORS AND REACTORS (600 VOLTS AND UNDER)

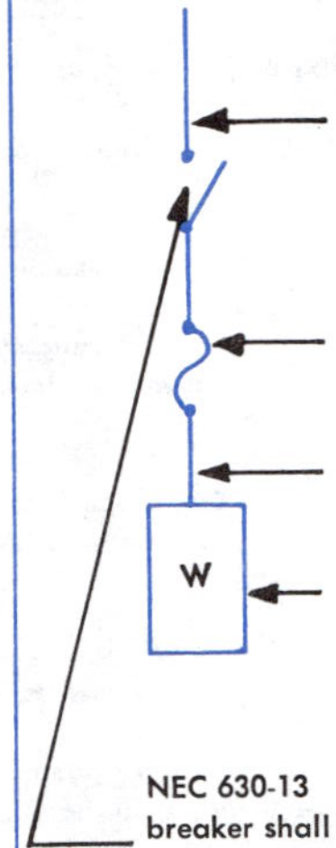

NEC 630-11 (a) Ampacity of supply conductors shall not be less than rated primary current times the duty-cycle factor stated in accompanying table.

NEC 630-12 (a) Welder overcurrent protection not over 200% primary current.

NEC 630-12 (b) Conductor protected at not over 200% of rating.

Transformer Arc Welder

NEC 630-13 Disconnect switch or circuit breaker shall have ampere rating sufficient to accommodate required overcurrent protection, as per 630-12.

TRANSFORMER ARC WELDER

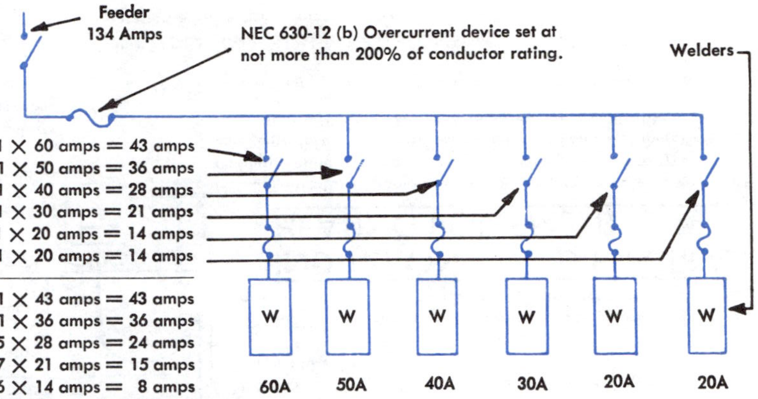

.71 × 60 amps = 43 amps
.71 × 50 amps = 36 amps
.71 × 40 amps = 28 amps
.71 × 30 amps = 21 amps
.71 × 20 amps = 14 amps
.71 × 20 amps = 14 amps

1 × 43 amps = 43 amps
1 × 36 amps = 36 amps
.85 × 28 amps = 24 amps
.7 × 21 amps = 15 amps
.6 × 14 amps = 8 amps
.6 × 14 amps = 8 amps

TOTAL 134 amps

Based on 50% duty-cycle; multiplying factor .71 as per NEC 630-11 (a).

NEC 630-11 (b) Conductor rating for a group of welders shall be determined in each case by the nature of use. It is suggested that adequate conductor size will be obtained if the following percentages of current values determined as per NEC 630-11 (a) are applied: 100 percent for the two largest welders, plus 85% for the third largest, plus 70% for the fourth largest, plus 60% for all others. See FPN.

GROUP OF TRANSFORMER ARC WELDERS

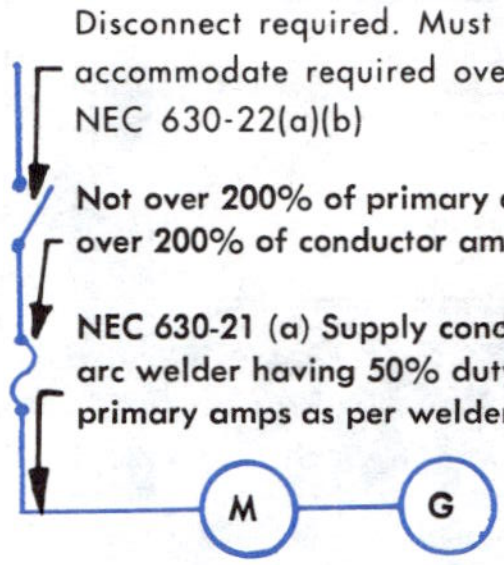

Disconnect required. Must be large enough to accommodate required overcurrent protection. NEC 630-22(a)(b)

Not over 200% of primary amps of welder and not over 200% of conductor amps.

NEC 630-21 (a) Supply conductor rating for M-G arc welder having 50% duty cycle is .75 of rated primary amps as per welder nameplate.

NEC 630-22 (a) Overcurrent protection of welder not over 200% of rated primary current. (Protection may also protect supply conductors.)

NEC 630-22 (b) Welder supply conductors to be protected at not over 200% of conductor rating.

NEC 630-23 Each welder shall have disconnect. Rating not less than required to accommodate overcurrent protection required by 630-22.

MOTOR-GENERATOR WELDER

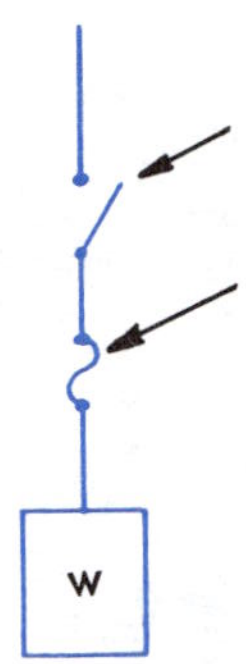

NEC 630-33 Rating of disconnect not less than that of supply conductors as required by 630-31.

NEC 630-32 (a) Maximum rating of overcurrent device 300% rated primary current.

NEC 630-31 (a-1) Minimum rating of supply conductors, varying duty 70% rated primary current automatic; 50% manual.

NEC 630-31 (a-2) Rating of supply conductors for specific cooperation as per duty-cycle table: Multiplier 0.71 for 50% duty, 0.45 for 20% duty.

RESISTANCE WELDER

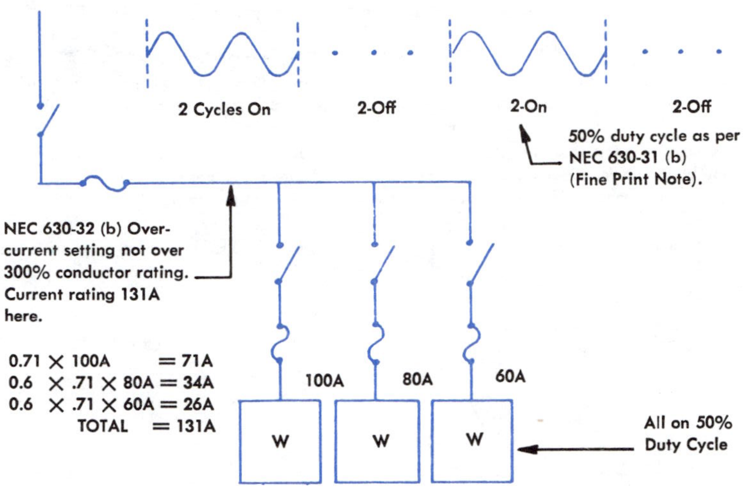

NEC 630-32 (b) Overcurrent setting not over 300% conductor rating. Current rating 131A here.

0.71 × 100A = 71A
0.6 × .71 × 80A = 34A
0.6 × .71 × 60A = 26A
TOTAL = 131A

NEC 630-31 (a)(1)(2)(b) Conductor rating based on 630-31(a)(2) value obtained for largest welder plus 60% of that for the others.

NEC 630-33 Rating of disconnect not less than that of supply conductors.

GROUP OF RESISTANCE WELDERS

X-RAY EQUIPMENT REQUIREMENTS

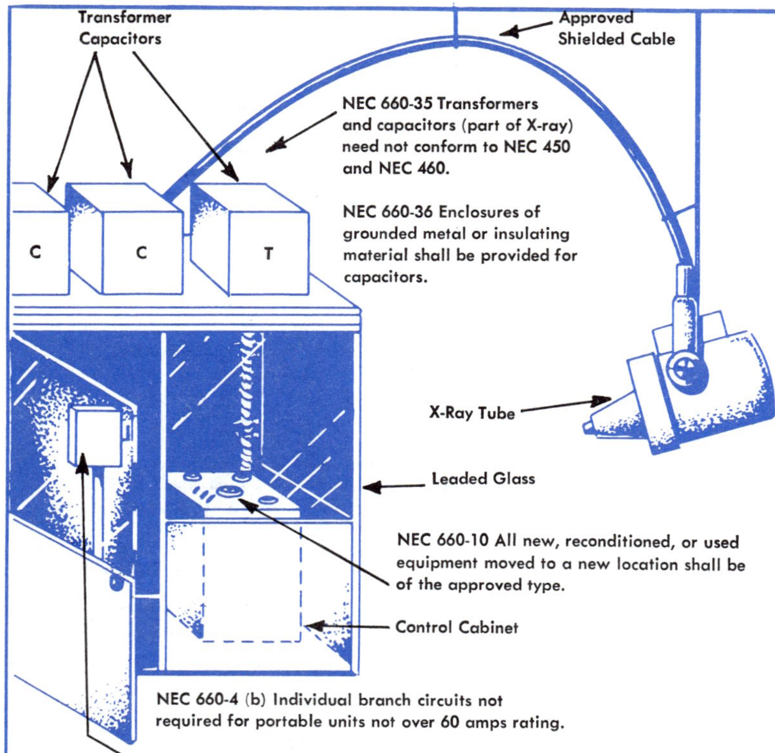

NEC 660-35 Transformers and capacitors (part of X-ray) need not conform to NEC 450 and NEC 460.

NEC 660-36 Enclosures of grounded metal or insulating material shall be provided for capacitors.

NEC 660-10 All new, reconditioned, or used equipment moved to a new location shall be of the approved type.

NEC 660-4 (b) Individual branch circuits not required for portable units not over 60 amps rating.

NEC 660-5 A switch or circuit breaker at least 50% capacity of momentary rating or 100% of long time rating, whichever is greater, shall be readily accessible from X-ray control station. Where 120V, 30 amps or less, plug and receptacle may serve as disconnect.

NEC 660-21 Manually controlled device located in or on equipment required for portable X-ray, and shall comply with 660-20.

NEC 660-4 (a) Hard-service cord and plug may be used to supply fixed and stationary unit not over 30 amps. Over 30 amps requires wiring meeting general code requirements.

NEC 660-6 (a) Ampacity of conductors and overcurrent devices shall be 50% of momentary rating or 100% of long time rating, whichever is greater.

NEC 660-24 Where two or more pieces of equipment use same high-voltage devices, each must have separate disconnect.

NEC 660-47 (a) Connection from high-voltage equipment to X-ray tubes to be made with high-voltage shielded cables.

NEC 560-48 Noncurrent-carrying metal parts to be grounded.

NEC 660-48 Portable equipment shall have approved grounding-type plug.

MOTOR-GENERATOR EQUIPMENT

OTHER THAN MOTOR-GENERATOR EQUIPMENT

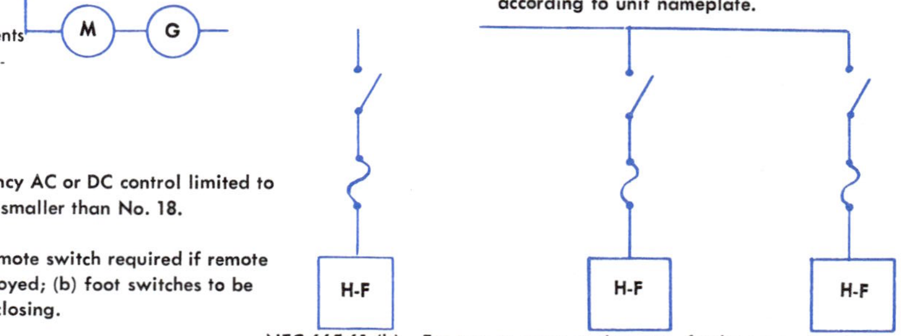

NEC 665-44 (a) Output circuit shall be isolated from ground. If over 500 volts, DC ground protector required, DC impressed on output circuit not to exceed 30 volts, 5 mils. (See Ex 1, 2.)

NEC 665-41, 42, 43 Motor Rules, Article 430, apply.

NEC 665-44 (b) Components shall be connected by multi-conductor or coaxial cable or busbar. Conduit shall be nonferrous.

NEC 665-28 Low frequency AC or DC control limited to 150 volts. Conductors not smaller than No. 18.

NEC 665-47 (a) Local remote switch required if remote application controls employed; (b) foot switches to be shielded from accidental closing.

NEC 665-61 (a) For single unit, ampacity of supply conductors according to unit nameplate.

NEC 665-61 (b) For two or more equipments, feeder ampacity shall not be less than sum of nameplate currents, except where simultaneous operation not possible. Nameplate current for largest unit plus sum of standby currents of the others.

NEC 665-64 (b) If more than 2 ft long, to be enclosed in nonferrous, noncombustible material.

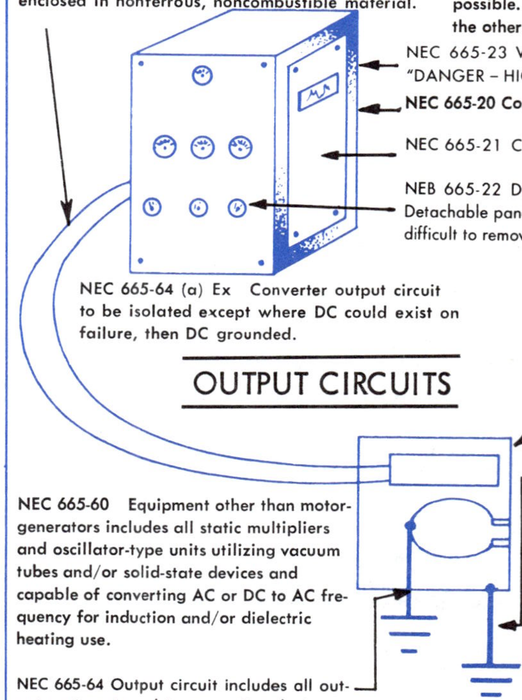

NEC 665-23 Warning labels where necessary that read "DANGER – HIGH VOLTAGE – KEEP OUT".

NEC 665-20 Converting apparatus to be enclosed.

NEC 665-21 Control panels to be of dead-front construction.

NEB 665-22 Doors or panels required for internal access. Detachable panels (not normally used for access) to be difficult to remove.

NEC 665-67 Block oscillator peak voltage not over 100 volts in using R-F converters.

NEC 665-64 (a) Ex Converter output circuit to be isolated except where DC could exist on failure, then DC grounded.

OUTPUT CIRCUITS

NEC 665-24 (General Rule) Capacitors to have bleeder resistors as per 460-6 (a).

NEC 665-25 Work applicator to be guarded.

NEC 665-26 Grounds or bonds to limit stray R-F potentials.

NEC 665-66 "Coupled" low-voltage control circuits limited to 150 volts.

NEC 665-60 Equipment other than motor-generators includes all static multipliers and oscillator-type units utilizing vacuum tubes and/or solid-state devices and capable of converting AC or DC to AC frequency for induction and/or dielectric heating use.

NEC 665-68 (a) "Local-Remote" switches to be interlocked; (b) foot switches to be shielded from accidental contact.

INDUCTION AND DIELECTRIC HEATING EQUIPMENT

NEC 665-64 Output circuit includes all output parts external to converting device including transformers, contactors, busbars, and other conductors, and shall comply with 665-64 (a) and (b).

NEC 517-71(a)(b) Equipment to be permanently installed unless portability essential. Where portable, a 3-wire hard-service type cord with insulated, green, or green with yellow stripe(s) equipment grounding wire required. Plug to be as per 250-59 (b).

NEC 665-20 Converting apparatus to be enclosed with noncombustible material.

NEC 665-22 Doors and detachable panels giving access to more than 500V to be interlocked with circuit-interrupting device.

NEC 550-2 A mobile home is a transportable factory-assembled structure or structures, constructed on a permanent chassis, for use as a dwelling. A mobile home is not constructed on a permanent foundation but is connected to the required utilities. The term "mobile home" includes manufactured homes.

NEC 550-8(b)(e) At least one bathroom and one outdoor receptacle required. Outdoor receptacle permitted in compartment of home if accessible from outside. All 15- and 20-amp, 120-volt, outdoor and bathroom receptacles to be on GFCI.

Attachment-plug-cap configurations used for mobile home supply cords and mobile home parks. See Figure 550-5(c).

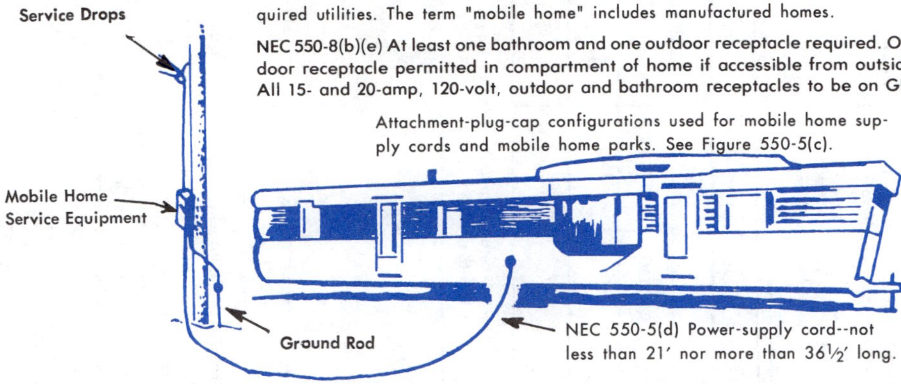

NEC 550-5(i) Where load exceeds 50 amps, or permanent feeder used, then supply shall be (1) mast weatherhead installed, as per 230, and contain 4 continuous, insulated, color-coded feeder conductors, or (2) raceway from home disconnect to under home with provisions to attach suitable junction box or fitting on underside of house.

NEC 550-4(c) Rules of Article 550 apply when 120/240-volt, 3 wire used for service. (Also see 550-4(a))

NEC 550-6(a) Disconnecting means required inside home; neutral strip to be insulated.

NEC 550-7(a) Outside dimensions to be used in calculating minimum lighting load (3 VA per square foot).

NEC 550-10 (a–k) Nonmetallic-sheathed cable to be protected if exposed less than 15″ from floor; if rigid conduit, double-locknuts required. Slack to be left in range cable. Under-chassis wiring is to be in rigid conduit or IMC. EMT or PVC if in protected area.

NEC 550-13 See 550-13(a) for calculating lighting and small appliance load. See 550-13(b) for total power supply load.

NEC 550-13(c) This section permits use of 220-30 and Table 220-30 for calculating lighting and appliance load in mobile homes.

NEC 550-5 (a through i) Service equipment shall be located outside but adjacent to home. Power supply to home shall be one approved 50-amp mobile home power-supply cord, or a permanently installed circuit, except if home has oil or gas-fired heating and cooking, then 40-amp cord acceptable. Cord to be secured to distribution panelboard with suitable cord clamp and have 4 wires, one to be the grounding conductor and to be green or green with yellow stripe(s). Cord 21 to 36½ feet and have a molded attachment plug cap, 125/250 V, 50 A, 3-pole, 4-wire grounding type, as per Figure 550-5(c). Cord shall enter home in exterior wall, floor, or roof and have suitable protection where entering.

NEC 550-8 (a)(c)(d) Plug receptacles shall be of grounding type. Receptacles shall be installed in all rooms other than the bath, closet, and hall areas.

NEC 550-11(a, b, c) No equipment shall be grounded to the neutral wire inside the mobile home. All exposed metal parts shall be bonded to grounding terminal or panel enclosure. Bond required (No. 8 copper or equal not exposed to physical damage) between panel and accessible terminal on chassis. Panel ground connected from service ground. Metallic (gas, water, etc.) pipes may be grounded by connecting to terminal on chassis.

550-12(a)(b) Wiring of each mobile home shall be tested for (1) dielectric strength, (2) continuity and operational tests, (3) polarity checks.

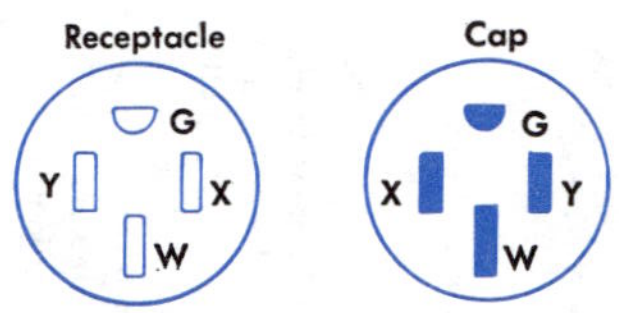

125/250-Volt, 50-Amp.
3-Pole, 4-Wie, Grounding Type

The receptacle and attachment-plug-cap configurations shown here are used for mobile home supply cords and mobile home parks and are required by 550-5(c). Details can be found in ANSI/NEMA WD6-1989, Figure 14-50.

MOBILE HOMES

NEC 551-3 When requirements … requirements of 551 shall prevail.

NEC 551-2 Definition of RECREATIONAL VEHICLE: A vehicular type unit which is self-propelled or is mounted on or pulled by another vehicle and is primarily designed as temporary living quarters for camping, travel, or recreational use. Normally identified as Travel Trailer, Camping Trailer, Truck Trailer, Truck Camper, and Motor Home.

NEC 551-2 Travel trailer is: mounted on wheels, has a living area less than 320 square feet (excluding wardrobes, closets, cabinets, kitchen units, fixtures, etc.) is of such size and weight so that a special highway use permit is not required, and is designed as temporary living quarters while camping or traveling.

NEC 551-46(e) Power-supply assembly located with 15′ of rear, left (driver or road) side, or at rear, left within 18″ of outside wall. See Ex. 1 and 2.

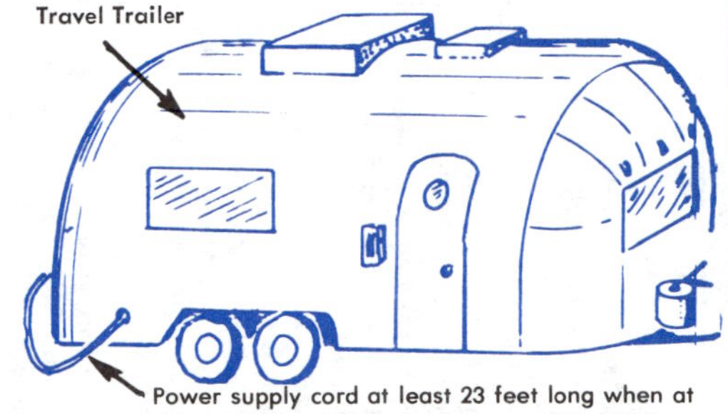

NEC 551-41(c) Each 120 V, 1ϕ, 15 A or 20 A receptacle outlet shall have GFCI where installed in following locations: (1) adjacent to bathroom lavatory (2) within 6′ of any lavatory or sink; (3) area used for toilet, toilet and/or shower, or toilet and tub-shower enclosure; (4) on exterior of vehicle except if located in access panel. See Exception.

NEC 551-10(a) Low-voltage circuits installed by manufacturer subject to NEC® except automotive vehicle circuit lights subject to Federal and State regulations but not lower than code.
NEC 551-10(b) Low-voltage wiring shall be: (1) copper except metal chassis or frame permitted as return path if connections are accessible and are copper and mechanically secure; (2) Type HDT, SGT, SGR, SLX or insulated as per Table 310-13 or equal. Conductors No. 6 through 18 shall be listed; (3) single wires to be stranded; (4) wires to be marked on surface every 4 ft. and be listed.
NEC 551-52 All receptacle outlets shall be grounding type and installed as per 210-7 and 210-21.
NEC 551-42 RV with 120 V system for lights, receptacles, and appliances: (a) 15 amp supply if one 15 amp circuit, (b) 20 amp supply if one 20 amp circuit, (c) 30 amp supply if two or more 15 amp or 20 amp circuits. [See 210-23(a) for permitted loads; (See 551-45(c) for disconnect and O.C. protection.)] 551-42(d) covers 40 or 50 amp supply.
NEC 551-54(a) Grounding conductor in power supply shall be connected to grounding bus or approved grounding means in distribution panelboard.
NEC 551-55(a) All exposed metal parts, enclosures, frames, light fixture canopies, etc., shall be bonded to grounding terminals or enclosure of distribution panelboard.
NEC 551-40(a) Equipment and material indicated for connection to system rated 120 volts, 2-wire with ground, or 120-240 volts, 3-wire with ground, shall be approved and installed as per NEC 551, Part A.

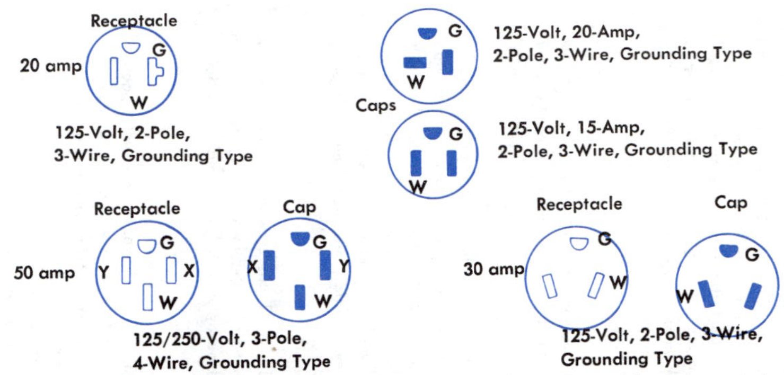

The receptacle and attachment-plug-cap configurations shown here are used for recreational vehicle supply cords and recreational vehicle parks and are required by NEC 551-46(c). Details can be found in ANSI/NEMA WD6-1989, Figure 14-50.

RECREATIONAL VEHICLES

NEC 680-6 (b) (2) Lighting outlets and fixtures, installed between 5 and 10 feet horizontally from walls of pool, to be protected by GFCI unless at least 5 feet above maximum water level and rigidly mounted.

NEC 680-6 (b) (3) Where cord-connected lighting fixtures are installed within 16 feet of nearest water surface, they shall comply with 680-7.

NEC 680, Sections 4, 11, 22 (a-5), 25 (c), and 26 give definitions and list specific requirements for electrically operated pool covers.

NEC 680-20 (a-1) Underwater fixture, during normal use (not relamping) shall be free from shock hazard by design. If operating at over 15 volts, GFCI required to eliminate shock hazard during relamping. If supplied by transformer, as per

NEC 680-20 (b-3) Noncurrent parts of wet-niche fixtures shall be grounded to forming shell by a positive locking device and requires tool to remove fixture from forming shell

NEC 680-20(b-1, 2) Wet-niche underwater fixtures shall have approved forming shell with threaded conduit entries. Rigid metal conduit or IMC of brass or similar metal, liquidtight flexible nonmetallic conduit, or rigid nonmetallic shall extend from forming shell to suitable junction box or enclosure located per 680-21. Where PVC used, a #8 insulated, Cu wire required in PVC with provisions to terminate in junction box or transformer or GFCI enclosure. Termination of #8 in forming shell to be protected by covering or encapsulating with listed potting compound. Where flexible cord is used, end of cord jacket and wire terminals in forming shell to be similarly protected.

680-2 All electric equipment installed in, on, or associated with pools, fountains, and similar installations shall conform to 680.

NEC 680-22(a) Bonding. The following parts shall be bonded: All metal parts of electric pool water handling equipment and other metallic elements within or attached to pool structure, including reinforcing rods, coping stones, deck, piping systems, ladders, rails, diving board platforms, drains, skimmers, refill pipes, forming shells and all other metal piping or fixed metal parts within 5 feet of inside walls of pool or 12 feet above the maximum water level of the pool or any observation stands, towers, or platforms (unless separated from pool by permanent barrier). Except: (1) tie wires acceptable as bond re-rod; (2) common bonding grid may be re-rod; (3) isolated metal parts, no more than 4 inches in any dimension and do not penetrate into pool structure more than 1 inch, need not be bonded.

NEC 680-6(b)(1) Lighting outlets and fixtures and ceiling fans not to be installed over pool or over area 5′ horizontally from inside of pool, unless at least 12′ above maximum water level. Ex. 1 Existing located less than 5′ horizontally from walls of pool and at least 5′ above maximum water level, rigidly mounted, and GFCI protected. See Ex. 2 for indoor pool areas.

NEC 680-22 (b) The parts required to be bonded by 680-22 (a) shall be connected to a common bonding grid, with not less than No. 8 solid copper wire, insulated, covered or bare, and connected by pressure connectors or clamps of brass, copper, or copper alloy. The common bonding grid is permitted to be: (1) reinforcing steel with usual tie wires; (2) wall of bolted or welded metal pool; (3) No. 8 or larger solid copper wire, insulated, covered or bare.

NEC 680-22 (c) Pool heaters over 50 amps to be grounded and bonded, as per specific instructions, as to which parts to be grounded and bonded.

NEC 680-20(a)(3) Pool lights in walls so top of lens at least 18″ below normal water level. Ex. lights identified for purpose need be only 4″ below normal water level.

NEC 680-6(a)(1), Ex. Receptacles on property required to be at least 10 feet from inside walls of pool. Where pool installed for existing dwelling, at least one receptacle required between 10 and 20 feet of pool. All receptacles between 10 and 20 feet of pool to be protected by GFCI. A single, locking and grounding type receptacle for a permanent water-pump motor as per 680-7 permitted between 5 and 10 ft. from pool. All 120-volt receptacles within 20 feet of pool to have GFCI. (See 210-8(a)(3).)

NEC 680-25(f) Other electrical equipment to be grounded as per Art. 250 and wiring as per Chapter 3.

NEC 680-21 (a-2, 3) Junction boxes to underwater pool-light forming shell shall be of copper, brass, suitable plastic, or other approved material, and provide continuity by means of copper, brass, or other approved corrosion-resistant metal that is integral with box.

NEC 680-20 (b-1) Metal parts of fixtures or forming shell shall be of brass or other corrosion-resistant metal.

NEC 680-25 (b-3) Flexible cords to wet-niche lighting fixtures shall contain a grounding conductor not smaller than No. 16.

NEC 680-20 (a-2) No underwater lighting fixture may operate at a voltage greater than 150V between conductors.

Lighting Panel

NEC 680-8 Proper clearances shall be maintained between open overhead wiring and swim pools, diving structures, observation stands, towers, and platforms. (See EX. 1, 2)

NEC 680-21 (d) Junction boxes shall have one more grounding terminal than conduit entries.

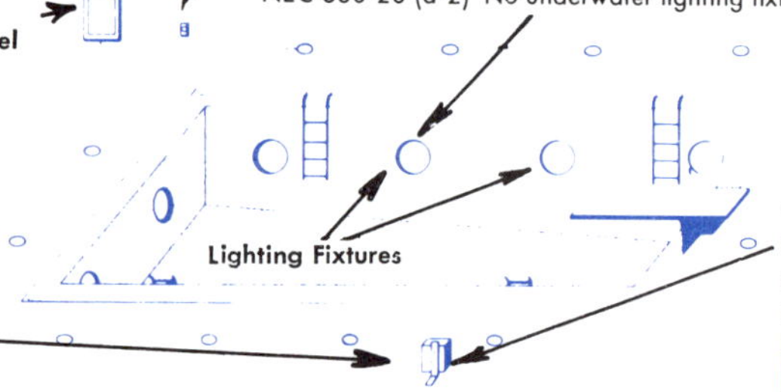

NEC 680-21 (b-4) Enclosures for electrical equipment that extends directly to underwater pool-light forming shell shall be not less than 4 feet from pool, unless separated by barrier and not less than 4 inches from the inside bottom of box above grade level or 8 inches above the maximum water level, whichever is greater.

NEC 680-5 (a) Transformers shall be of two-winding type, with metallic barrier between primary and secondary windings.

NEC 680-25 (b) (1) Conduit contains insulated copper equipment grounding conductor, sized as per Table 250-95; minimum of No. 12.

NEC 680-25 (d) Where equipment ground wire connected to subpanel, then equipment ground to continue to service ground.

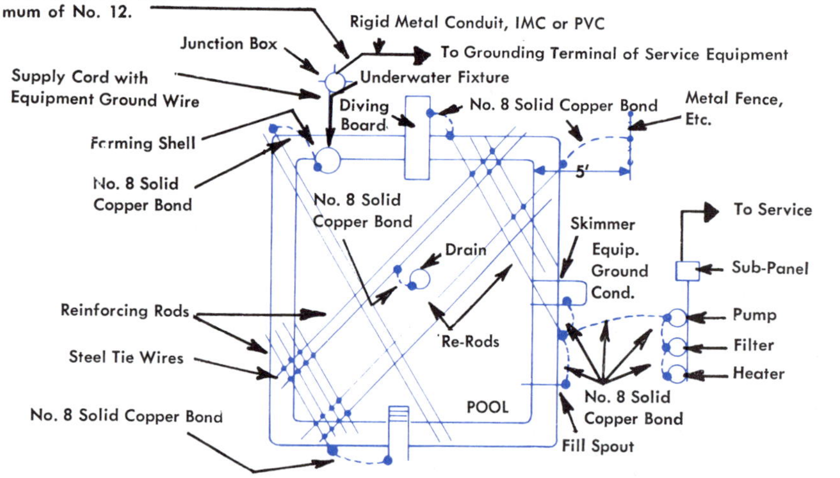

NEC 680-22 (b) Common Bonding Grid may consist of: (1) Structural reinforcing steel rods that are bonded together by usual tie wires; (2) Wall of bolted or welded metal pool; (3) Solid copper wire, insulated, covered or bare, and not smaller than No. 8.

NEC 680-24 Electrical equipment to be grounded. Wet-and-dry-nitche fixtures; equipment within 5 feet of pool; all associated recirculating equipment; junctions boxes; transformer enclosures; GFCI; panels not part of service equipment.

BONDING & GROUNDING PERMANENT SWIMMING POOLS

SWIMMING POOLS--ARTICLE 680

CLEARANCE	VOLTS		
Service Drop or Overhead Wiring	0-750 V	750-15 kV	15-50 kV
A = To Water	18′	25′	27′
B = To Diving Platform	14′	16′	18′
C = Horizontal from Inside Pool Walls	Limit Extends from Outer Edge of Above But Not Less Than 10 Feet		

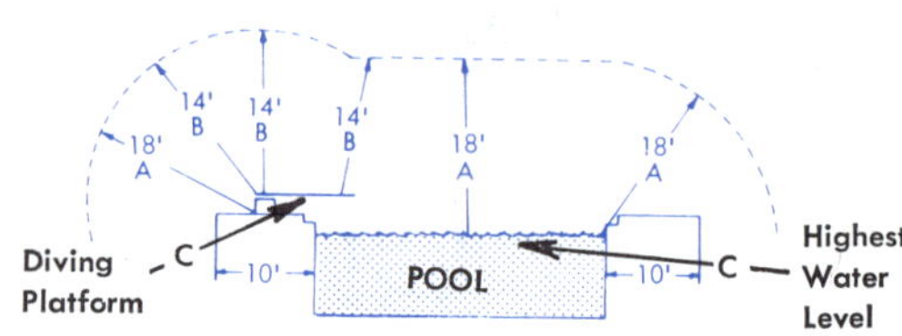

Overhead conductor clearances at 0-750 volts to ground.

NEC 680-8 Following parts of swim or wading pools not permitted under existing service drops or other overhead wiring, and new overhead wiring not permitted to be installed above the following: (1) swim or wading pools, including area extending 10 feet horizontally from walls of pool; (2) diving structure; (3) towers, platforms, or observation stands. Ex 1, 2 Utility power lines to have clearances as per above table. Utility communications and CATV to have 10 ft minimum clearance from above Table areas.

NEC 605-2 Partitions not permitted to extend from floor to ceiling. Wiring systems providing power for appliances and lighting accessories in wired partitions to be identified as suitable for such use.
NEC 605-2 (b) Code requirements to apply except where specifically modified by this article.

NEC 605-8(a) No. 12 extra-hard usage power supply cord with insulated grounding conductor not over 2 ft. long.

NEC 605-6 Partitions secured to building shall be permanently wired by wiring method per NEC Chapter 3.

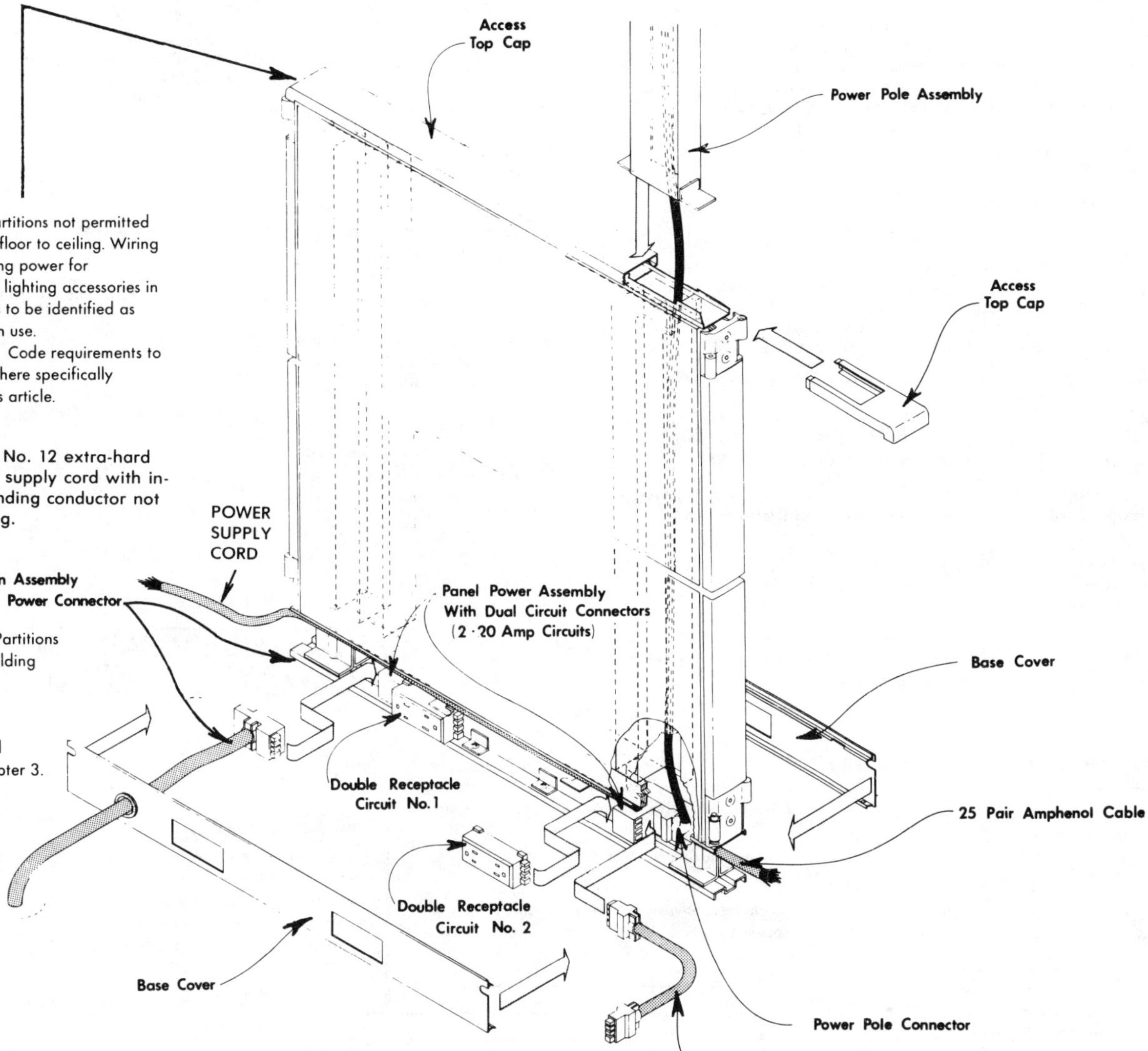

(All-Steel Inc.)

NEC 605-4 A flexible assembly, identified for use with wired partitions, shall be used for the electrical connection between partitions.
Exception: Flexible cord permitted to interconnect partitions when partitions are mechanically contiguous, cord is extra-hard usage type, cord is as short as possible, but not longer than 2 feet, and cord terminates in an attachment plug and cord-connector with strain relief.

NEC 605-2, Ex. The authority having jurisdiction can permit relocatable wired partitions to extend to the ceiling but not penetrate the ceiling.

NEC 605-1 This article deals with lighting accessories, electrical equipment, and wiring systems that are contained within or installed on relocatable wired partitions.

NEC 605-3 Wireways in partitions containing conductors and connections shall be identified as suitable for the condition of use. (FPN prohibits use of flexible cords).

NEC 605-5 Lighting equipment shall be identified for use with wired partitions and shall: (a) have adequate support; (b) where cord-and plug-connected shall not exceed 9 ft., not smaller than No. 18, contain an equipment grounding conductor and shall be of the hard usage type; (c) lighting accessories shall not contain receptacle outlets.

NEC 605-8 Free standing individual partitions that are not over 30 feet long and are mechanically contiguous are permitted to be cord and plug connected providing: (a) cord is at least No. 12 AWG with insulated grounding conductor of extra-hard usage type and not over 2 feet long, (b) permanent building wiring connection permitted by cord if receptacle is on separate circuit and located not more than 12 inches from partition (c) not more than 13, 15 Amps, 125 Volt receptacles permitted in individual or group of interconnected partitions, (d) multi-wire circuits not permitted.

NEC 770-1 Article 770 covers the installation of optical fiber cables and raceways. It does not cover construction of cables or installation in situations other than those covered in this article.

NEC 770-52(a) Optical fibers permitted in hybrid cables with electric light, power, or Class I circuits of 600 volts or less only where functions of all conductors are associated. Nonconductive cable conductors permitted in same raceway with light, power, or Class I circuits of 600 volts or less. Conductive and hybrid cables not permitted in same raceway with light, power, or Class I circuits. (See Exceptions)

NEC 300-17 Shall apply to conductors in raceways.

770-52(b) Optical fibers permitted in same cable and conductive and nonconductive cables permitted in same raceway or enclosure with any of the following: (1) Class 2 & 3 circuits per 725; (2) Fire protection systems per 760; (3) Communication circuits per 800; and (4) Community antenna TV and radio distribution systems per 820.

NEC 770-52(c) Grounding of conductive members of optical fiber cables per 250.

770-49 Optical fiber cables in buildings to be listed as resistant to the spread of fire in accordance with 770-50 and 770-51.

NEC 770-53(b) Vertical runs of cable in a shaft to be listed as having fire-resistance characteristics to prevent carrying of fire from floor to floor.

750-53(b), Ex. 1 Cables permitted in fireproof shaft having fire stops at each floor or cables enclosed in raceways.

NEC 770-53(a) Listed cables installed in ducts and plenums and other spaces used for environmental air have the same requirements as NEC 300-22 for other wiring methods. See Exception.

NEC 770-3 Conductive members transmit light for signaling, communication, and control through an optical fiber.

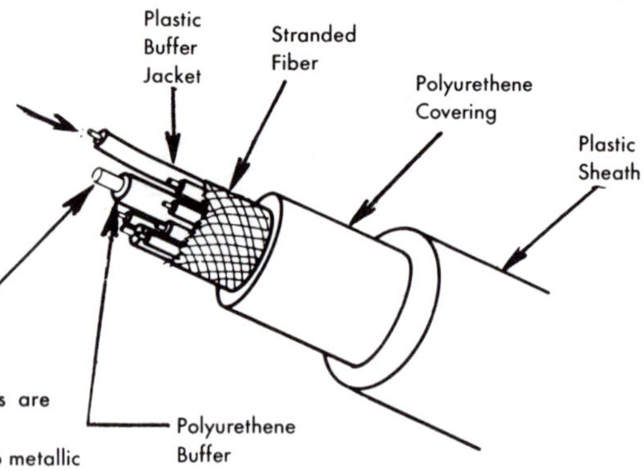

10 conductor with metallic strength member

770-4 Optical fiber cables are grouped into three types:

(a) NONCONDUCTIVE – no metallic members and no electrically conductive materials.

(b) CONDUCTIVE – contain noncurrent-carrying conductive members such as metallic vapor barriers and metallic strength members.

(c) COMPOSITE - contain both optical fibers and electrical conductors and are classified as electrical cables per the type of electrical conductors.

OPTICAL FIBER CABLES AND RACEWAYS — ARTICLE 770

NEC 690 Part H covers storage batteries. 690-71 deals with installation, 690-72 deals with state of charge and 690-73 deals with grounding.

NEC 690-5 Dwelling units with roof-mounted photovoltaic arrays shall be provided with ground-fault protection to reduce fire hazards.

Photovoltaic Source Circuits

NEC 690-2 The photovoltaic source circuit includes conductors between modules and from modules to common junction point(s) of the DC system.

Photovoltaic Output Circuit

NEC 690-2 The photovoltaic output circuits are conductors between photovoltaic source circuits and power conditioning units or DC utilization equipment.

Power Conditioning Unit

690-2 An inverter, also known as a power conditioning unit or power conversion system, is used to change DC input into AC output. Inverters on some systems may take AC from a generator and convert to DC for purpose of charging batteries.

NEC 690-2 An interactive solar photovoltaic system operates in parallel with and is permitted to deliver power to a normal utility service source connected to the same load.

NEC 690-61 When normal utility service is interrupted, the interactive power conditioning unit shall be automatically disconnected and shall not reconnect until the utility power is restored.

Module

NEC 690-2 A module is the smallest complete, environmentally protected assembly of solar cells, exclusive of tracking, designed to generate DC power under sunlight.

Panel

NEC 690-2 A panel is a collection of modules, secured together, wired, and designed to provide a field-installable unit.

Array

NEC 690-2 An array is an assembly of modules or panels mounted on a support structure.

NEC 690-18 A means is required to disable an array or portions of an array to protect persons against shock while installing, servicing, or replacing of array during daylight hours.

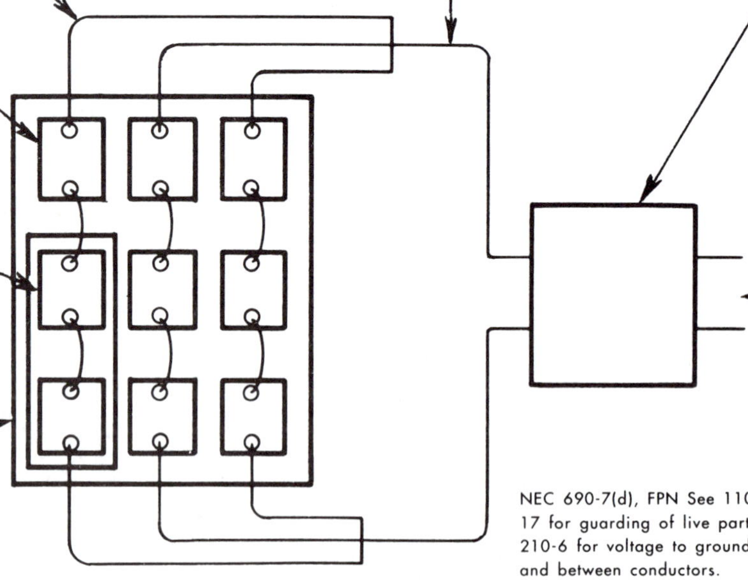

Power Conditioning Unit Output Circuit

NEC 690-1 Either AC or DC output for utilization is permitted.

NEC 690-2 Power conditioning unit output circuit-conductors between the power conditioning unit and the normal electrical service.

NEC 690-7(d), FPN See 110-17 for guarding of live parts, 210-6 for voltage to ground and between conductors.

Solar Photovoltaic Circuits

NEC 690-2 The photovoltaic power source is one or more arrays which generate DC power at system voltage and current.

NEC 690-2 The solar photovoltaic system is used to convert solar energy into electrical energy suitable for connection to a utilization load.

NEC 690-7 (d) In one-family and two-family dwellings, live parts in source and output circuits over 150 volts to ground shall not be accessible, while energized, to other than qualified persons.

SOLAR PHOTOVOLTAIC SYSTEMS — ARTICLE 690

Trade Competency Test 13

Name ______________________ Date ______________________

Section	Answer	
______	T F	**1.** The secondary circuits of 460 V wound-rotor induction motors are permitted to be protected against overload by the motor-overload device.
______	T F	**2.** The primary windings of voltage transformers installed indoors shall normally be protected with fuses.
______	______	**3.** The conductors in a pull box having any dimension over 6′ shall be cabled or ______ in an approved manner.
______	T F	**4.** A transformer is a polyphase or single transformer, identified by a single nameplate.
______	______	**5.** High fire point, liquid-insulated transformers installed indoors shall be installed in vaults if rated over ______ V. A. 5000 C. 25,000 B. 15,000 D. 35,000
______	______	**6.** In NEC 430, the term "in sight" refers to a distance of not more than ______′.
______	______	**7.** The motor disconnecting means shall disconnect both the motor and the ______.
______	______	**8.** The bottom of sign and outline enclosures when accessible to vehicles must have a clearance of ______′ from grade. A. 10 C. 14 B. 12 D. 16
______	______	**9.** Except for polyphase wound-rotor units, AC motors ______ hp and larger must have code letters or locked-rotor amps marked on their nameplates.
______	T F	**10.** Capacitors used with motors increase motor current.
______	______	**11.** The ampacity of supply conductors for a transformer arc welder with a 100% duty cycle and 30 A primary current is ______ A. A. 30 C. 60 B. 37.5 D. 90
______	T F	**12.** A $\frac{3}{4}$ hp, 115 V motor on an automatic air compressor does not need overload protection.
______	______	**13.** Separate running overcurrent protection for continuous-duty motors more than 1 hp shall be based on the FLC, which is listed on the ______ of the motor.
______	______	**14.** Wired office partitions may extend from floor to ______ with permission of the authority having jurisdiction.
______	T F	**15.** The rated load current of a sealed (hermetic) motor-compressor shall be marked on either the equipment or the nameplate.

__________ __________ **16.** The controller may be an attachment plug and receptacle for portable motors rated at _________ hp or less.

__________ __________ **17.** On an ungrounded circuit, _________ thermal overcurrent units are needed to protect a 3-phase induction motor from overload.

__________ __________ **18.** Branch-circuit conductors supplying a 20 hp, 3-phase, 230 V, 40°C induction motor shall be rated at not less than _________ A.

A. 50 | C. 67.5
B. 57.5 | D. 100

__________ __________ **19.** A suitable plug and hard-usage cord can supply a fixed or stationary nonmedical X-ray unit if the branch circuit is rated at no more than _________ A.

__________ __________ **20.** The maximum size of the overload relay used to protect a 50 A, 230 V, 40°C motor is _________ A.

A. 50 | C. 70
B. 57.5 | D. 100

__________ __________ **21.** Fuses used as motor overload protection devices shall be inserted in each _________ conductor.

__________ __________ **22.** The conductors of mineral-insulated, metal-sheathed cable (Type MI) are made of solid _________.

__________ __________ **23.** Flexible metal conduit used in a manufactured wiring system shall contain at least a No. _________ AWG copper grounding conductor.

__________ __________ **24.** The voltage considered in a solar photovoltaic power source is the rated _________ voltage.

__________ __________ **25.** The maximum setting of the primary overcurrent protection device for a 230/460 V, 1-phase, askarel-insulated transformer without secondary protection, with a 40 A primary FLC is _________ A.

A. 40 | C. 80
B. 50 | D. 100

__________ __________ **26.** A general-use snap switch suitable only for use on AC may be used to disconnect a 240 V, 2 hp motor if the FLC rating of the motor does not exceed _________% of the ampere rating of the switch.

__________ __________ **27.** A 3 hp, 3-phase, 240 V squirrel-cage motor with no code letter is started at full voltage. The size of a Type TW branch-circuit conductor shall be not less than No. _________.

A. 8 | C. 12
B. 10 | D. 14

__________ __________ **28.** Unless otherwise designed and marked, busways shall be supported at least every _________′.

__________ __________ **29.** The minimum length of a pull box for a straight-through pull of three No. 4/0 conductors in 2½″ conduit is _________″.

__________ __________ **30.** For stationary motors of 2 hp or less and 300 V or less, a general-use switch may be used as a controller if the rating is at least _________ times the full-load motor current.

__________ T F **31.** Transformers may have overcurrent protection devices in both primary and secondary circuits.

__________ __________ **32.** Optical fiber cables are used to transmit _________ for control, signaling, and communications.

Trade Competency Test 14

Name ______________________ Date ______________

Section	Answer	
______	T F	**1.** An overload in electrical apparatus includes short circuits or ground faults.
______	______	**2.** Control circuit transformers rated less than ______ VA require no additional protection.
______	______	**3.** A 30 hp, 3-phase, 230 V squirrel-cage motor with Code letter B is started with an autotransformer compensator. The initial rating of a dual element fuse for branch circuit protection shall not exceed ______ A. A. 100 C. 150 B. 125 D. 175
______	______	**4.** No underground lighting fixture shall operate at a voltage greater than ______ V between conductors.
______	T F	**5.** An attended self-service station shall have emergency controls installed 150′ or less from dispensers.
______	______	**6.** A remote No. 14 control circuit tapped from the line side of a magnetic starter requires no additional protection if the branch circuit is protected by a(n) ______ A device or less.
______	______	**7.** Sealed refrigeration units have ______ refrigerant motor compressors.
______	______	**8.** In general, a controller used to stop and start a motor shall have a rating equivalent to the ______ of the motor.
______	______	**9.** The maximum overcurrent protection of a No. 6 THHN feeder that supplies two 30 A transformer arc welders rated for continuous duty shall be ______ A.
______	T F	**10.** A nonautomatically started, not permanently installed, 1 hp motor within sight of the controller location shall be permitted to be protected against overload by the branch-circuit, short-circuit, and ground-fault protective device.
______	______	**11.** Overhead 120/240 V conductors crossing a swimming pool shall have a clearance of ______′ above water level.
______	T F	**12.** The overcurrent protection device for a 30 A transformer arc welder shall not exceed 30 A.
______	______	**13.** To satisfy Code requirements, conductors supplying a capacitor with a nameplate current of 50 A shall be rated not less than ______ A. A. 57.5 C. 67.5 B. 62.5 D. 72.5
______	T F	**14.** A 277 V-to-ground circuit is permitted to supply lighting fixtures incorporating mogul-base screw-shell lampholders.
______	______	**15.** The basic rule requires an individual disconnecting means for ______ motor.

__________ __________ **16.** A manually-operated resistance welder draws 50 A according to the nameplate. Circuit conductors are permitted to be chosen on the basis of _________ A.

A. 25
B. 50
C. 75
D. 100

__________ __________ **17.** A busway may be reduced in size provided the smaller busway has a current rating of at least one third of the rating of the main busway overcurrent protection device and does not extend more than _________'.

__________ __________ **18.** Pool heaters rated over _________ A are to be grounded and bonded.

__________ T F **19.** An adjustable overload protection device protecting a 50°C motor with an FLC rating of 20 A shall be selected to trip at not more than 23 A.

__________ __________ **20.** Junction boxes housing conductors for swimming pool equipment shall have _________ more grounding terminal(s) than the number of conduit entries.

__________ T F **21.** The disconnecting means for 460 V capacitors need not open all ungrounded conductors simultaneously.

__________ __________ **22.** The conductors to an adjustable speed drive shall be based on the rated _________ of the power conversion equipment.

__________ __________ **23.** The three types of listed optical fiber cables are the nonconductive, the conductive, and the _________.

__________ __________ **24.** A No. 14 motor-control circuit conductor requires separate overcurrent protection when the control circuit extends beyond the control equipment enclosure and the motor branch-circuit protective device exceeds _________ A.

A. 30
B. 45
C. 60
D. 75

__________ __________ **25.** The disconnecting means for a motor shall have an ampacity of at least _________% of the FLC rating of the motor.

__________ __________ **26.** The minimum height of the door sill or curb in an approved transformer vault is _________".

__________ __________ **27.** The controller can be an attachment plug and receptacle for a(n) _________ motor rated at ⅓ hp or less.

__________ __________ **28.** The FLC rating for a 10 hp, 230 V, 2-phase AC motor is _________ A.

__________ __________ **29.** The minimum size of conductors permitted in a cablebus is No. _________.

A. 1
B. 2
C. 1/0
D. 2/0

__________ __________ **30.** It is mandatory that one of the disconnecting means to a motor be _________ accessible.

__________ __________ **31.** Transformers used near swimming pools shall be provided with a metallic _________ between primary and secondary windings.

__________ __________ **32.** Overcurrent protection for a manually operated resistance welder that draws 50 A shall not exceed a maximum value of _________ A.

A. 150
B. 175
C. 200
D. 250

__________ __________ **33.** The allowable ampacities of wires used in crane circuits is _________ than for general electrical construction work.

__________ __________ **34.** The metal shell of wet niche fixtures shall be _________.

Trade Competency Test 15

Name ______________________ Date ______________________

Section	Answer	
______	______	**1.** Flexible cords used to wire wet niche lighting fixtures shall be equipped with equipment grounding conductor not smaller than No. ______.
______	______	**2.** A stop-and-start station that can be locked open ______ be used as a disconnecting means.
______	T F	**3.** A 1/40 hp, split-phase, automatically started motor does not require running overcurrent protection if protected by the branch-circuit, short-circuit, and ground-fault protective device.
______	______	**4.** Conductors rated over 15,000 V shall enter metal-enclosed switchgear or be installed in a transformer ______.
______	T F	**5.** Motor overload protection shall never be shunted during the starting period.
______	______	**6.** Motors with no starting current problems and with a temperature rise of 50°C shall have overload protection provided at ______% of nameplate current.
______	T F	**7.** Replacement parts may be stored in transformer vaults if at least 3′ from operating equipment.
______	______	**8.** Lightning arresters installed indoors shall be located so they are ______ only to qualified personnel.
______	T F	**9.** The second largest X-ray unit on a feeder, in a health care facility, has a demand factor of 150% of the momentary rating.
______	______	**10.** Bare conductors and busbars rated at over 600 V are permitted to be accessible where only ______ persons have access.
______	T F	**11.** Mineral-insulated, metal-sheathed cable is not approved for continuously moist locations.
______	______	**12.** The disconnecting means for a stationary AC motor need not be horsepower rated if the motor is larger than ______ hp. A. 25 C. 75 B. 50 D. 100
______	T F	**13.** A motor branch circuit, short-circuit protection device consisting of an NTD fuse shall never have a rating in excess of 400% of the full-load motor current.
______	______	**14.** The smallest stationary motor rated 300 V or less requiring a horsepower-rated switch is ______ hp. A. ½ C. 2 B. 1 D. 3
______	______	**15.** Foreign pipes or ducts shall not pass ______ a transformer vault.

_______ T F **16.** Infrared heating units may be supplied from a 50 A branch circuit in an industrial occupancy.

_______ _______ **17.** Raceways routed in _______″ of concrete or brick are considered to be outside the building.

_______ _______ **18.** The secondary conductors of a 2-wire to _______ transformer can be protected by the primary overcurrent protection device.

_______ T F **19.** Single-phase conductors supplying a phase converter shall have an ampacity of not less than 125% of the 1-phase input FLC rating.

_______ _______ **20.** A feeder circuit supplying power to a mobile home shall be provided with a(n) _______ conductor.

_______ T F **21.** An inherently short-time intermittent duty motor usually does not require separate running overload protection.

_______ _______ **22.** The grounded ends to a midpoint grounded transformer shall have conductors not smaller than No. _______.

A. 14 C. 18
B. 16 D. none of the above

_______ _______ **23.** The feeder circuit to a number of resistance welders shall not exceed _______% of the conductors' rating.

_______ T F **24.** In some cases, overload protection for a 50 A continuous duty motor may be set at 70 A.

_______ _______ **25.** The ampacity of branch-circuit conductors supplying power to X-ray equipment shall be _______% of the momentary rating or 100% of the long time rating, whichever is greater.

_______ _______ **26.** Enclosed overcurrent protection devices for services rated over 600 V are not required to be derated _______% for continuous operation.

_______ _______ **27.** Secondary open circuit voltage of 7500 V or less shall have an output not exceeding _______ mA.

A. 240 C. 280
B. 260 D. 320

_______ _______ **28.** All metal enclosures within the walls of a swimming pool shall be bonded to the grid with a common conductor No. _______ or larger in size.

_______ _______ **29.** A resistance welder with a 50% duty cycle has a multiplier of _______%.

_______ _______ **30.** Each welder is required to have a disconnect either at the welder or in the _______.

_______ _______ **31.** Dry-type transformers rated over _______ V, primary or secondary, are required to be installed in a vault.

_______ _______ **32.** For capacitors rated over 600 V, the discharge means is required to be reduced to 50 V or less in _______ minutes or less.

_______ _______ **33.** The metallic shielding of high-voltage cables shall be _______.

_______ _______ **34.** Control conductors shall be used with ampacities rated for _______°C.

A. 60 C. 90
B. 75 D. 120

_______ _______ **35.** Conductors from lightning arresters to line and ground should be kept as _______ as possible.

Hazardous Locations 5

CLASSES AND DIVISIONS

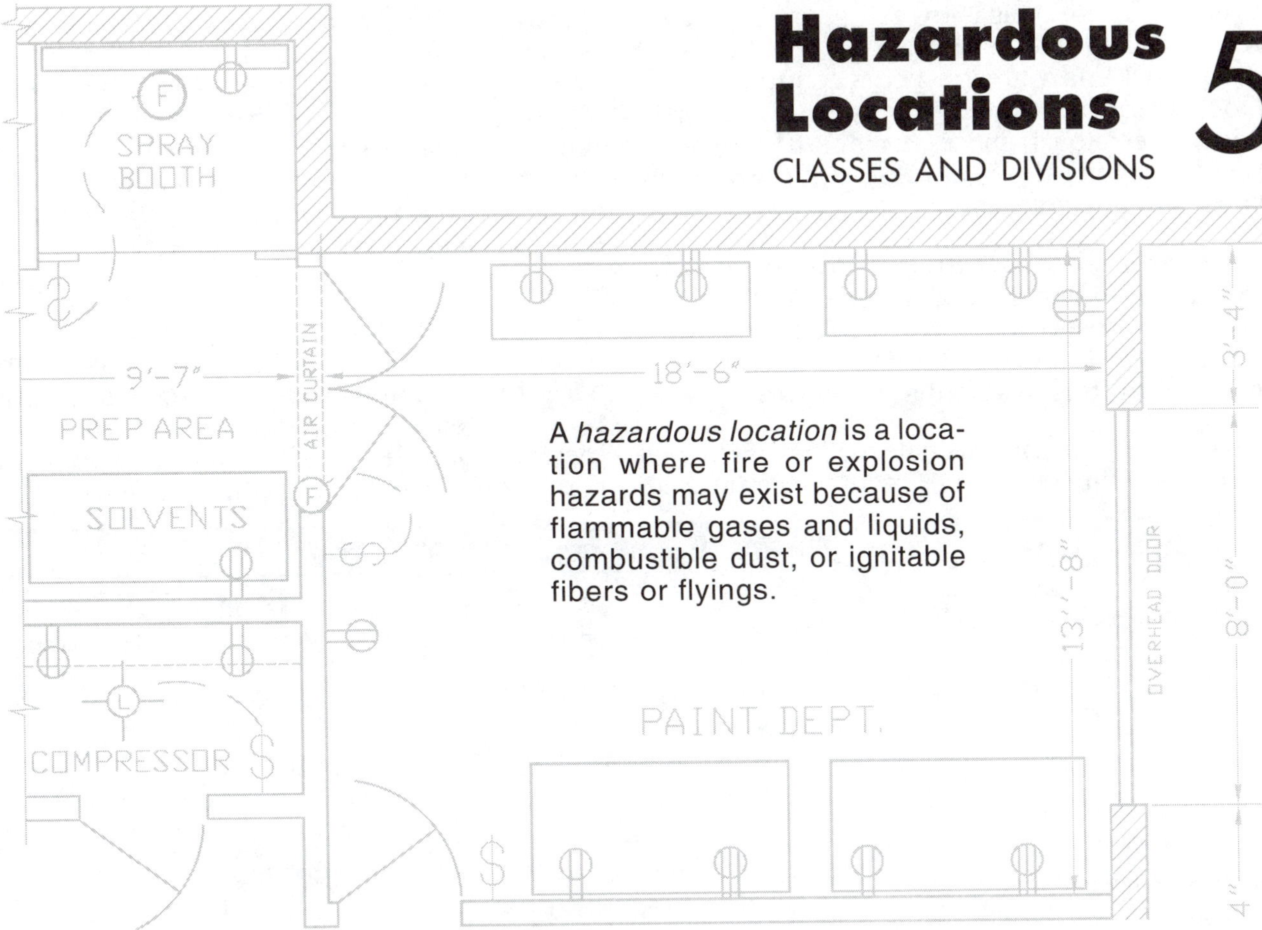

A *hazardous location* is a location where fire or explosion hazards may exist because of flammable gases and liquids, combustible dust, or ignitable fibers or flyings.

Classes

This chapter is concerned with locations that, because of their nature, call for higher standards with respect to materials or for methods of installation, than are thought necessary in the general run of electrical wiring. The NEC classifies hazardous locations under three headings: Class I, Class II, and Class III. Class I concerns activities that may cause fire or explosion from gases or vapors. Class II concerns the fire or explosion risk from dusts. Class III concerns the dangers of combustible fibers.

Each class has two divisions. In Division 1 locations, the particular danger is imminent at any or all times during the normal course of operation. In Division 2 locations, the danger is not present under normal conditions, but is likely to arise from a reasonable, foreseeable, accidental occurrence. For example, a building that houses a machine for compressing acetylene gas falls under Class I, Division 1. If high-pressure mains from this machine pass into another building where shut-off valves and pressure-reducing instruments are inserted in the lines, the adjacent building falls under Class I, Division 2.

Plants where combustible or electrically conductive dusts are present are in Class II, Division 1. Locations where readily ignitible fibers are merely stored or handled, other than in the actual process of manufacture, are in Class III, Division 2. This classification applies mostly to warehouses where manufactured goods, in bales, are held for shipment to consumers.

Methods for Reducing Hazards

The NEC requirements for proper seals in Class I, Division 1 and 2 locations is of utmost importance. Seals, when required, are installed within 18″ of an enclosure for switches, relays, circuit breakers, fuses, or other equipment that could produce arcs, sparks, or high temperatures. The purpose of these seals is to restrict an explosion to one enclosure. Seals are also required in Class I, Division 1 and 2 areas when a conduit run of 2″ or larger enters an enclosure or fitting that contains terminals, splices, or taps. This seal is provided to limit the total volume that could be exposed to an explosion.

Seals are also required in conduits passing from a hazardous to a nonhazardous location. The purpose of this seal is to prevent an explosive mixture from being communicated, through the conduit, from a hazardous to a nonhazardous area.

In Class II areas, an alternate method of sealing is provided that permits a horizontal raceway not less than 10′ long or a vertical raceway, extending not less than 5′ downward from a dust-ignition proof enclosure. All sealing fittings and compound shall be approved for the purpose. At the present time, Underwriter's Laboratories lists a sealing fitting, but does not list a compound except when a compound is used with the explosionproof seal fittings of a specific manufacturer.

Intrinsically safe equipment and associated wiring that is approved for a specific location is permitted in hazardous areas. This equipment is such that under normal or abnormal conditions, it is not capable of releasing suf-

HAZARDOUS LOCATIONS

MATERIAL	CLASS I		CLASS II			CLASS III	
	Division 1	Division 2	Division 1	Division 2	Metal dust	Division 1	Division 2
RACEWAY							
Rigid	✓	✓	✓	✓	✓	✓	✓
IMC	✓	✓	✓	✓	✓	✓	✓
EMT				✓			
Open wiring							✓[e]
BOXES—Fittings							
Threaded	✓	✓	✓	✓	✓	✓	✓
Explosionproof	✓						
Dust-ignition proof			✓[c]		✓		
Dust-tight			✓[d]				
Dust-minimal				✓		✓	✓
FLEXIBLE CONDUIT							
Explosionproof	✓						
Standard							
Extra-hard-usage cord	✓	✓	✓	✓	✓	✓	✓
SWITCHES							
Explosionproof	✓	✓					
General purpose—hermetic[a]		✓					
General purpose—oil[b]		✓					
Dust-ignitionproof			✓		✓		
Dust-minimal				✓		✓	✓

[a]General purpose enclosure—contacts hermetically sealed
[b]General purpose enclosure—contacts immersed in oil
[c]Where joints or terminals
[d]Where no joints or terminals
[e]Storage areas only

ficient electrical or thermal energy to cause ignition of a specific explosive mixture.

The NEC recommends that wherever possible, electrical equipment for hazardous locations be located in less hazardous areas. It also suggests that by adequate, positive-pressure ventilation from a clean source of outdoor air, the hazards may be reduced, or hazardous locations may be limited or eliminated. In many cases, the installation of dust collection systems can greatly reduce the hazards in a Class II area.

Rigid metal conduit, IMC, or Type MI cable are prescribed as the only acceptable wiring methods in the majority of hazardous locations. Grounding continuity must be assured by means of bonding and not by double lockouts or locknut and bushing. Fittings must be explosionproof, dust-ignition proof, or dust-tight, depending on the nature of the hazard.

Motors in the scope of Class I, Division 1 must be explosionproof. Under Class I, Division 2, standard motors can be used if they have no sliding contacts. Motors in Classes II and III must be totally enclosed or pipe-ventilated. Flexible cords of the three class areas shall be approved for extra hard use.

Specially Listed Occupancies

The NEC considers all areas up to a level of 18″ in a commercial garage as Class I, Division 2 locations. A commercial garage where only exchange of parts and routine maintenance work are performed is considered a nonhazardous location. Aircraft hangars are given stricter, but similar protection. Space in the immediate vicinity of a gasoline dispensing island is Class I, Division 2 to the top of the dispenser enclosure extending 18″ horizontally to finished grade. Surrounding territory within a radius of 20′ of the island is Class I, Division 2 to a height of 18″ above grade. Bulk storage plants for gasoline are subject to comparable restrictions.

Finishing processes, where paint, lacquer, or other flammable coatings are applied by spraying, dipping, brushing, or similar means, must be carried out under prescribed safeguards for electrical wiring and equipment. Of particular interest here are electrostatic units that incorporate high-voltage transformers.

Anestheizing locations of hospitals are Class I, Division 1 to a height of 5′ above floor level. Special rules apply to lighting and surgical equipment in the operating room, isolating transformers being generally required. Gas storage rooms are designated as Class I, Division 1 throughout.

Theaters and movie studios are given adequate consideration in the NEC. Unless theaters are provided with sufficient emergency lighting, there is a likelihood of panic in case of fire. Fires may occur suddenly from temporary wiring used in connection with stage scenery or from projection room equipment. Movie studios also have a threat of fire from temporary wiring, arc lamps, and portable devices.

Radio or television studios and receiving stations must be guarded from the danger of high-voltage electronic wiring systems or equipment.

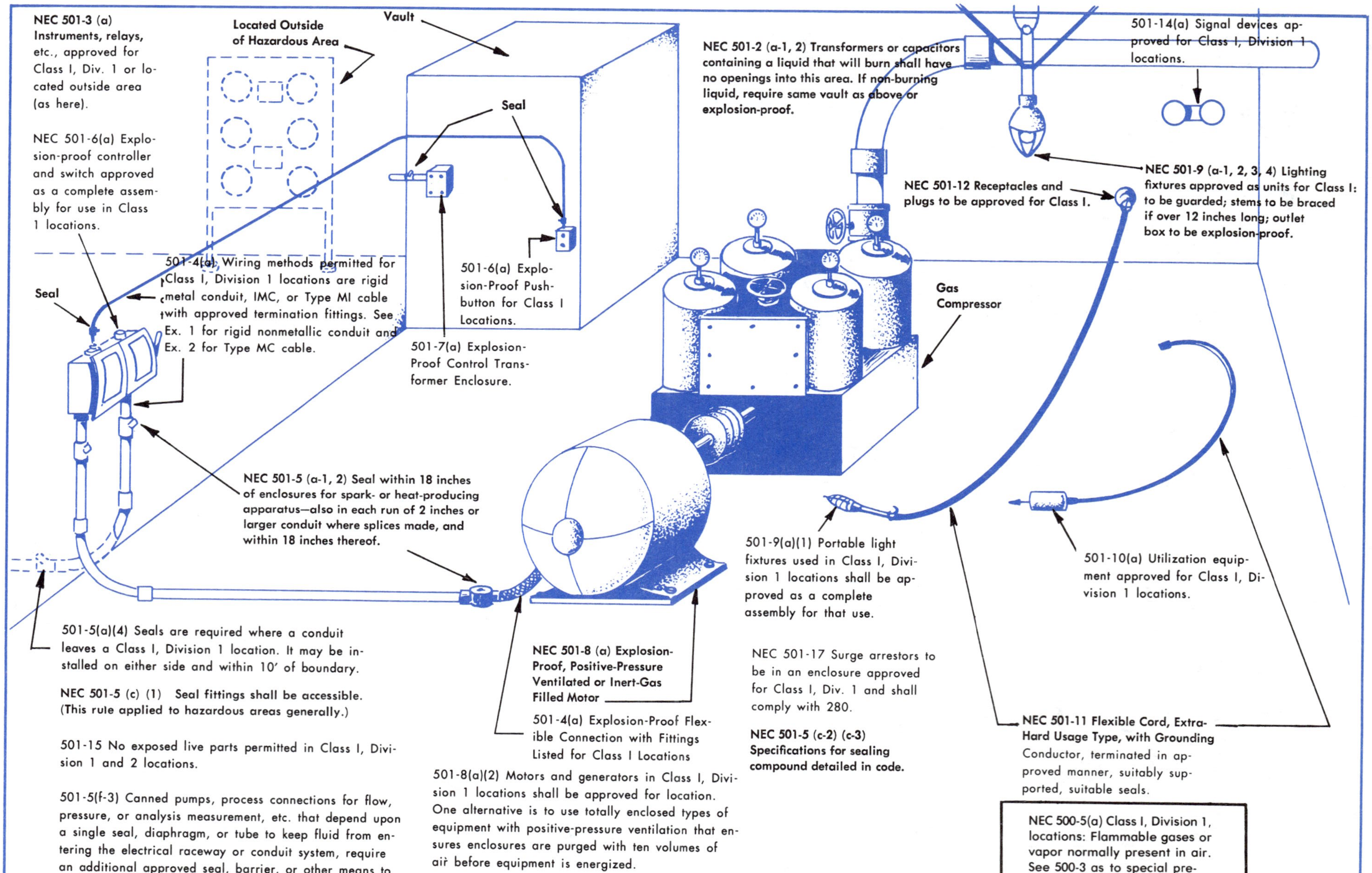

CLASS I, DIVISION 1, LOCATIONS

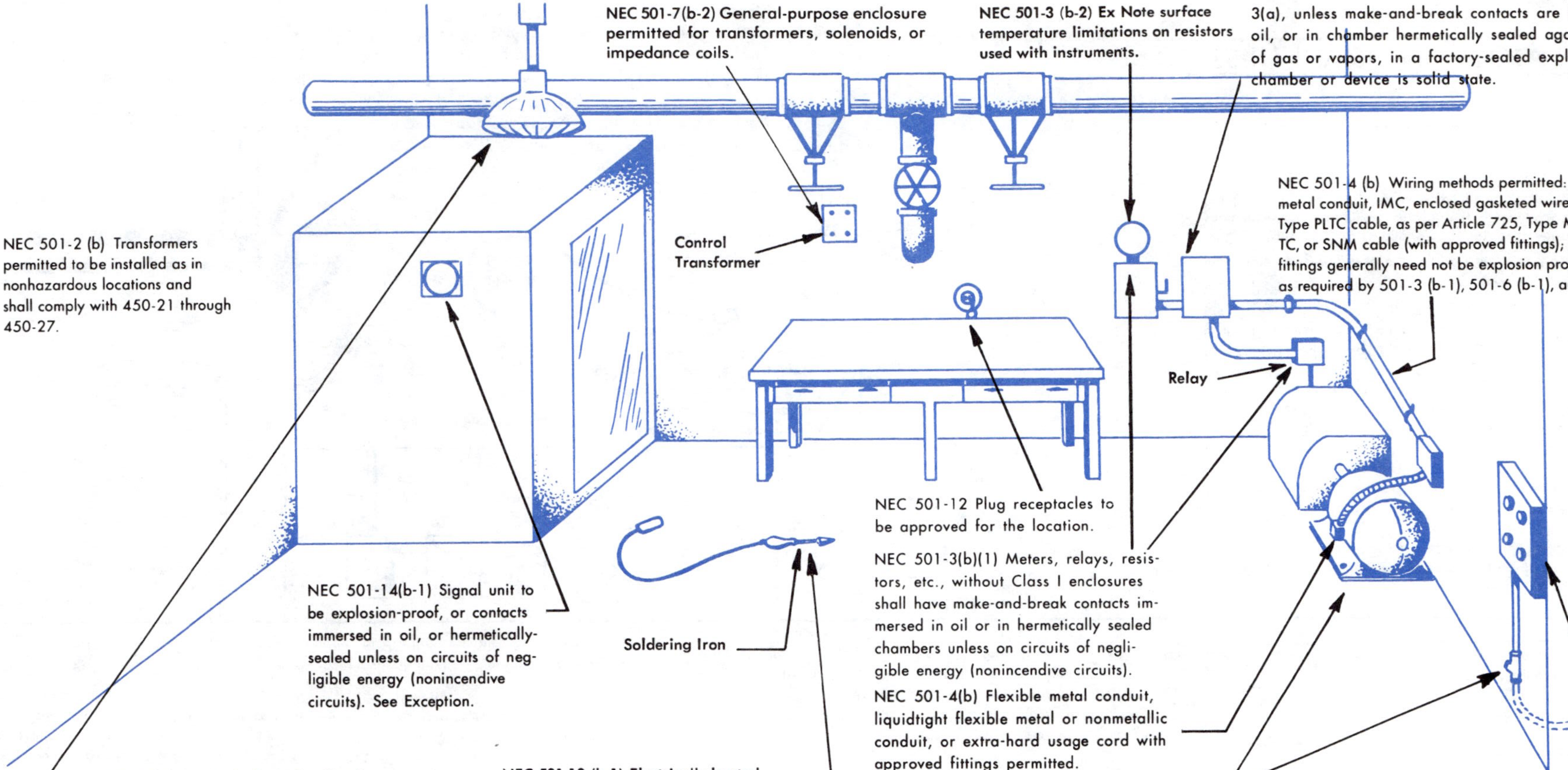

NEC 501-6(b-1) Motor controllers, switches, and CBs shall be listed as Class I, Division 1 equipment, per 501-3(a), unless make-and-break contacts are immersed in oil, or in chamber hermetically sealed against entrance of gas or vapors, in a factory-sealed explosion proof chamber or device is solid state.

NEC 501-7(b-2) General-purpose enclosure permitted for transformers, solenoids, or impedance coils.

NEC 501-3 (b-2) Ex Note surface temperature limitations on resistors used with instruments.

NEC 501-4 (b) Wiring methods permitted: threaded rigid metal conduit, IMC, enclosed gasketed wireways or busways, Type PLTC cable, as per Article 725, Type MI, MC, MV, TC, or SNM cable (with approved fittings); boxes and fittings generally need not be explosion proof, except as required by 501-3 (b-1), 501-6 (b-1), and 501-14 (b-1).

NEC 501-2 (b) Transformers permitted to be installed as in nonhazardous locations and shall comply with 450-21 through 450-27.

NEC 501-12 Plug receptacles to be approved for the location.

NEC 501-3(b)(1) Meters, relays, resistors, etc., without Class I enclosures shall have make-and-break contacts immersed in oil or in hermetically sealed chambers unless on circuits of negligible energy (nonincendive circuits).

NEC 501-4(b) Flexible metal conduit, liquidtight flexible metal or nonmetallic conduit, or extra-hard usage cord with approved fittings permitted.

NEC 501-14(b-1) Signal unit to be explosion-proof, or contacts immersed in oil, or hermetically-sealed unless on circuits of negligible energy (nonincendive circuits). See Exception.

NEC 501-9 (b) (1, 2, 3, 4, 5) Fixed fixtures to have protection from physical damage. Note surface temperature limitation. Portable lamps to be specifically approved for purpose. Pendants to have reinforcing supports if over 12 inches long. Fixture switches to comply with 501-6 (b) (1). Electric discharge starting and control equipment to be as per 501-7 (b), except potted thermal protector in ballast of fluorescent approved for hazardous locations.

NEC 501-10 (b-1) Electrically-heated appliances to be approved for Class I except where low operating temperature on exposed surface.

NEC 501-15 No exposed live parts permitted in Class I, Division 2 locations.

NEC 501-16(b) Locknut-bushing type contacts are not approved for general bonding purposes in hazardous locations. Bonding jumpers are required for flexible metal conduit. External bonding jumpers per 250-79. See Exception.

NEC 501-5 (b-2) Sealing fitting required in conduit run passing to nonhazardous area.

See 501-16 for the rules of bonding and grounding.

NEC 501-8 (b) Standard squirrel cage induction motors may be employed. If motor has sliding contacts, it must be approved for Class I, Division 1, unless contact arrangements especially approved for location.

NEC 501-6 (b) (4) Lighting fuses or circuit breaker (maximum 10 sets), may be general-purpose type (no switching) to protect lamps in fixed positions only.

NEC 501-3 (b-5) Instrument fuses permitted in general-purpose enclosures per 501-3 (b-1, 2, 3, & 4) if not subject to overloading and preceded by a switch conforming to 501-3 (b-1).

NEC 500-5(b) Areas where flammable liquids or gases are handled, processed or used, but are normally contained, and where positive mechanical ventilation is provided or areas where a barrier is provided, but which could become hazardous upon a malfunction shall be classed as a Class I, Div. 2 area.

NEC 500-5(b) Flammable mixtures not normally present, but may appear through mechanical failure or accident.

CLASS I, DIVISION 2, LOCATIONS

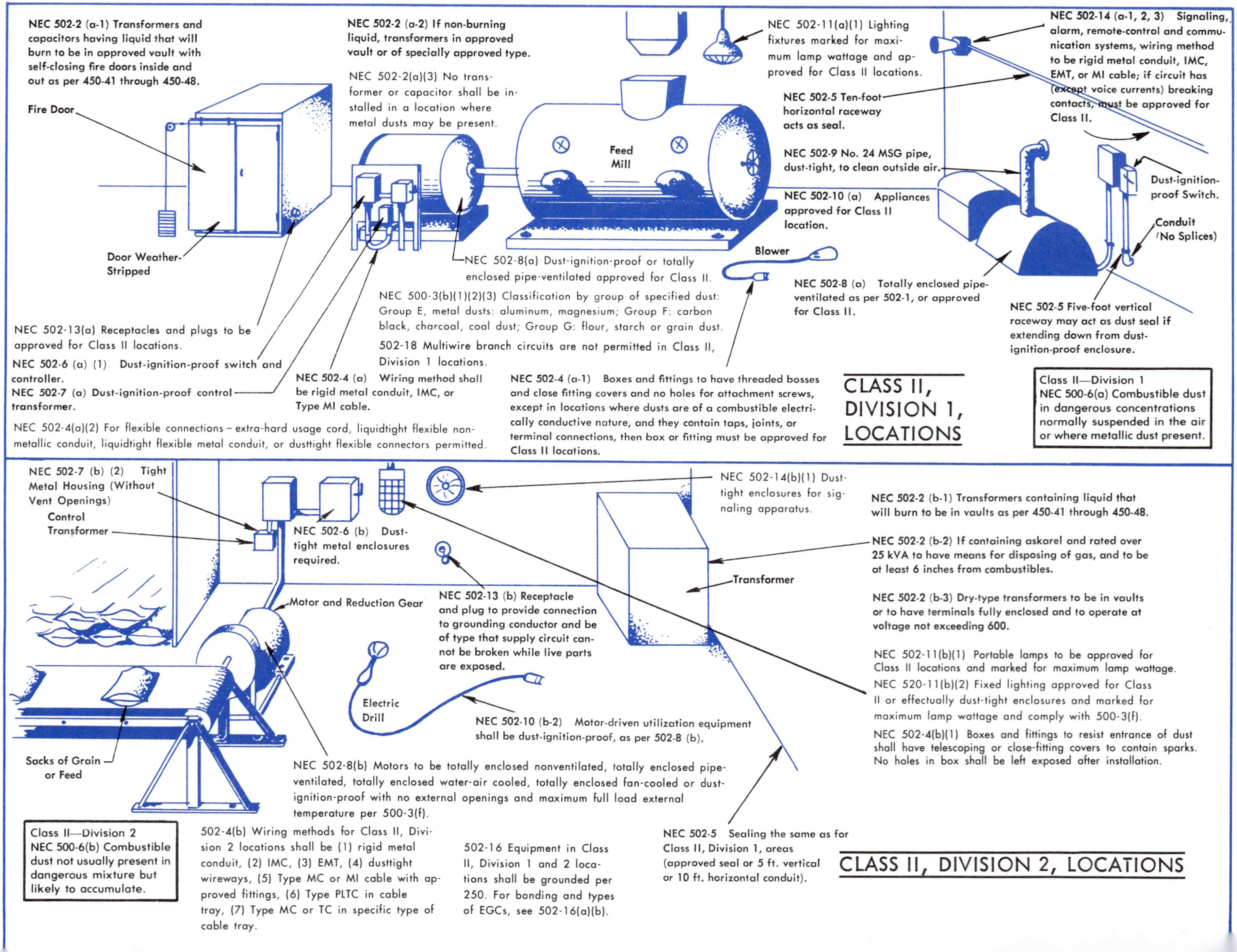
NEC 502-2 (a-1) Transformers and capacitors having liquid that will burn to be in approved vault with self-closing fire doors inside and out as per 450-41 through 450-48.
Fire Door
Door Weather-Stripped
NEC 502-2 (a-2) If non-burning liquid, transformers in approved vault or of specially approved type.
NEC 502-2(a)(3) No transformer or capacitor shall be installed in a location where metal dusts may be present.
Feed Mill
NEC 502-11(a)(1) Lighting fixtures marked for maximum lamp wattage and approved for Class II locations.
NEC 502-5 Ten-foot horizontal raceway acts as seal.
NEC 502-9 No. 24 MSG pipe, dust-tight, to clean outside air.
NEC 502-10 (a) Appliances approved for Class II location.
NEC 502-14 (a-1, 2, 3) Signaling, alarm, remote-control and communication systems, wiring method to be rigid metal conduit, IMC, EMT, or MI cable; if circuit has (except voice currents) breaking contacts, must be approved for Class II.
Dust-ignition-proof Switch.
Conduit (No Splices)
Blower
NEC 502-8(a) Dust-ignition-proof or totally enclosed pipe-ventilated approved for Class II.
NEC 502-8 (a) Totally enclosed pipe-ventilated as per 502-1, or approved for Class II.
NEC 502-5 Five-foot vertical raceway may act as dust seal if extending down from dust-ignition-proof enclosure.
NEC 500-3(b)(1)(2)(3) Classification by group of specified dust: Group E, metal dusts: aluminum, magnesium; Group F: carbon black, charcoal, coal dust; Group G: flour, starch or grain dust.
502-18 Multiwire branch circuits are not permitted in Class II, Division 1 locations.
NEC 502-13(a) Receptacles and plugs to be approved for Class II locations.
NEC 502-6 (a) (1) Dust-ignition-proof switch and controller.
NEC 502-7 (a) Dust-ignition-proof control transformer.
NEC 502-4 (a) Wiring method shall be rigid metal conduit, IMC, or Type MI cable.
NEC 502-4 (a-1) Boxes and fittings to have threaded bosses and close fitting covers and no holes for attachment screws, except in locations where dusts are of a combustible electrically conductive nature, and they contain taps, joints, or terminal connections, then box or fitting must be approved for Class II locations.
NEC 502-4(a)(2) For flexible connections – extra-hard usage cord, liquidtight flexible nonmetallic conduit, liquidtight flexible metal conduit, or dusttight flexible connectors permitted.
CLASS II, DIVISION 1, LOCATIONS
Class II—Division 1
NEC 500-6(a) Combustible dust in dangerous concentrations normally suspended in the air or where metallic dust present.
NEC 502-7 (b) (2) Tight Metal Housing (Without Vent Openings)
Control Transformer
NEC 502-6 (b) Dust-tight metal enclosures required.
NEC 502-14(b)(1) Dust-tight enclosures for signaling apparatus.
NEC 502-2 (b-1) Transformers containing liquid that will burn to be in vaults as per 450-41 through 450-48.
NEC 502-2 (b-2) If containing askarel and rated over 25 kVA to have means for disposing of gas, and to be at least 6 inches from combustibles.
Transformer
NEC 502-2 (b-3) Dry-type transformers to be in vaults or to have terminals fully enclosed and to operate at voltage not exceeding 600.
Motor and Reduction Gear
NEC 502-13 (b) Receptacle and plug to provide connection to grounding conductor and be of type that supply circuit cannot be broken while live parts are exposed.
NEC 502-11(b)(1) Portable lamps to be approved for Class II locations and marked for maximum lamp wattage.
NEC 520-11(b)(2) Fixed lighting approved for Class II or effectually dust-tight enclosures and marked for maximum lamp wattage and comply with 500-3(f).
NEC 502-4(b)(1) Boxes and fittings to resist entrance of dust shall have telescoping or close-fitting covers to contain sparks. No holes in box shall be left exposed after installation.
Electric Drill
NEC 502-10 (b-2) Motor-driven utilization equipment shall be dust-ignition-proof, as per 502-8 (b).
Sacks of Grain or Feed
NEC 502-8(b) Motors to be totally enclosed nonventilated, totally enclosed pipe-ventilated, totally enclosed water-air cooled, totally enclosed fan-cooled or dust-ignition-proof with no external openings and maximum full load external temperature per 500-3(f).
Class II—Division 2
NEC 500-6(b) Combustible dust not usually present in dangerous mixture but likely to accumulate.
502-4(b) Wiring methods for Class II, Division 2 locations shall be (1) rigid metal conduit, (2) IMC, (3) EMT, (4) dusttight wireways, (5) Type MC or MI cable with approved fittings, (6) Type PLTC in cable tray, (7) Type MC or TC in specific type of cable tray.
502-16 Equipment in Class II, Division 1 and 2 locations shall be grounded per 250. For bonding and types of EGCs, see 502-16(a)(b).
NEC 502-5 Sealing the same as for Class II, Division 1, areas (approved seal or 5 ft. vertical or 10 ft. horizontal conduit).
CLASS II, DIVISION 2, LOCATIONS

CLASS III, DIVISION 1, LOCATIONS

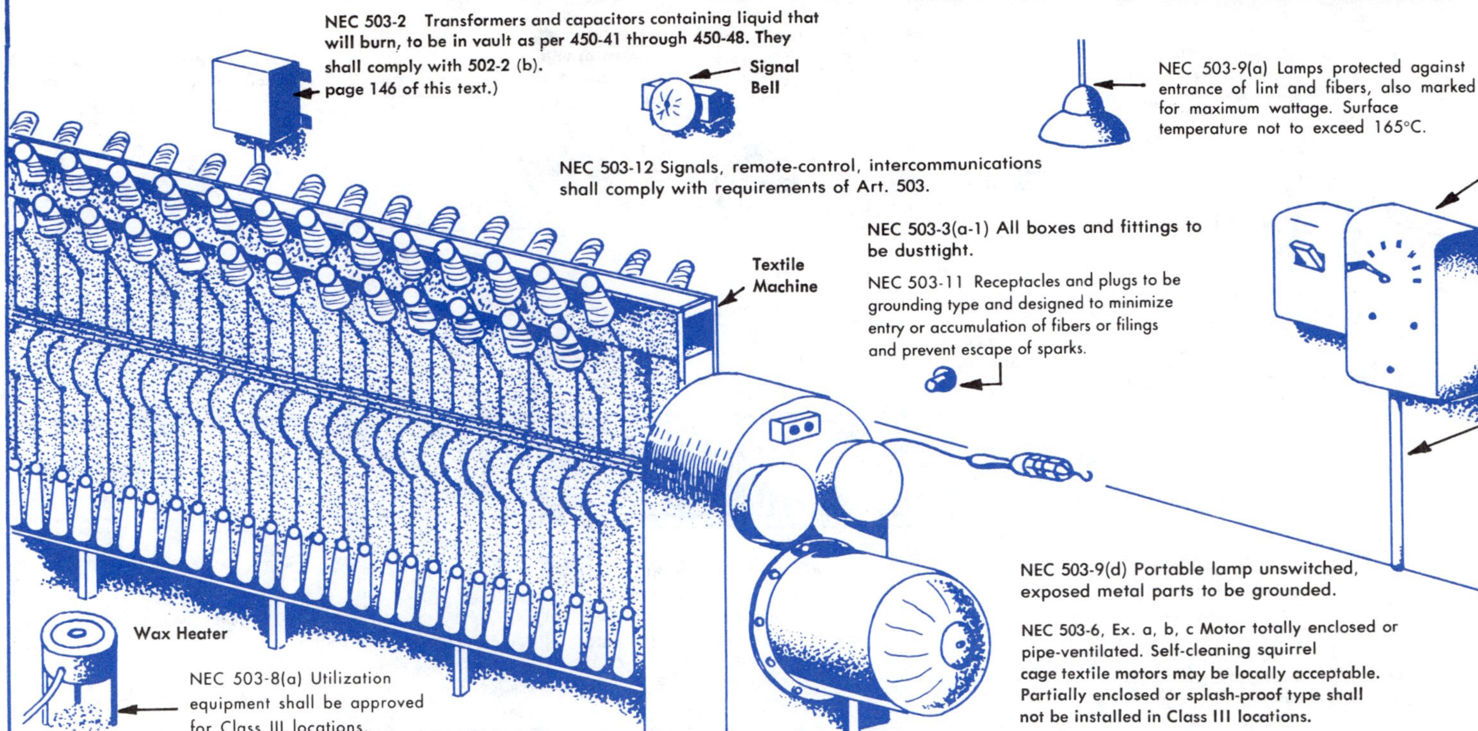

NEC 503-2 Transformers and capacitors containing liquid that will burn, to be in vault as per 450-41 through 450-48. They shall comply with 502-2 (b). page 146 of this text.)

NEC 503-12 Signals, remote-control, intercommunications shall comply with requirements of Art. 503.

NEC 503-9(a) Lamps protected against entrance of lint and fibers, also marked for maximum wattage. Surface temperature not to exceed 165°C.

NEC 503-4 Switches and controllers shall be in dusttight enclosures.

NEC 503-3(a-1) All boxes and fittings to be dusttight.

NEC 503-11 Receptacles and plugs to be grounding type and designed to minimize entry or accumulation of fibers or filings and prevent escape of sparks.

NEC 503-3(a) Wiring method to be: rigid metal conduit, IMC, approved Types MI or MC cables, or rigid nonmetallic conduit, EMT, or dusttight wireways.

Class III—Division 1.
NEC 500-7(a) Location where easily ignited fibers or materials producing combustible flyings are handled, manufactured, or used.

NEC 503-9(d) Portable lamp unswitched, exposed metal parts to be grounded.

NEC 503-6, Ex. a, b, c Motor totally enclosed or pipe-ventilated. Self-cleaning squirrel cage textile motors may be locally acceptable. Partially enclosed or splash-proof type shall not be installed in Class III locations.

NEC 503-8(a) Utilization equipment shall be approved for Class III locations.

CLASS III, DIVISION 2, LOCATIONS

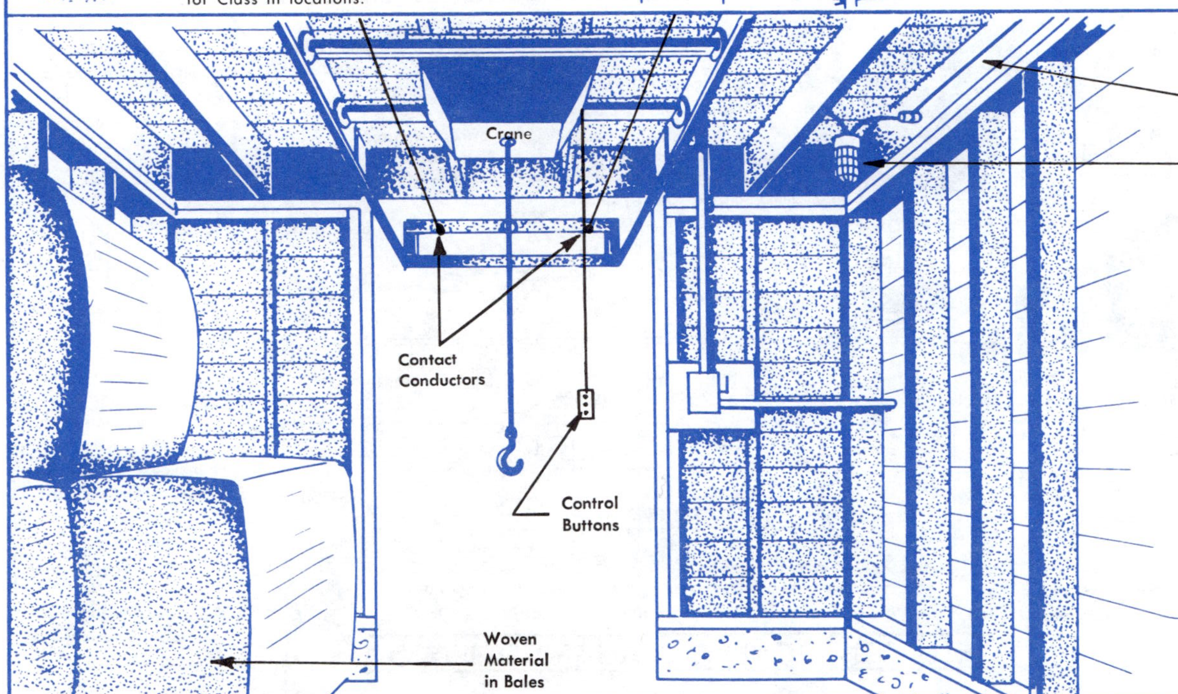

NEC 503-3 (b) Same wiring methods as required for Class III, Division 1, except open wiring permissible in storage areas which have no machinery. See Exception.

NEC 503-9 (a) Lamps protected against entrance of lint and fibers, also marked for maximum wattage. Surface temperature not to exceed 165°C.

NEC 503-13 Cranes operating over combustible fibers, etc., to be specially wired: (a) power supply to contact conductors isolated and ungrounded; (b) contact conductors to be guarded against accidental contact; (c) current collectors to be guarded against escape of hot particles or sparks.

NEC 503-15 No exposed live parts except as per 503-13.

NEC 503-16 Equipment and wiring to be grounded per 250. Bonding and EGCs to conform to 503-16(a)(b).

Class III—Division 2
NEC 500-7(b) Locations in which easily ignited fibers are stored or handled.

(NOTE: In general, provisions for wiring in Class III, Division 1, apply here.)

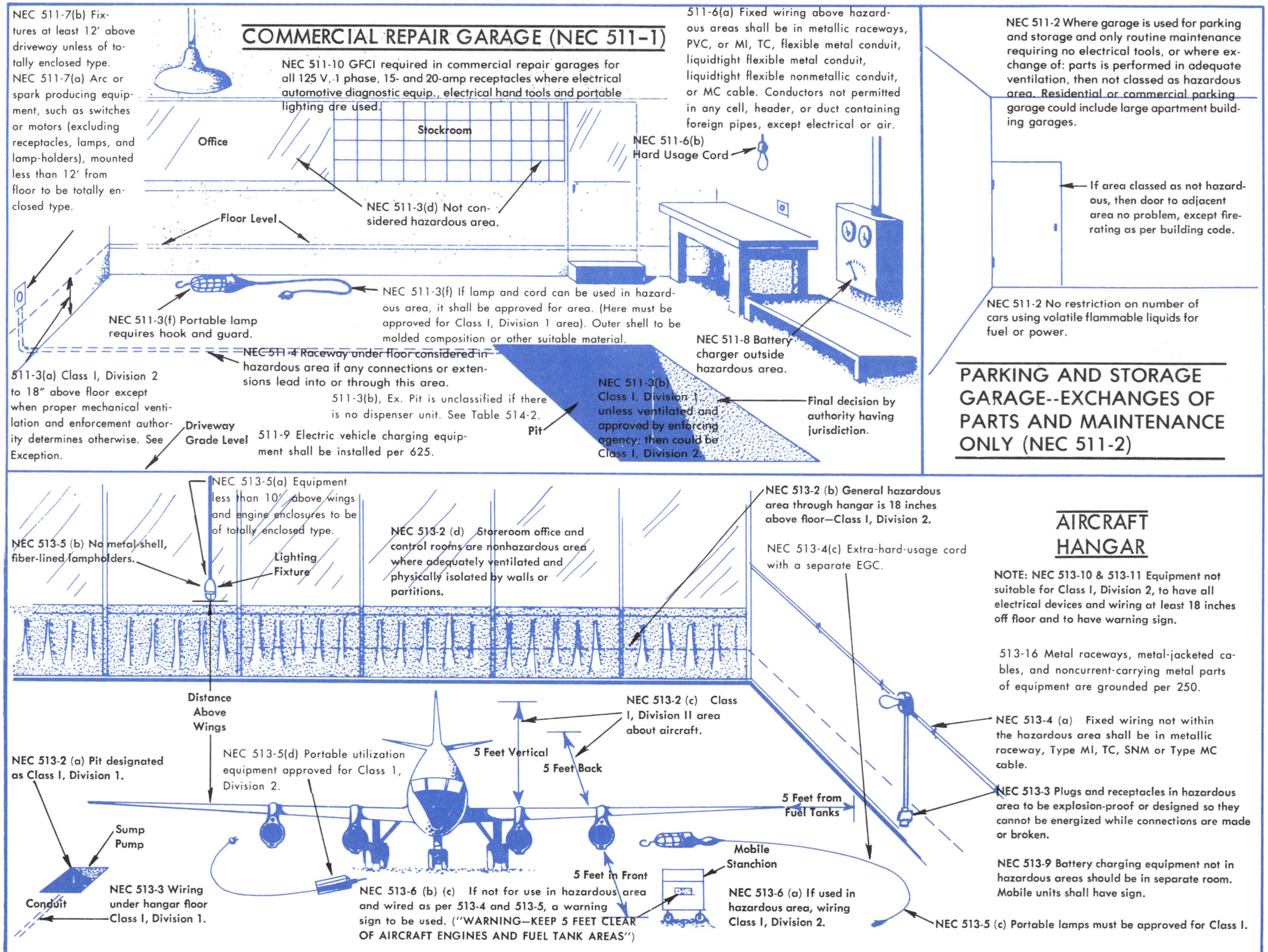
COMMERCIAL REPAIR GARAGE (NEC 511-1)
NEC 511-10 GFCI required in commercial repair garages for all 125 V, 1 phase, 15- and 20-amp receptacles where electrical automotive diagnostic equip., electrical hand tools and portable lighting are used.
NEC 511-7(b) Fixtures at least 12′ above driveway unless of totally enclosed type.
NEC 511-7(a) Arc or spark producing equipment, such as switches or motors (excluding receptacles, lamps, and lamp-holders), mounted less than 12′ from floor to be totally enclosed type.
Office
Stockroom
NEC 511-3(d) Not considered hazardous area.
Floor Level
511-6(a) Fixed wiring above hazardous areas shall be in metallic raceways, PVC, or MI, TC, flexible metal conduit, liquidtight flexible metal conduit, liquidtight flexible nonmetallic conduit, or MC cable. Conductors not permitted in any cell, header, or duct containing foreign pipes, except electrical or air.
NEC 511-6(b) Hard Usage Cord
NEC 511-3(f) Portable lamp requires hook and guard.
NEC 511-3(f) If lamp and cord can be used in hazardous area, it shall be approved for area. (Here must be approved for Class I, Division 1 area). Outer shell to be molded composition or other suitable material.
NEC 511-4 Raceway under floor considered in hazardous area if any connections or extensions lead into or through this area.
511-3(b), Ex. Pit is unclassified if there is no dispenser unit. See Table 514-2.
511-3(a) Class I, Division 2 to 18″ above floor except when proper mechanical ventilation and enforcement authority determines otherwise. See Exception.
Driveway Grade Level
511-9 Electric vehicle charging equipment shall be installed per 625.
Pit
NEC 511-3(b) Class I, Division 1 unless ventilated and approved by enforcing agency; then could be Class I, Division 2.
NEC 511-8 Battery charger outside hazardous area.
Final decision by authority having jurisdiction.
NEC 511-2 Where garage is used for parking and storage and only routine maintenance requiring no electrical tools, or where exchange of: parts is performed in adequate ventilation, then not classed as hazardous area. Residential or commercial parking garage could include large apartment building garages.
If area classed as not hazardous, then door to adjacent area no problem, except fire-rating as per building code.
NEC 511-2 No restriction on number of cars using volatile flammable liquids for fuel or power.
PARKING AND STORAGE GARAGE--EXCHANGES OF PARTS AND MAINTENANCE ONLY (NEC 511-2)
NEC 513-5(a) Equipment less than 10′ above wings and engine enclosures to be of totally enclosed type.
NEC 513-5 (b) No metal-shell, fiber-lined lampholders.
Lighting Fixture
NEC 513-2 (d) Storeroom office and control rooms are nonhazardous area where adequately ventilated and physically isolated by walls or partitions.
NEC 513-2 (b) General hazardous area through hangar is 18 inches above floor—Class I, Division 2.
NEC 513-4(c) Extra-hard-usage cord with a separate EGC.
AIRCRAFT HANGAR
NOTE: NEC 513-10 & 513-11 Equipment not suitable for Class I, Division 2, to have all electrical devices and wiring at least 18 inches off floor and to have warning sign.
513-16 Metal raceways, metal-jacketed cables, and noncurrent-carrying metal parts of equipment are grounded per 250.
Distance Above Wings
NEC 513-2 (c) Class I, Division II area about aircraft.
5 Feet Vertical
5 Feet Back
NEC 513-2 (a) Pit designated as Class I, Division 1.
NEC 513-5(d) Portable utilization equipment approved for Class 1, Division 2.
NEC 513-4 (a) Fixed wiring not within the hazardous area shall be in metallic raceway, Type MI, TC, SNM or Type MC cable.
5 Feet from Fuel Tanks
NEC 513-3 Plugs and receptacles in hazardous area to be explosion-proof or designed so they cannot be energized while connections are made or broken.
Sump Pump
Mobile Stanchion
5 Feet in Front
NEC 513-3 Wiring under hangar floor Class I, Division 1.
Conduit
NEC 513-6 (b) (c) If not for use in hazardous area and wired as per 513-4 and 513-5, a warning sign to be used. ("WARNING—KEEP 5 FEET CLEAR OF AIRCRAFT ENGINES AND FUEL TANK AREAS")
NEC 513-6 (a) If used in hazardous area, wiring Class I, Division 2.
NEC 513-9 Battery charging equipment not in hazardous areas should be in separate room. Mobile units shall have sign.
NEC 513-5 (c) Portable lamps must be approved for Class I.

GASOLINE DISPENSING AND SERVICE STATION

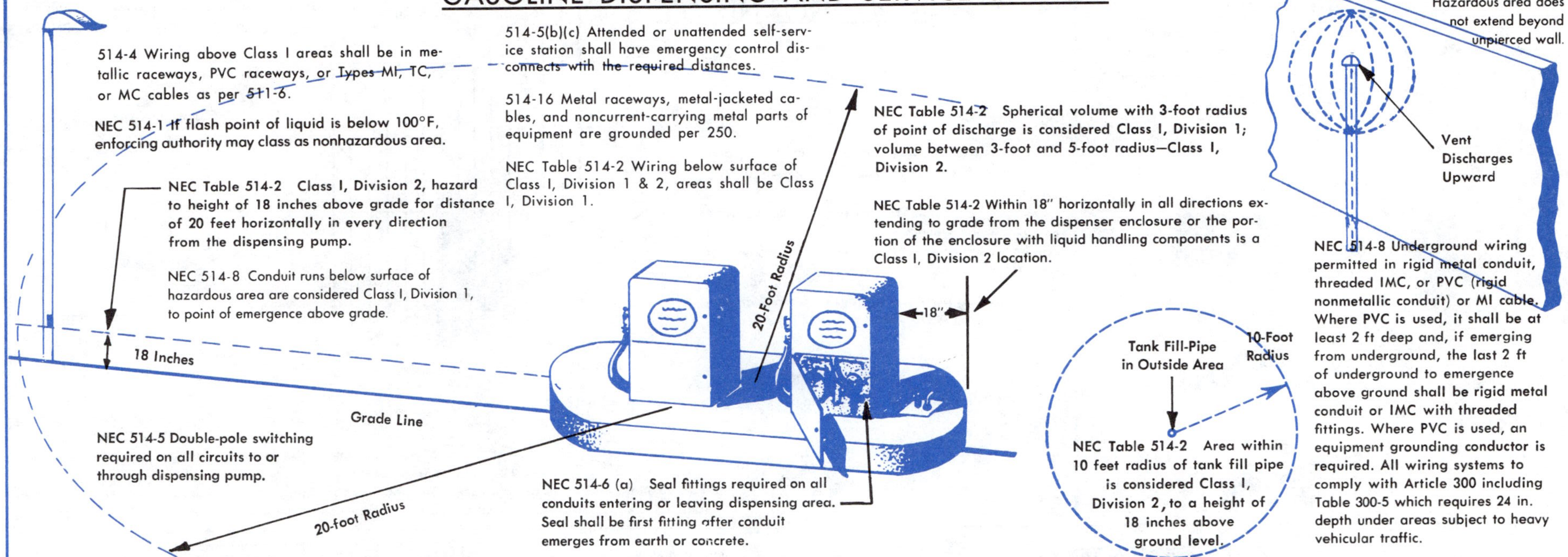

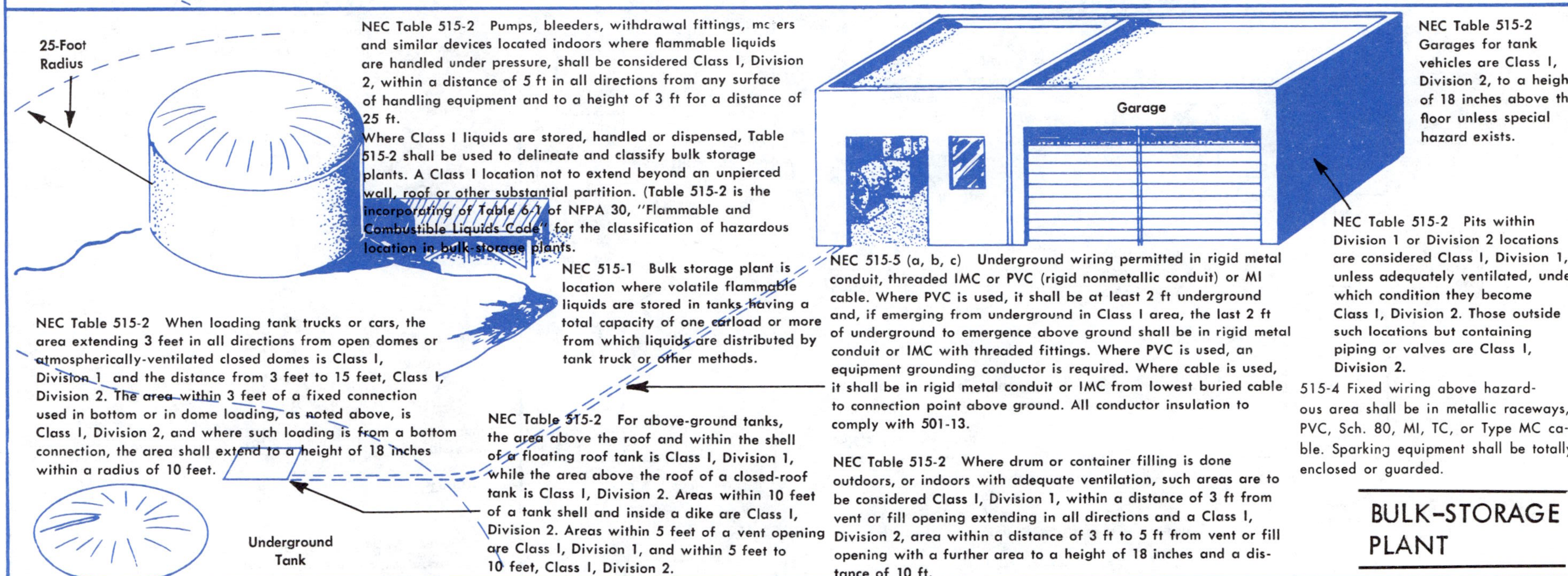

BULK-STORAGE PLANT

NEC 516-7(a) Fixed wiring above hazardous areas in metal raceways, ENT, PVC, or Type MI, TC, or MC cable.

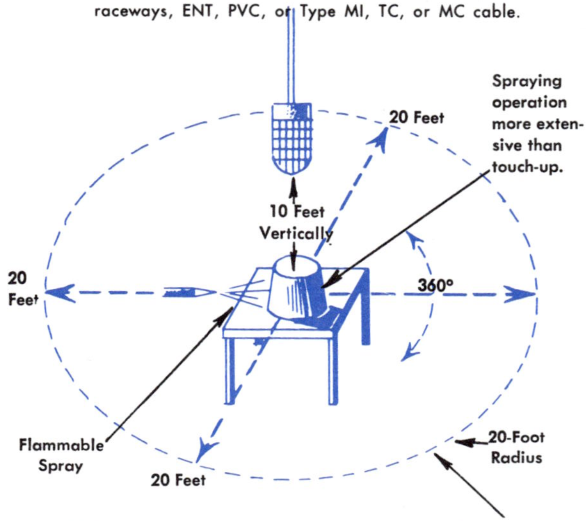

NEC 516-2 (a-3) Area containing paint spray considered Class I, Division I hazardous area.

NEC 516-2 (b-1) Area outside of the Class I, Division I (as per 516-2 (a-3) classed as Class I, Division II hazardous area.

OPEN SPRAYING

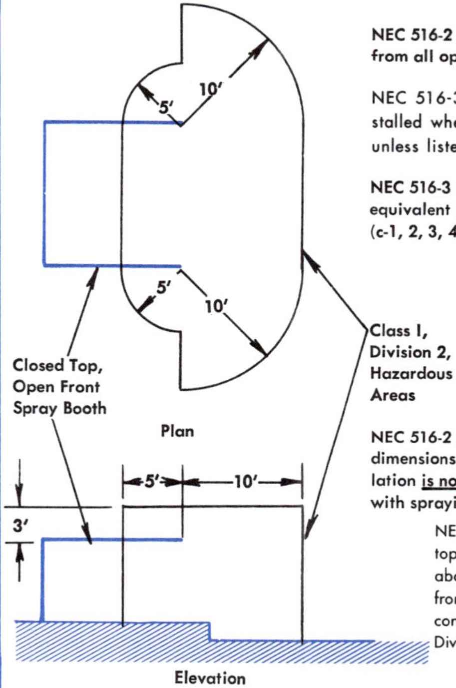

NEC 516-2 (b-4) If enclosed booth, 3 feet in all directions from all openings are Class I, Division 2, area.

NEC 516-3(b) Electrical equipment not to be installed where subject to vapor or deposit of residue unless listed for purpose.

NEC 516-3 (c) Illumination through glass panels or equivalent permitted if installed as per 516-3 (c-1, 2, 3, 4, 5).

NEC 516-2 (b-2 (b)) These dimensions apply if ventilation is not interlocked with spraying equipment.

NEC 516-2 (b-3) If open top spray booth, the 3 ft. above the booth and the 3 ft. from the other openings are considered as Class I, Division 2.

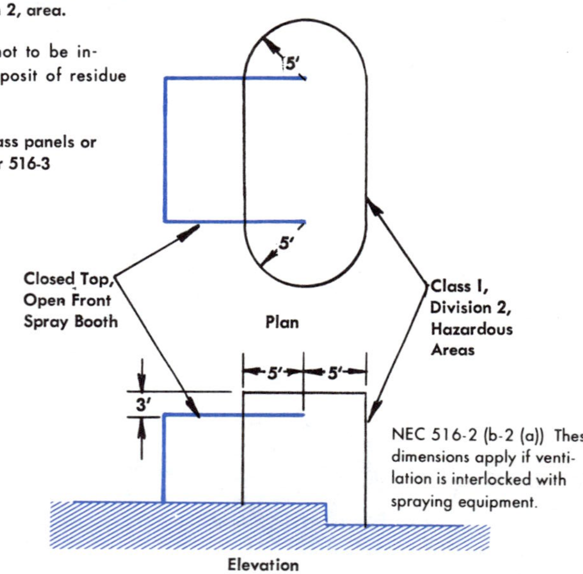

NEC 516-2 (b-2 (a)) These dimensions apply if ventilation is interlocked with spraying equipment.

SPRAY BOOTH

NEC 516-5 Electrostatic hand-spraying equipment shall be approved for location. Transformers, etc. shall be located outside the hazardous area. Handle of spray gun and other conductive objects shall be grounded. Target objects shall maintain contact with grounded support. Area shall be properly ventilated and interlocked with spraying equipment. Hand gun and connections to power supply permitted in hazardous area.

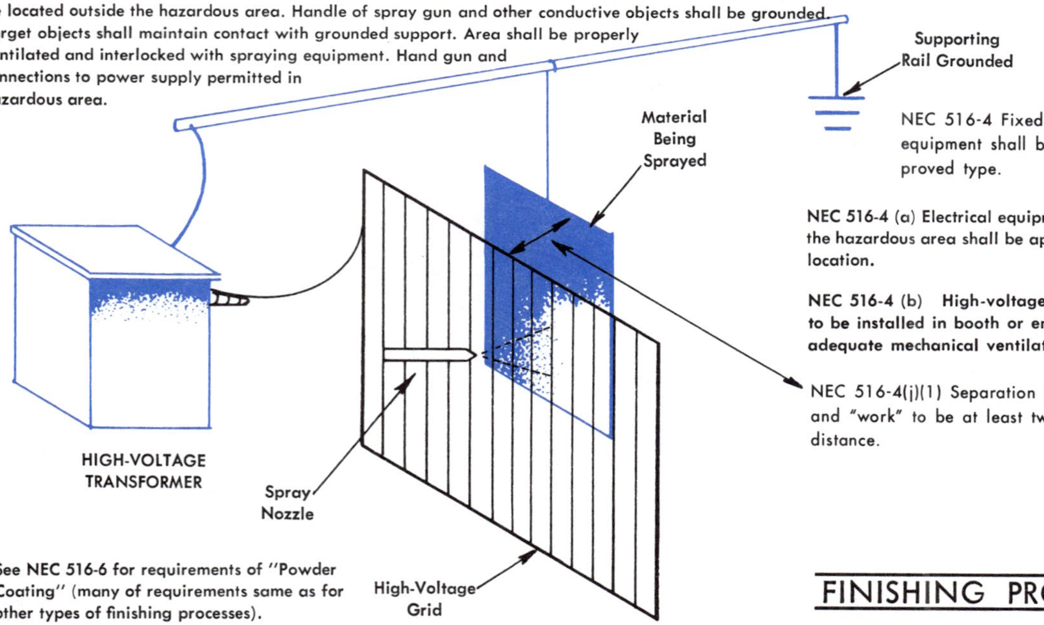

NEC 516-4 Fixed electrostatic spray equipment shall be listed or an approved type.

NEC 516-4 (a) Electrical equipment within the hazardous area shall be approved for the location.

NEC 516-4 (b) High-voltage grid or electrodes to be installed in booth or enclosure which has adequate mechanical ventilation.

NEC 516-4(j)(1) Separation between grid and "work" to be at least twice sparking distance.

See **NEC 516-6** for requirements of "Powder Coating" (many of requirements same as for other types of finishing processes).

FINISHING PROCESSES

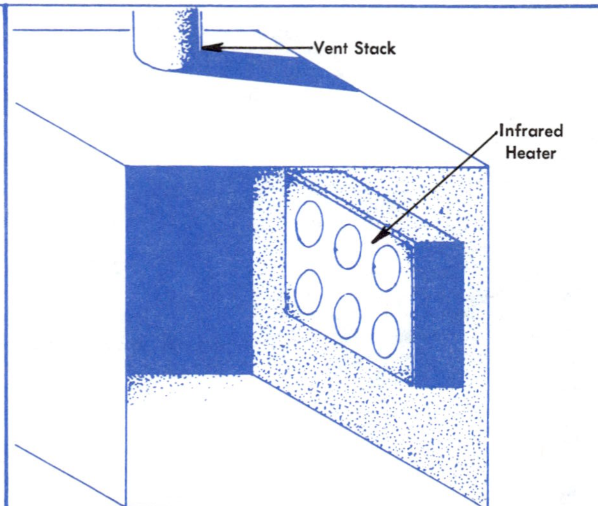

NEC 516-2(d) Drying and baking areas generally considered nonhazardous if adequately ventilated, and protective devices of ventilating system interlocked with heating circuit, and acceptable to authority having jurisdiction.

PAINT DRYING

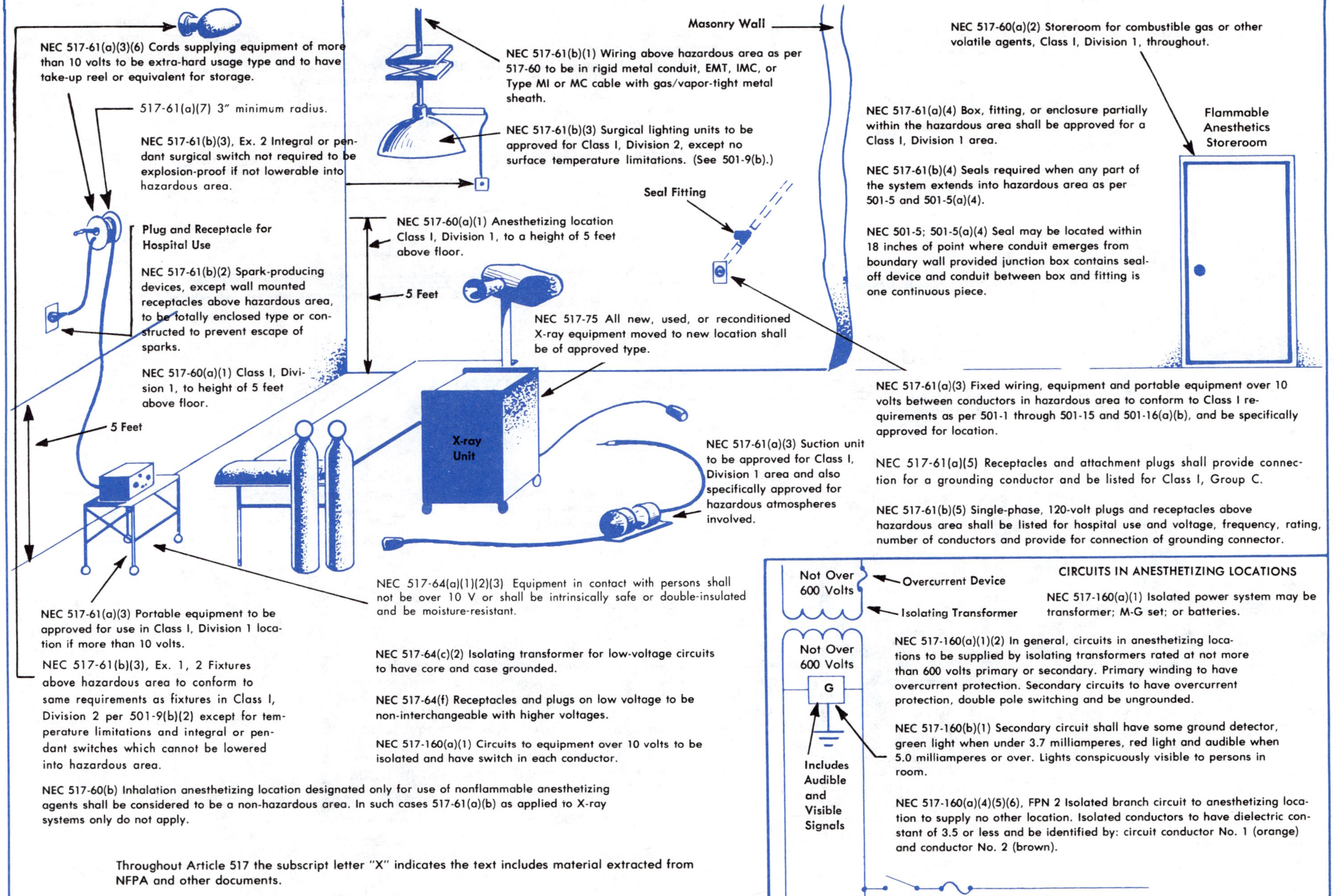

INHALATION ANESTHETIZING LOCATIONS

THEATERS AND SIMILAR LOCATIONS

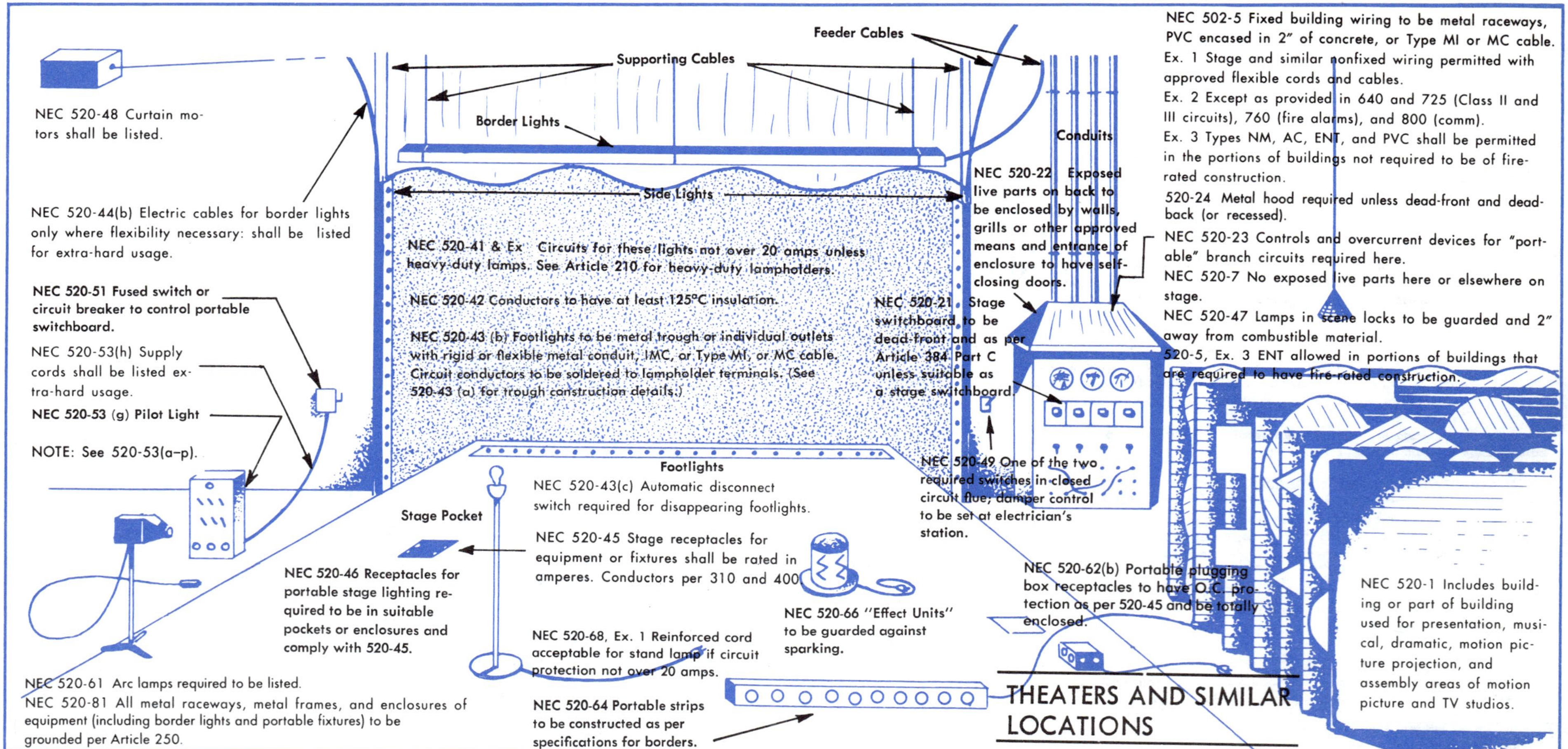

CABLE TRAYS--ARTICLE 318

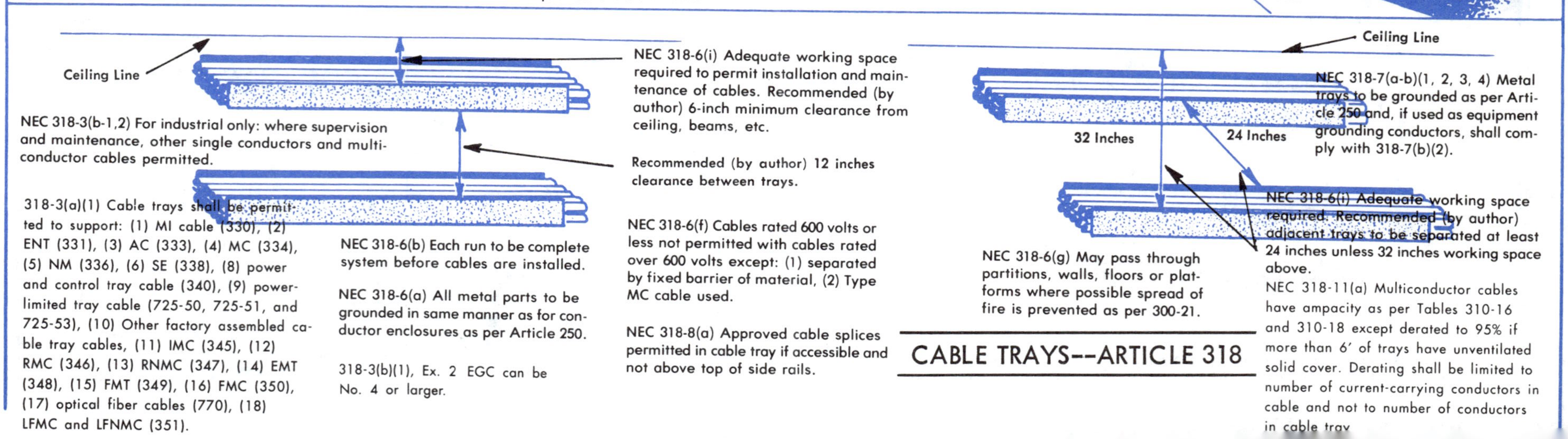

BRACKET FIXTURE WIRING -- STAGE

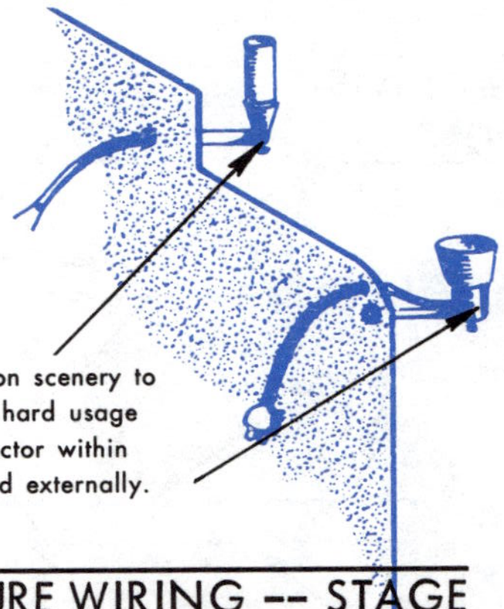

NEC 520-63(a) Brackets on scenery to be wired internally or, if hard usage cord terminating in connector within 18" is used, may be wired externally. See Exception.

FESTOONS -- STAGE WIRING

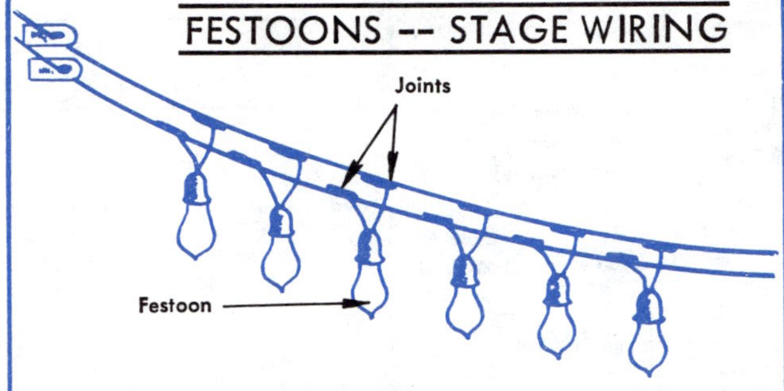

NEC 520-65 Joints in festoon lighting shall be staggered. Lamp, such as a lantern, shall have guard if enclosed in combustible material.

CABLE CONNECTORS -- STAGE WIRING

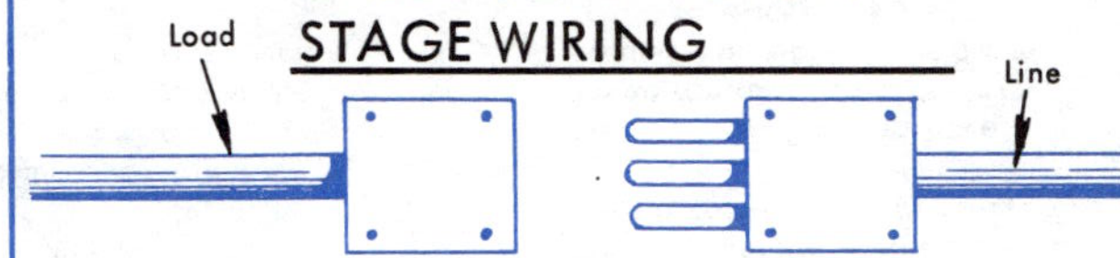

520-67 Female half of the connector shall be attached to the load side of cord or cable. Cable connectors constructed so that tension on cord or cable will not be transmitted to connections. Connectors to be rated in amps and designed so that different rated males and females cannot be properly connected and be polarized. OCPD shall not be greater than feeder ampacity.

RESISTANCE- OR REACTOR-TYPE DIMMERS -- STAGE WIRING

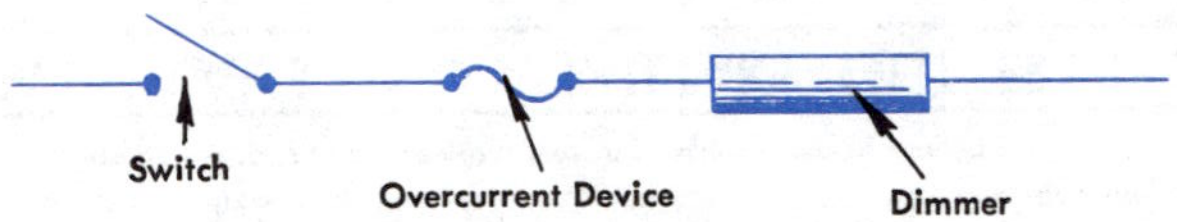

NEC 520-25 (a) Where installed in ungrounded conductors, each dimmer shall be protected by an overcurrent device set at a value not exceeding 125% of its rating.

NEC 520-25 (b) Resistance or reactor (series) dimmers may be placed in either line wire and shall comply with 380-1. Dimmers in grounded neutral shall not open the circuit.

AUTOTRANSFORMER-TYPE DIMMERS

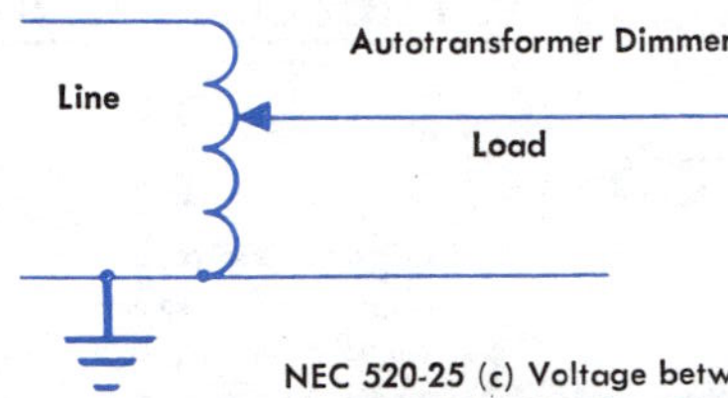

NEC 520-25 (c) Voltage between conductors not to exceed 150 volts. Grounded conductor to be common to both load and supply.

STAGE DETAILS

FLUE DAMPER CONTROL -- STAGE

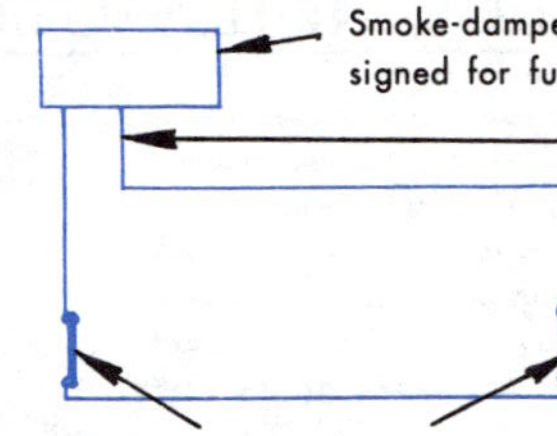

Smoke-damper control device designed for full circuit voltage.

Control circuit normally closed.

Smoke-damper device to be located in loft above scenery and enclosed in suitable metal box with tight, self-closing door.

At least two switches in series, one at electrician's station.

NEC 520-53 (e) Dimmer terminals to have enclosures and faceplate and be so arranged that accidental contact between live parts and faceplate is improbable.
NEC 520-49 Requirements for smoke ventilator control device and its circuit.

THEATER DRESSING ROOMS

NEC 520-72 Lamps within 8 feet of floor to have riveted, open-end guards or otherwise locked in place.

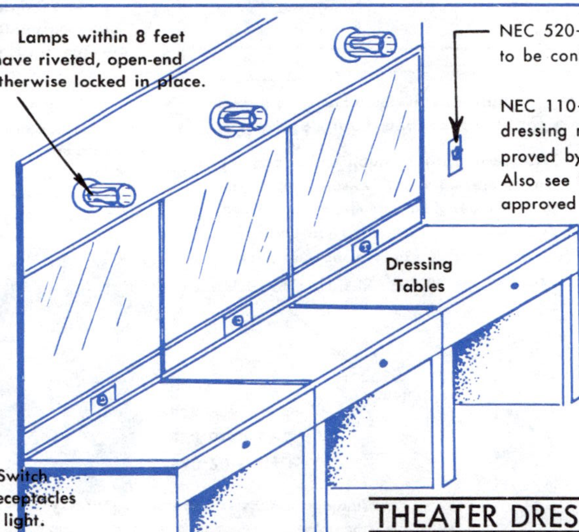

Wall Switch with Pilot Light

NEC 520-73 Switch controlling receptacles to have pilot light.

NEC 520-73 Lamps and receptacles to be controlled by wall switches.

NEC 110-2 Equipment (including dressing room lighting) shall be approved by authority enforcing code. Also see 90-4 and 100 – Definitions, approved for purpose.

NEC 520-71 Pendant lampholders not permitted in dressing rooms.

MOTION PICTURE AND TV STUDIOS

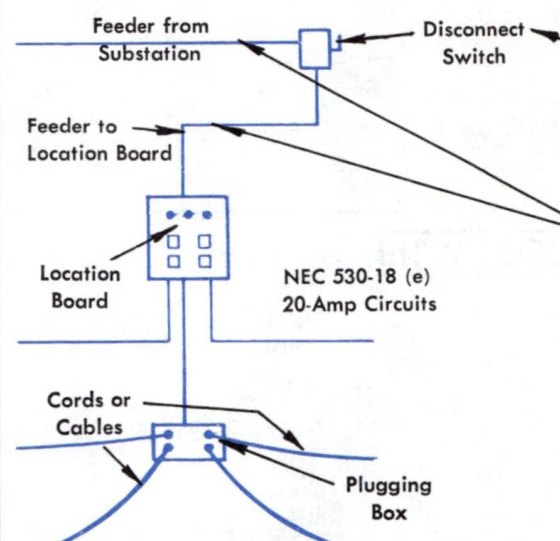

NEC 530-11, 530-12, 530-13 Studio wiring requirements, generally, follow those for theater stage.

NEC 530-13 Ex A single switch may be used to disconnect all contactors on a location board which is not more than 6 feet away.

NEC 530-19 See Table 530-19 (a) for feeder demand factors for stage set lighting.

NEC 530-18 (b) Overcurrent devices for feeders may be set at not over 400% of conductor rating.

NEC 530-20 Pendant and portable lamps and special portable equipment need not be grounded if operating supply voltage not over 150 V DC to ground.

NEC 530-18 (d) Where plugging box does not have overcurrent devices, every cord or cable smaller than No. 8 supplied through it shall have its own overcurrent device.

NEC 540-12 Clear working space of at least 30 inches on each side and at rear of projector, floodlight, spotlight, or similar equipment.

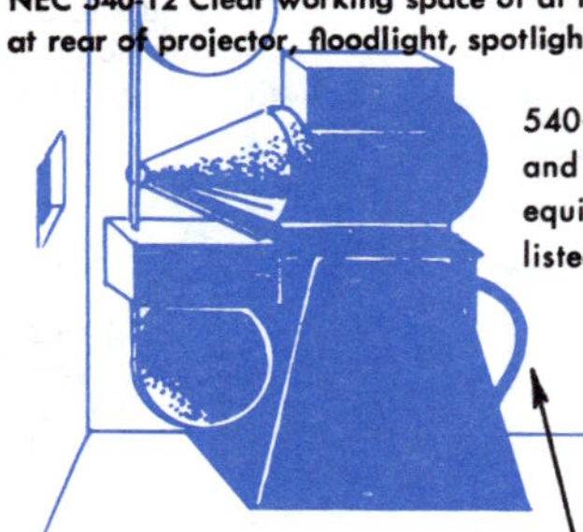

540-20 Projectors and associated equipment to be listed.

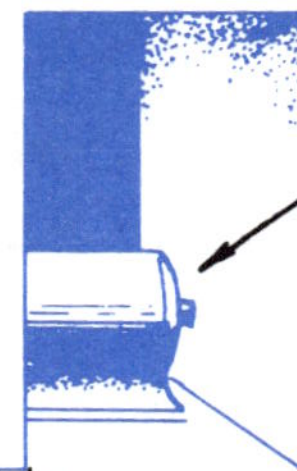

NEC 540-11(a) Motor-generators and similar equipment to be located in a separate room where possible.

NEC 540-13 Circuit conductors for arc projectors shall not be smaller than No. 8. Conductors for incandescent-type projectors shall conform to general provisions of this code. (See 210-24.)

NEC 540-11(b) In general, switches or overcurrent devices foreign to projector operation not permitted. Ex. 1 Rooms approved for safety film only require signs inside and outside. Ex. 2 Remote control for auditorium lights or curtain motors.

NEC 540-14 Wiring on lamps to have 200°C insulation where ambient temperature over 50°C.

Article 540 covers both professional and nonprofessional equipment.

MOTION PICTURE PROJECTOR

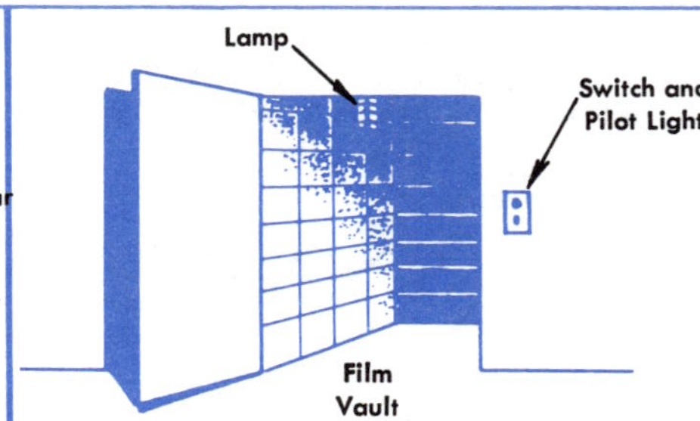

NEC 530-51 Lamps in cellulose nitrate film storage vaults must be rigid, glass-enclosed, gasketed. A switch with pilot light shall be located outside the vault.

NEC 530-52 No electrical equipment other than permanent lighting fixtures shall be located in the vault.

FILM STORAGE VAULTS

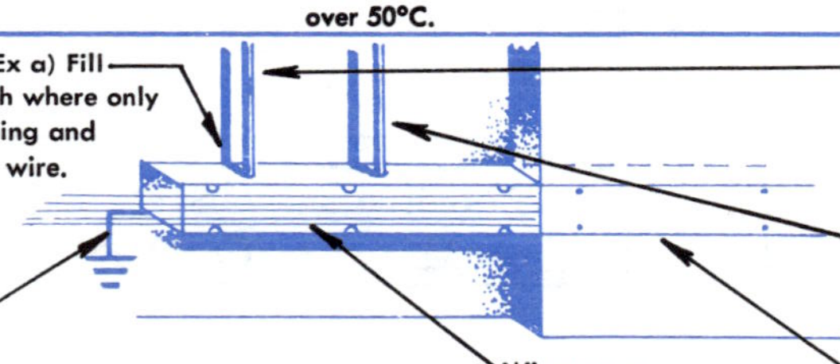

NEC 640-4 (Ex a) Fill to 75% depth where only sound-recording and reproduction wire.

640-3, Ex. 1 Two 2-wire lead cables in 3/4" conduit if area of each cable not over 11 sq in. and special permission is granted.

640-3, Ex. 2 Two 2-wire #19 leads, 1/2" conduit (not over 32% fill) and special permission is granted.

NEC 640-4, Ex.(d) If no power wires, ground wire may be #14 Cu or equal.

640-4, Ex. (c) Troughing may be run in concealed places if in straight lines and covers accessible.

NEC 640-6 Where conductors of different systems grouped: (a) Power supply conductors only for associated equipment; (b) Supply leads of generating or converting units to run separately from output leads; (c) Conductors to have individual or collective insulation equal to that of power wires unless lead covered or equal.

WIREWAYS AND AUXILIARY GUTTERS (SOUND RECORDING)

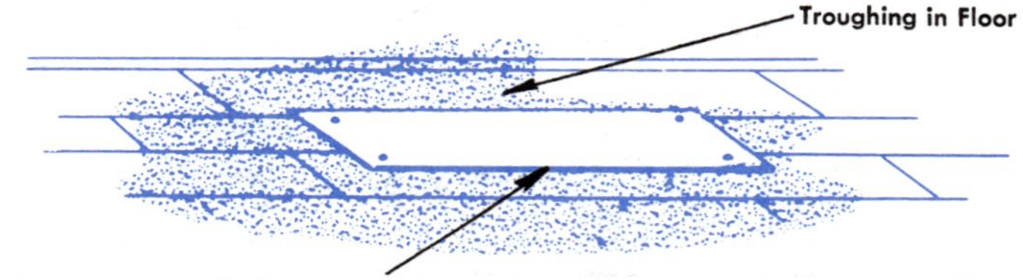

NEC 640-4, Ex. (b) 1/4" steel cover where heavy object moved over floor. No. 10 MSG sufficient where no heavy objects involved.

AUXILIARY GUTTERS -- RECESSED

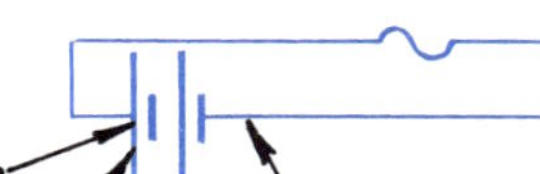

NEC 640-9 (b) Leads shall be rubber or thermoplastic insulated.

NEC 640-10 Overcurrent protection: (a) If cathode supplied by branch circuit or battery of more than 20 ampere-hour capacity, overcurrent protection shall not exceed 15 amps; (b) Overcurrent for plate and screen grid circuit not to exceed 1 amp and placed in each positive lead; (c) If control grid circuit supplied in same manner, overcurrent protection not to exceed 1 amp; (d) Overcurrent devices near as practicable to supply.

OVERCURRENT PROTECTION (SOUND EQUIPMENT)

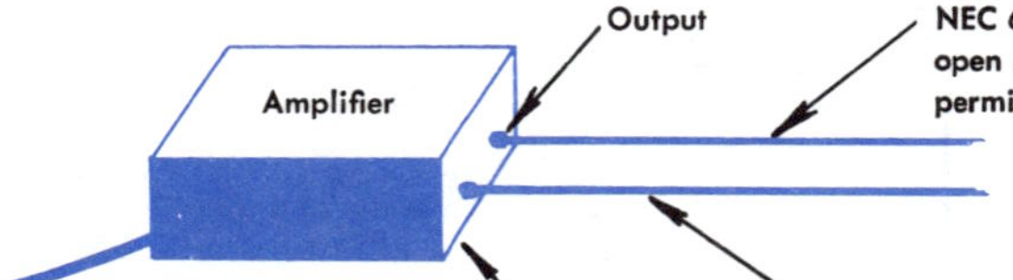

NEC 640-5 If output signal not over 70 volts or open circuit 100 volts, Class II or Class III wiring permitted, as per 725.

NEC 640-11 (a) (b) (c) Amplifiers and rectifiers to be suitably housed, readily accessible and not subjected to high temperatures.

NEC 640-7 Flexible cords not connected to power supply and output limited to 150 VA may be not smaller than No. 26. Flexible cords to be Type S, SJ, ST, SJO or SJT, or other specifically approved types.

AMPLIFIERS (SOUND EQUIPMENT)

SOUND-RECORDING AND SIMILAR EQUIPMENT

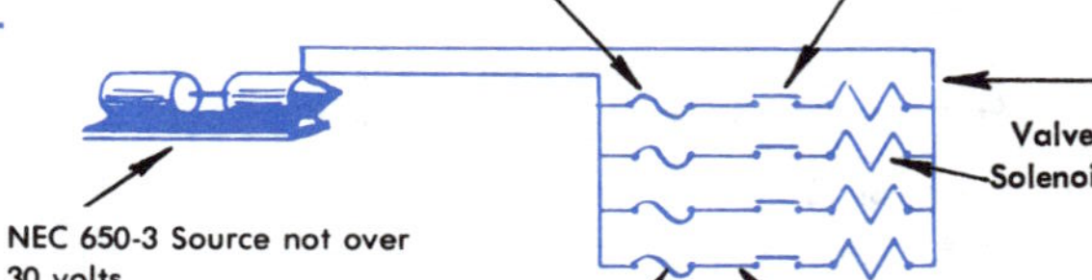

NEC 650-3 Source not over 30 volts.

650-4 Rectifiers used as source of supply for organs shall be grounded per 250.

NEC 650-7 Overcurrent protection not over 6 amps except for main supply and common return conductors.

650-5(a) Conductors for electro-magnetic valve supply shall be a minimum of #26 AWG.

ORGANS

NEC 650-5(a) Common return not smaller than No. 14.

NEC 650-5(b) Insulation: conductors shall have thermoplastic or thermosetting insulation.

NEC 650-5(c) Wires to be cabled except common return conductor.

NEC 650-5(d) Cable to have flame-retardant braid or close-wound fireproof tape unless in metal raceways.

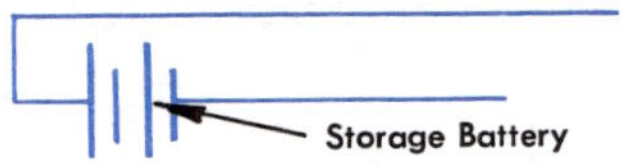

NEC 700-12 (a) Storage batteries permitted for emergency service if capacity is adequate to supply and maintain 87½% of system voltage when supplying full emergency load for 1½ hours. (See Article 480 for installation of storage batteries.)

STORAGE BATTERY

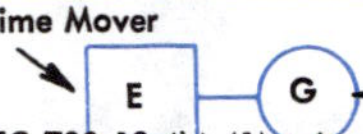

NEC 700-12 (b) (1) A generator with automatic transfer and automatically-started prime mover with 15-minute time delay before restoring normal power is permitted for emergency power.
NEC 700-12 (b) (2) Where internal combustion engine is prime mover, a 2-hour fuel supply is required.
NEC 700-12 (b) (3) Where prime mover fuel is natural gas from utility, a second fuel supply is required with automatic transfer to second fuel supply. (See 700-12 (b) (3) Ex.)
NEC 700-12 (b) (4) Where battery is used to start prime mover, it shall be suitable for the purpose and have automatic charging means.

GENERATOR SET

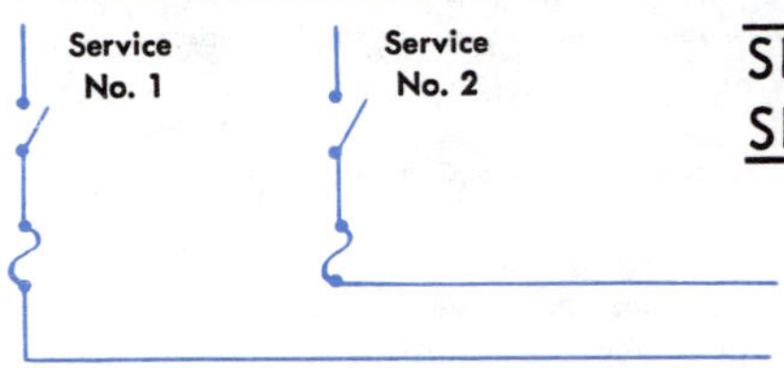

SEPARATE SERVICE

NEC 700-12 (d) Two distinct services, with separate service drops or laterals, widely separated physically and electrically, is permitted. Each service to be treated as a separate service and installed as per Article 230 shall be approved by authority having jurisdiction.

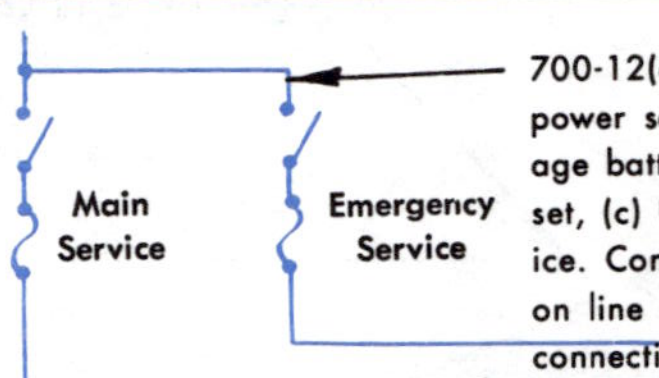

700-12(a–d) Emergency power sources can be (a) storage batteries, (b) generator set, (c) UPS, (d) separate service. Connections ahead of or on line side of a service disconnecting means are not permitted to serve as an emergency power source.

CONNECTION AHEAD OF DISCONNECTING MEANS

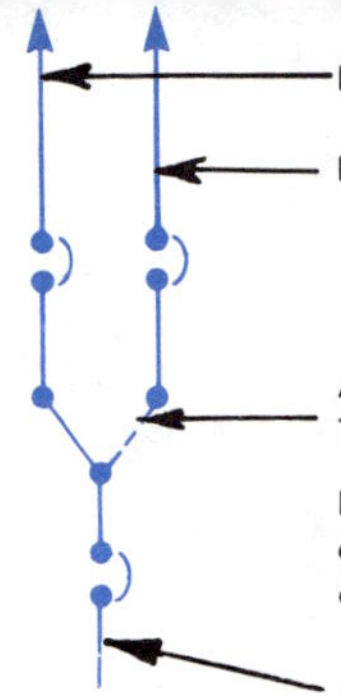

NEC 230-83 All ungrounded conductors of one source to be disconnected before conductors of second source is connected.

NEC 700-6 Transfer equipment with automatic transfer switches to be automatic and identified for use or approved by authority having jurisdiction. To be designed and installed so that connection of both normal and emergency power is not possible.

See 250-5 (d) for grounding of separately derived systems.

NEC 700-18 Branch circuits for emergency use to be automatically transferred upon failure of normal power supply.

TRANSFER SWITCHES

UNIT EQUIPMENT

700-12(e) Unit equipment to be listed for emergency use.

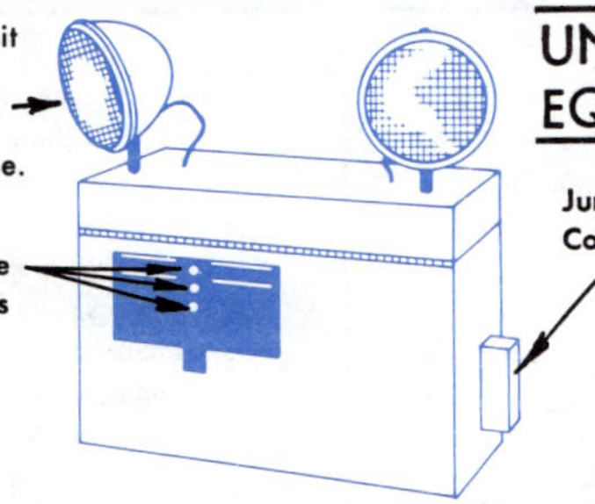

700-12(e) Unit equipment for emergency lighting shall consist of: (1) rechargeable battery; (2) means to charge battery; (3) one or more lamps on unit (terminals are permitted for remote lamp location); (4) a relay automatically energized by failure of general lighting system. Battery shall be able to maintain at least 87½% of nominal voltage at full lamp load for at least 1½ hours or maintain at least 60% of rated illumination for at least 1½ hours. Unit must be approved for emergency service.

Unit not to be portable and to be wired as per Chapter 3. Flexible cord and plug permitted if cord not over 3 feet long. Unit wired to same branch circuit as other normal lighting in area and connected ahead of any local switches. Remote location unit equipment lamps shall be wired per Chapter 3. Branch circuit feeding unit equipment shall be clearly identified at distribution panel.

EMERGENCY SYSTEMS

NEC 700-1 Emergency systems are normally installed in buildings where large numbers of persons assemble or occupy such buildings, such as theaters, assembly halls, hotels, sports arenas, hospitals, and similar institutions. Emergency systems also provide power for hospital equipment that is essential to maintain life and for illumination and power for operating rooms, fire alarm systems, fire pumps, critical public address systems, industrial processes where interruption would cause hazards, and similar functions. See Life Safety Code, NFPA 101, 1991 (ANSI) for critical emergency illumination and exit lights and lighting.

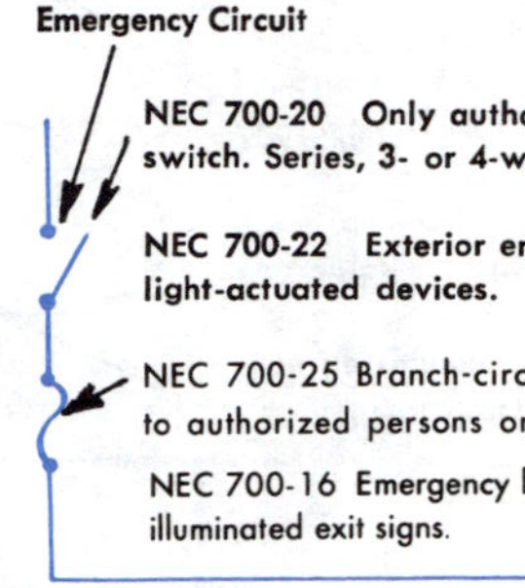

NEC 700-20 Only authorized persons have access to emergency switch. Series, 3- or 4-way switches not permitted. (See Ex 1, 2.)

NEC 700-22 Exterior emergency light may be switched by light-actuated devices.

NEC 700-25 Branch-circuit overcurrent devices accessible to authorized persons only.

NEC 700-16 Emergency lighting includes exit lights and illuminated exit signs.

NEC 700-1 Applies when emergency systems or circuits are legally required by Municipal, State, Federal, or other codes or by any governmental agency having jurisdiction.
NEC 700-7 Audible and visual derangement signals recommended to insure constant readiness of emergency system. See 700-7(a)(b)(c)(d).
NEC 700-9 Wiring to be entirely separate from all other wiring and boxes and raceways to be identified. See NEC 700-9, Ex. 1–5.
NEC 700-15 Only emergency light outlet may be served by emergency lighting system.

EMERGENCY SYSTEMS

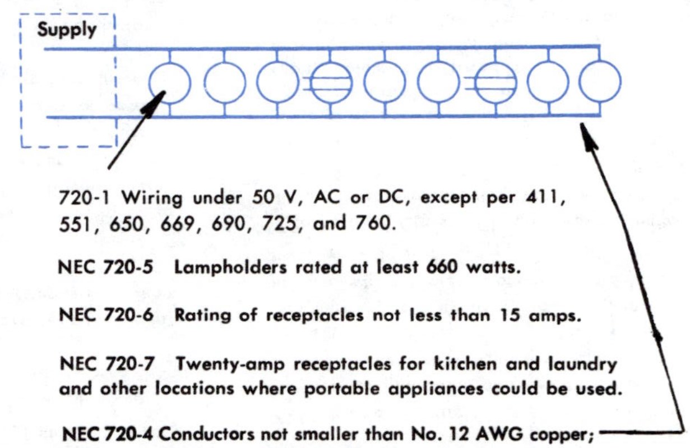

720-1 Wiring under 50 V, AC or DC, except per 411, 551, 650, 669, 690, 725, and 760.

NEC 720-5 Lampholders rated at least 660 watts.

NEC 720-6 Rating of receptacles not less than 15 amps.

NEC 720-7 Twenty-amp receptacles for kitchen and laundry and other locations where portable appliances could be used.

NEC 720-4 Conductors not smaller than No. 12 AWG copper; not smaller than No. 10 AWG copper if supplying more than one appliance receptacle.

720-10 Grounding per 250.

CIRCUITS AND EQUIPMENT LESS THAN 50 VOLTS

CLASS 1, CLASS 2, AND CLASS 3 REMOTE-CONTROL, SIGNALING, AND POWER-LIMITED CIRCUITS

CLASS 2 AND CLASS 3 CIRCUITS

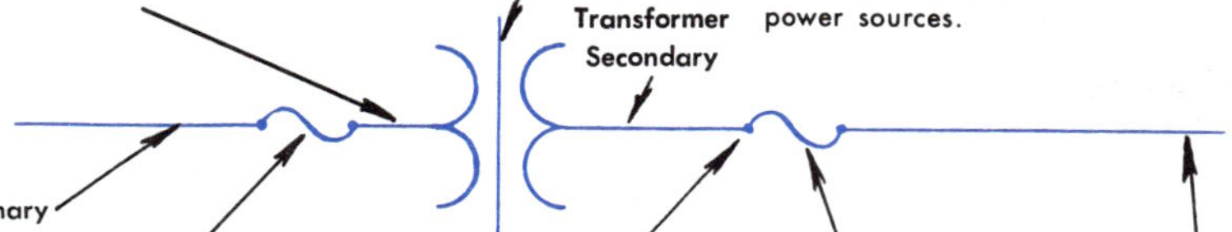

725-51, Ex. Input leads not smaller than #14 except may be #18 if not over 12″ long and insulation per 725-27(b).

Chapter 9, Table 11 A provides required power source limitations for Class 2 and Class 3 AC power sources.

725-51 Transformer protected (maximum) by 20 A overcurrent device.

Notes to Chapter 9, Table 11 A Overcurrent devices shall not be interchangeable with devices of higher ratings and they may be an integral part of power supply.

725-6 Class 2 and 3 circuits and equipment to be grounded per 250.

725-41(b) Power supplies not to be interconnected or paralleled unless listed for such interconnection.

Notes to Chapter 9, Table 11 A Overcurrent devices shall be located at point where conductor to be protected receives its supply.
725-54(a)(1), Ex. Open conductors to be separated from Class 1 or light power wiring not in raceway, or in metal-clad, metal-sheathed, nonmetallic-sheathed or UF cables, a distance of 2″, and to be separated from light, power, or Class 1 wires in raceways or boxes by effective partitions. Conductors in vertical runs to be installed with a view to inhibiting spread of fire.
725-54(b)(4) Class 2 and Class 3 may be run in same cable as communication circuits, but must be classed as communication circuits and conform to 800.

CLASS 1 CIRCUITS

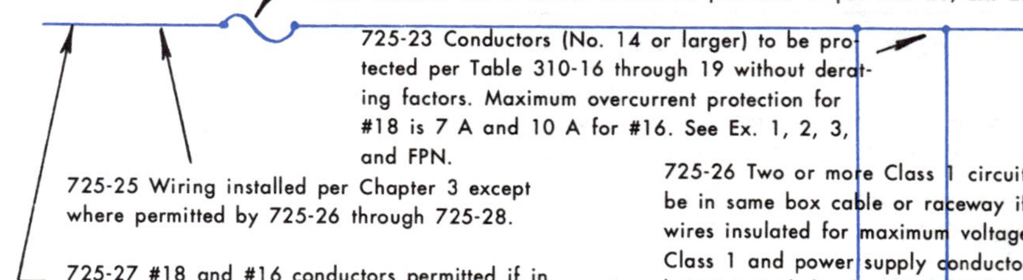

725-24 Overcurrent device at point where conductor receives supply. Ex. 1–Where overcurrent device protecting larger conductor also protects smaller. Ex. 2–Where overcurent provided as per 725-23, Ex. 2.

725-23 Conductors (No. 14 or larger) to be protected per Table 310-16 through 19 without derating factors. Maximum overcurrent protection for #18 is 7 A and 10 A for #16. See Ex. 1, 2, 3, and FPN.

725-25 Wiring installed per Chapter 3 except where permitted by 725-26 through 725-28.

725-27 #18 and #16 conductors permitted if in raceways or listed cables and if loads do not exceed ampacity per 402-5.

725-28(a)(b) Ampacity of Class 1 wires to be reduced for over 3 conductors if continuous load. Ampacity of all reduced when power supply and Class 1 in raceway and more than three Class 1 carry continuous loads or ampacity of power supply conductors only when more than three.

725-27(b) Conductors rated 600 V. Conductors, #16 and larger per 310. #16 and #18 shall be TF or similar. See Types listed in 725-27(b).

725-6 Grounding of Class 1 circuits and equipment per 250.

725-26 Two or more Class 1 circuits may be in same box cable or raceway if all wires insulated for maximum voltage. Also Class 1 and power supply conductors may be intermingled only when supplying equipment to which Class 1 conductors are functionally associated. See Ex. 1, 2.

726-21(a)(b) Class 1 Circuits– Power Limitations: Source rated output not over 30 V and 1000 VA and having overcurrent protection not over 167% of VA rating of source divided by source rated volts.
Overcurrent devices to be noninterchangeable and approved for purpose.
May be integral part of power supply.
(1) Article 450 applies where transformers are used to supply Class I circuits. (2) If source is other than transformer it shall have maximum output of 2500 VA and maximum amps times maximum volts shall not exceed 10,000 VA with overcurrent protection by-passed. (b) Class I remote-control and signaling circuits not over 600 volts but power output need not be limited.

COMBINING CLASS 2 AND CLASS 3 CIRCUITS

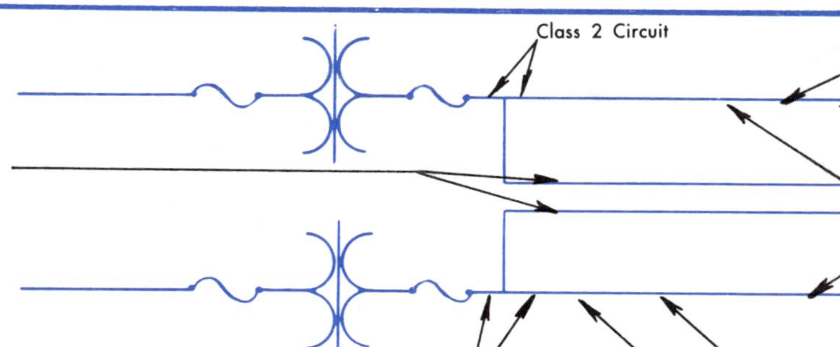

725-54(b)(3) One or more Class 2 circuits are permitted in same raceway, cable, or enclosure with Class 3 circuits if Class 2 conductor insulation is at least equal to that required for Class 3 circuits.

725-71(a–h) Class 2 circuit conductor insulation and size shall be suitable for application.

725-54(b)(1) Two or more Class 2 circuits are permitted in same raceway, cable, or enclosure.

725-54(c) Where Class 2 or Class 3 circuits extend beyond building and are subject to possible contact with over 300 V to ground, light, or power circuits they shall conform to 800-30, 800-10, 800-13, 800-12, 800-31, 800-32, 800-33, and 800-40.

725-54(b)(2) Two or more Class 3 circuits are permitted in same raceway, cable, or enclosure.

725-71(a–h) Class 3 circuit conductors per Table 725-71.

SIGNS -- GENERAL RULES -- 600 VOLTS OR LESS

600-3 All signs to be listed and installed per listing unless otherwise approved by special permission.

NEC 600-8 See specified details for sign enclosures (metal galvanized or equal).

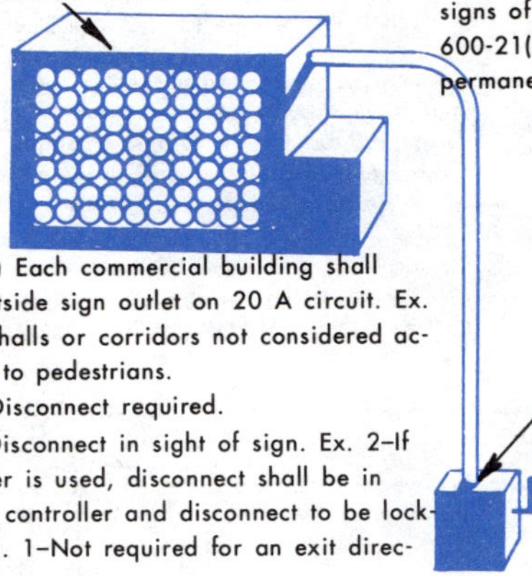

600-5(a) Each commercial building shall have outside sign outlet on 20 A circuit. Ex. Interior halls or corridors not considered accessible to pedestrians.

600-2 Disconnect required.

600-6 Disconnect in sight of sign. Ex. 2–If controller is used, disconnect shall be in sight of controller and disconnect to be lockable. Ex. 1–Not required for an exit directional sign.

600-6(b) Devices such as switches or flashers to have twice ampere rating of transformer, or be approved for purpose, except AC general-use snap switches permitted on AC inductive loads not exceeding switch rating.

NEC 600-21(b) Conductors not smaller than #14.

600-21(b), Ex. 1(a–f) If listed in Table 402-3, #18 permitted for: portable signs, short leads to lampholders or ballasts, permanent leads not over 8′ long to discharge lampholders or ballasts if in channels, multiple conductor incandescent signs of not over 250 VA.

600-21(b), Ex. 2 Not smaller than #20 permitted as short, permanent leads to synchronous motors.

600-7 All metal equipment of outline lighting systems and signs shall be grounded. Listed flexible metal conduit is permitted to serve as equipment grounding means in lengths not exceeding 100′ if it supplies secondary wiring of a transformer or power supply for electrical discharge tubing.

600-5(b)(1) Rating of branch circuits that supply signs and other outline lighting systems with either fluorescent, incandescent, or high-intensity-discharge illumination shall not exceed 20 A.

NEON SIGNS – OVER 1000 V

600-32(a) Conductors over 1000 V shall be installed in flexible metal conduit, liquidtight flexible metal conduit, EMT, rigid nonmetallic conduit, rigid or intermediate metal conduit, or as concealed conductors on insulators, or as Type MC cable.

600-32(d) Sharp bends in conductors are not permitted.

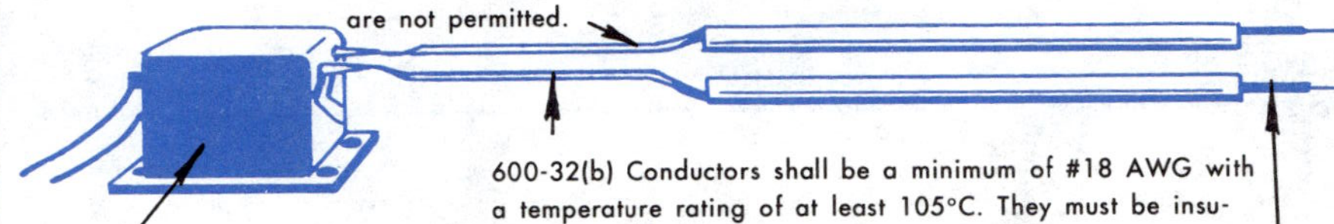

600-32(b) Conductors shall be a minimum of #18 AWG with a temperature rating of at least 105°C. They must be insulated and listed for purpose.

600-32(l) In dwellings, open circuit voltage for neon conductors shall not exceed 1000 V.

600-32(c) Conductors shall be installed so they are not subject to physical damage.

600-32(h) If listed for purpose, and provided with terminals at midpoint, conductors shall be permitted to run from neon tubing ends to grounded midpoint of a transformer.

OUTSIDE SIGN OR OUTLINE LIGHTING

600-9(a)(b) Signs shall be installed so they are suitable for weather conditions they may encounter. Unless protected from physical damage, signs and outline lighting equipment shall be installed a minimum of 14′ above ground where accessible to vehicles.

100 Outline lighting is an arrangement of either incandescent or electric-discharge lighting components designed to highlight certain elements of a building such as a window.

100 Electric signs can be either fixed, stationary, or portable, self-contained utilization equipment that are electrically illuminated. Signs call attention to or provide information about something.

SKELETON TUBING

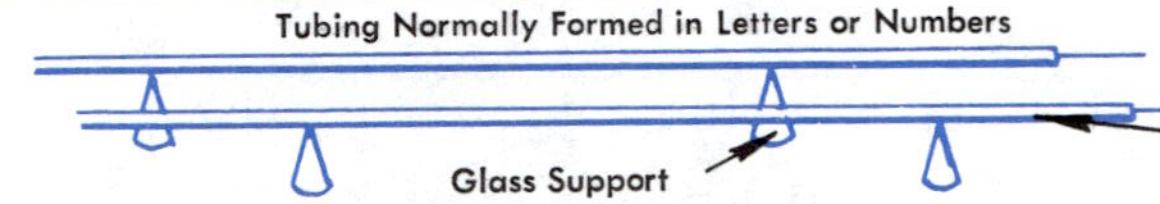

600-2 Skeleton tubing is neon tubing that is the sign of outline lighting. It does not contain, nor is attached to a sign body or enclosure.

600-31(a) Conductors for skeleton tubing shall be installed per Ch. 3.

ELECTRIC SIGNS & OUTLINE LIGHTING

600-10(c)(1)(2) Portable signs installed in wet or damp locations shall be equipped with cords rated for junior-hard service or hard service per 400-4. All portable or mobile signs shall have factory-installed GFCI protection.

PORTABLE SIGNS SHALL BE EQUIPPED WITH GFCI IN OR WITHIN 12″ OF ATTACHMENT PLUG

NEON TUBING

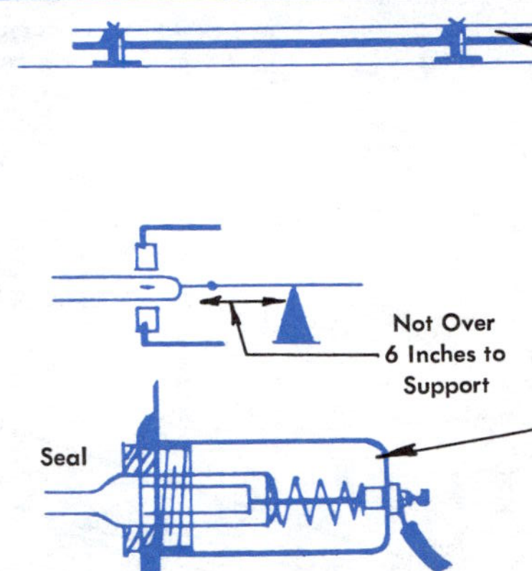

600-41(a) Neon tubing shall be designed so that length of the tubing shall not cause a continuous overcurrent of transformer or electronic power supply.

600-41(b) Neon tubing shall be supported by listed tube supports.

600-41(c) A minimum of 1/4″ spacing shall be provided between neon tubing and any surface other than its' support.

600-42(a) Terminal connections of neon tubing electrode shall be accessible only to qualified personnel.

600-42(b) Neon tubing terminal connections shall be made by: (1) a connection device, (2) twisting conductors together, (3) a listed electrode receptacle. Connections shall be made in an enclosure listed for purpose.

600-42(c) Neon tubing shall be supported no more than 6″ from electrode termination.

600-42(e) Bushings listed for purpose shall be used where electrodes penetrate an enclosure, unless receptacles are used.

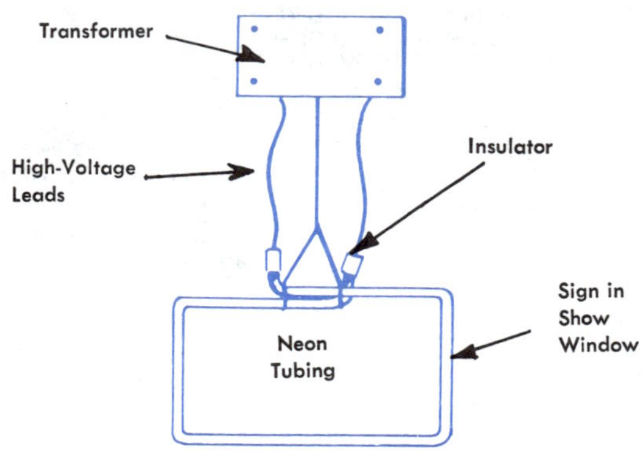

NEC 600-31 (f) Conductors that hang freely in air away from combustible material and not subject to physical damage need not be otherwise protected provided they are insulated for the voltage.

NEC 600-4 All electric signs shall be listed and installed as per listing unless excepted by special permission.

NEC 600-32(d) EX. 2. Transformers with permanent secondary leads may use No. 18 leads to tubing if not over 8 ft. long.

SHOW WINDOW SIGNS -- OVER 600 VOLTS

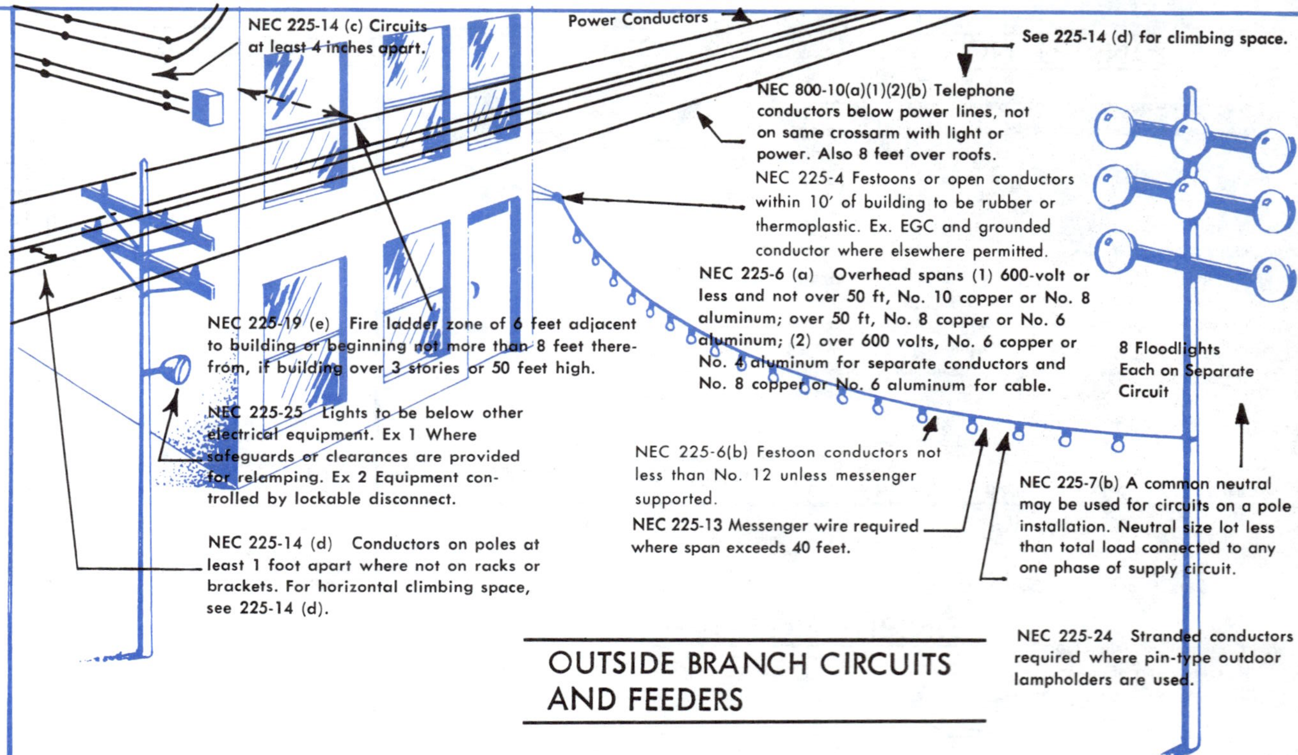

NEC 225-14 (c) Circuits at least 4 inches apart.

See 225-14 (d) for climbing space.

NEC 800-10(a)(1)(2)(b) Telephone conductors below power lines, not on same crossarm with light or power. Also 8 feet over roofs.

NEC 225-4 Festoons or open conductors within 10′ of building to be rubber or thermoplastic. Ex. EGC and grounded conductor where elsewhere permitted.

NEC 225-6 (a) Overhead spans (1) 600-volt or less and not over 50 ft, No. 10 copper or No. 8 aluminum; over 50 ft, No. 8 copper or No. 6 aluminum; (2) over 600 volts, No. 6 copper or No. 4 aluminum for separate conductors and No. 8 copper or No. 6 aluminum for cable.

NEC 225-19 (e) Fire ladder zone of 6 feet adjacent to building or beginning not more than 8 feet therefrom, if building over 3 stories or 50 feet high.

NEC 225-25 Lights to be below other electrical equipment. Ex 1 Where safeguards or clearances are provided for relamping. Ex 2 Equipment controlled by lockable disconnect.

NEC 225-6(b) Festoon conductors not less than No. 12 unless messenger supported.

NEC 225-13 Messenger wire required where span exceeds 40 feet.

NEC 225-7(b) A common neutral may be used for circuits on a pole installation. Neutral size lot less than total load connected to any one phase of supply circuit.

NEC 225-14 (d) Conductors on poles at least 1 foot apart where not on racks or brackets. For horizontal climbing space, see 225-14 (d).

NEC 225-24 Stranded conductors required where pin-type outdoor lampholders are used.

OUTSIDE BRANCH CIRCUITS AND FEEDERS

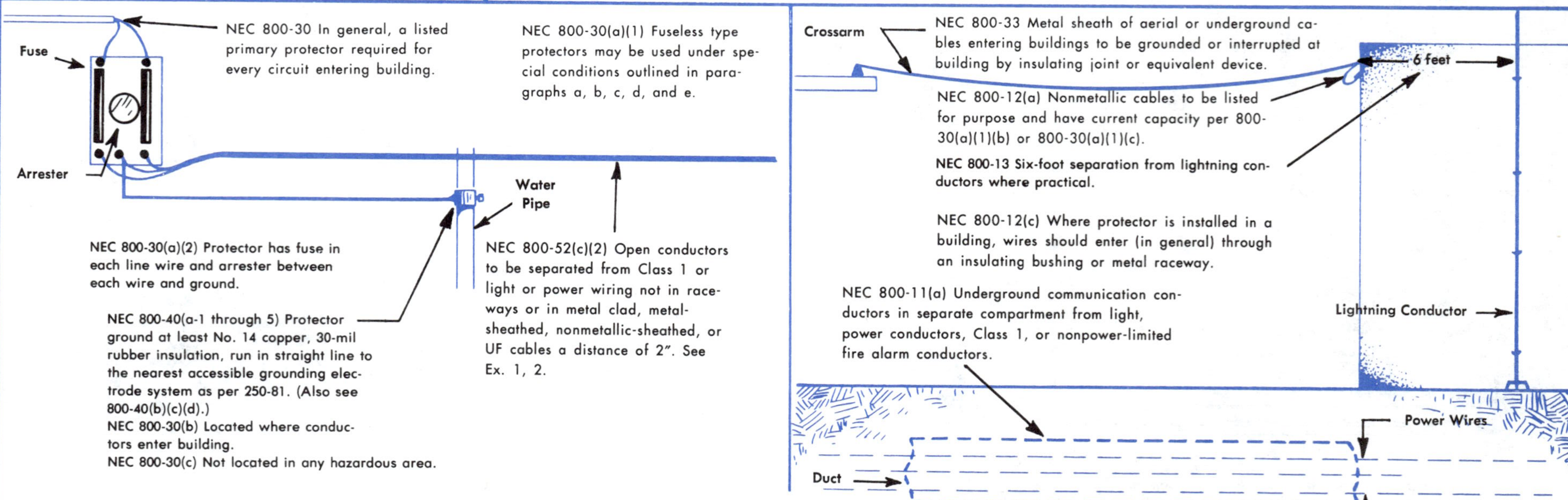

NEC 800-30 In general, a listed primary protector required for every circuit entering building.

NEC 800-30(a)(1) Fuseless type protectors may be used under special conditions outlined in paragraphs a, b, c, d, and e.

NEC 800-30(a)(2) Protector has fuse in each line wire and arrester between each wire and ground.

NEC 800-40(a-1 through 5) Protector ground at least No. 14 copper, 30-mil rubber insulation, run in straight line to the nearest accessible grounding electrode system as per 250-81. (Also see 800-40(b)(c)(d).)
NEC 800-30(b) Located where conductors enter building.
NEC 800-30(c) Not located in any hazardous area.

NEC 800-52(c)(2) Open conductors to be separated from Class 1 or light or power wiring not in raceways or in metal clad, metal-sheathed, nonmetallic-sheathed, or UF cables a distance of 2″. See Ex. 1, 2.

NEC 800-33 Metal sheath of aerial or underground cables entering buildings to be grounded or interrupted at building by insulating joint or equivalent device.

NEC 800-12(a) Nonmetallic cables to be listed for purpose and have current capacity per 800-30(a)(1)(b) or 800-30(a)(1)(c).

NEC 800-13 Six-foot separation from lightning conductors where practical.

NEC 800-12(c) Where protector is installed in a building, wires should enter (in general) through an insulating bushing or metal raceway.

NEC 800-11(a) Underground communication conductors in separate compartment from light, power conductors, Class 1, or nonpower-limited fire alarm conductors.

COMMUNICATION CIRCUITS

NEC 810--RADIO AND TELEVISION EQUIPMENT

RECEIVING EQUIPMENT--ANTENNA SYSTEMS

Power Wires Not Over 250 Volts

NEC 810-11 Antenna and lead-in conductors to be copper, bronze, or other corrosion-resistant material.

NEC 810-11 Ex Soft- or medium-drawn copper for lead-in if span not over 35 feet.

NEC 810-12 Antenna and associated conductors not to be attached to pole carrying light and power wires over 250 volts between conductors or to electric service masts.

NEC 810-13 Not to cross over light and power conductors or to come within 2 feet of them if of 250 volts or less.

NOTE: NEC 810-18 (b) Indoor antennas and lead-ins not closer than 2 inches to other wiring systems. Ex 1 Unless in metal raceways or cable armor. Ex 2 Effectively separated therefrom by continuous and firmly secured nonconductor such as porcelain tubes or flexible tubing.

NEC 810-21 (i) A single grounding conductor may be used for both protective and operating purposes.

NEC 810-16, 810-17 See Table 810-16 (a) for sizes of antenna and lead-in conductors; Table 810-52 for amateur stations.

NEC 810-18(a) Lead-in not closer than 6 feet to lightning rod system unless bonded as per 250-86. Underground conductors shall be separated at least 12 inches from light or power on Class I circuits.

NEC 810-15 Masts to be effectively grounded; grounding conductor to be installed per 810-21.

NEC 810-21 (e) Grounding conductor to run in straight line from mast or discharge unit as practicable.
NEC 810-21 (f) Grounding conductor shall be connected to grounding means as per 810-21 [f (1, 2, & 3)]. Lightning rod conductors, or steam or hot water pipes, not permitted as electrodes for protectors.

NEC 810-21 (b) Grounding conductor may be uninsulated.

NEC 810-21 (g) Grounding conductor for receiving station may be inside or outside of building.

NEC 810-21 (h) Not smaller than No. 10 copper, No. 8 aluminum, or No. 17 copper-clad steel or bronze.

NEC 810-18 (a) Lead-in conductors to be supported so they cannot swing closer than 2 feet to 250-volt wires or 10 feet to conductors over 250 volts. Exception: If not over 250 volts, at 150 volts between conductors, clearance may be 4 inches.

NEC 810-20 (a) Lead-in to have listed discharge unit. Ex Listing not required where lead-in wires enclosed in continuous metal shield that is effectively grounded or protected by antenna discharge unit.

AMATEUR TRANSMITTING AND RECEIVING STATIONS--ANTENNA SYSTEMS

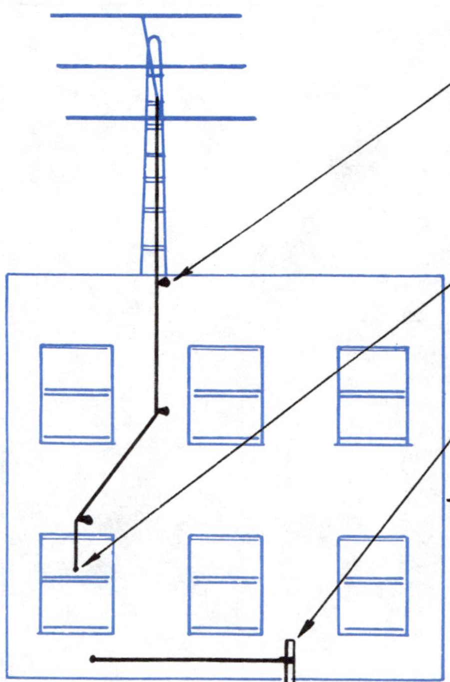

NEC 810-54 Antenna and lead-in wires to clear surface 3 inches. Ex Lead-in may be attached to surface if encased in grounded shield.

NEC 810-57 Antenna discharge unit required in transmitting stations or other suitable means to drain static charges. Ex 1 Where continuous metal shield is effectively ground. Ex 2 Where antenna is effectively grounded.

NEC 810-55 Except when shielded, lead-in conductors shall enter buildings through: (1) an approved insulating tube, or (2) a hole giving 2 inches of clearance all around, or (3) a drilled window pane.

NEC 810-58 (b) Protective ground for transmitting stations as large as lead-in but not less than No. 10 copper, bronze, or copper-clad steel.

NEC 810-58 (c) Operating ground not less than No. 14 copper or equal.

Amateur Transmitting or Receiving Station

INTERIOR INSTALLATION--TRANSMITTING STATION

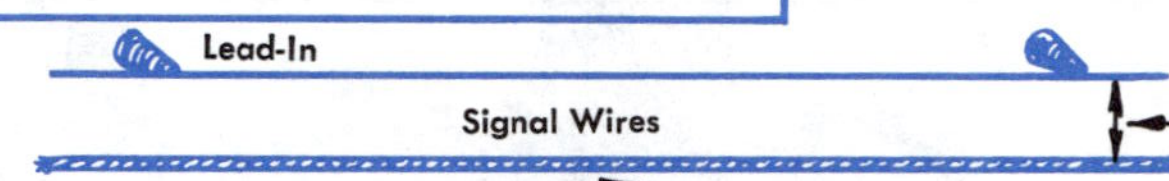

NEC 810-70 Conductors inside building to be separated from other open conductors at least 4 inches except (1) as provided in Article 640, (2) where conductors are separated by fixed nonconductor or are in conduit.

NEC 810-71 (a) Transmitter enclosed by metal or by installing grounded barrier to separate transmitter from operating space.

NEC 810-71 (a), (b) All noncurrent metal parts including handles to be grounded.

NEC 810-71 (c) All doors giving access to voltages in excess of 350 to have interlocks.

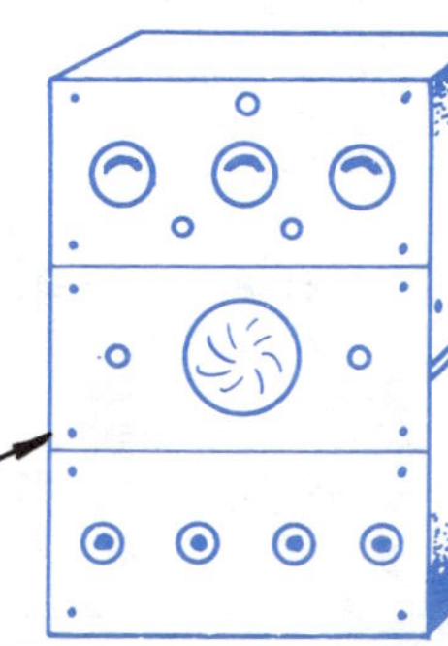

CAPACITORS

C = capacitance in farads
V = voltage
A = current

k = 1000
Xc = capacitive reactance
F = Hertz (cycles per second)

Capacitors in parallel

$C_t = C_1 + C_2 + C_3$ etc.

Capacitors in series

$\frac{1}{C_t} = \frac{1}{C_1} + \frac{1}{C_2} + \frac{1}{C_3}$ etc.

Capacitive Reactance (Xc)

$X_c = \frac{1}{2\pi FC}$

Capacitance (C)

$C = \frac{\text{kVAR } 10^3}{(2\pi F)\ (\text{kV})^2} = \frac{10^6}{(2\pi F)\ Xc}$

Capacitive kVAR's

$\text{kVAR} = \frac{(2\pi F)C\ (\text{kV})^2}{10^3}$ or $\frac{10^3\ (\text{kV})^2}{Xc}$

To determine capacitor rating when capacitor is operated at voltage or frequency below the rated value.
kVAR = Actual burden in kVA
E = Working voltage C = Rated capacity in micro-farads F = Working Hertz

$\text{kVAR} = \frac{CE^2\ 2\pi F}{10^3}$

Capacitor rating decreases with the frequency and as the square of voltage if the frequency or voltage used is lower than that listed on the nameplate.

$\text{kVAR} = \text{Rated kVAR} \times \left(\frac{\text{applied volts}}{\text{rated volts}}\right)^2$

$\text{kVAR} = \text{Rated kVAR} \times \left(\frac{\text{applied frequency}}{\text{rated frequency}}\right)$

TRANSFORMER CONNECTIONS AND RATINGS

$\text{Efficiency} = \frac{\text{kVA (secondary)}}{\text{kVA (primary)}} = \frac{\text{Output}}{\text{Input}}$

Neglecting primary losses:
kVA (primary) = kVA (secondary)

Three single-phase transformers connected and operating as a three-phase unit:
Total kVA capacity = kVA of each x 3 or the total of the three units.

Two single-phase transformers connected as a three-phase open-delta bank:
Total kVA capacity = Total kVA of both units x .866

Three single-phase transformers operating as three-phase open-delta and one single-phase transformer is lost:
Total kVA = Total kVA of all units x .577

TRANSFORMER FORMULAS

Tp = Turns in primary
Ep = Primary voltage
Ip = Primary current

Ts = Turns in secondary
Es = Secondary voltage
Is = Secondary current

Neglecting transformer losses:
kVA primary (input) = kVA secondary (output) and
Ip Tp (amp turns of primary = Is Ts (amp turns of secondary).

To find voltage, current and turns:

$Ep \times Ip = Es \times Is$ $\quad Ep = \frac{Es \times Is}{Ip}$ $\quad Ip = \frac{Es \times Is}{Ep}$

$Es = \frac{Ep \times Ip}{Is}$ $\quad Is = \frac{Ep \times Ip}{Es}$

$Ip \times Tp = Is \times Ts$ $\quad Tp = \frac{Is \times Ts}{Ip}$ $\quad Ip = \frac{Is \times Ts}{Tp}$

$Ts = \frac{Ip \times Tp}{Is}$ $\quad Is = \frac{Ip \times Tp}{Ts}$

$Ep \times Ts = Es \times Tp$ $\quad Ep = \frac{Es \times Tp}{Ts}$ $\quad Tp = \frac{Ep \times Ts}{Es}$

$Es = \frac{Ep \times Ts}{Tp}$ $\quad Ts = \frac{Es \times Tp}{Ep}$

AUTOTRANSFORMERS

Buck and Boost autotransformers (used as per manufacturer's specifications and installed as per NEC)
CHECK ALL LOCAL AND NATIONAL CODES RELATING TO THE USE OF AUTOTRANSFORMERS.
The basic transformer formula applies:
kVA input = kVA output (load)
Single-phase buck and boost transformer connected as an autotransformer to boost 208 volt to 240 volt or to buck 240 volt to 208 volt.
Total kVA capacity = 7.5 x kVA nameplate rating.
Two single-phase transformers connected as a buck or boost autotransformer to boost three-phase 208 volt primary to 230 volt secondary or to buck three-phase 230 volt primary to 208 volt secondary.
Total three-phase kVA capacity = (Total kVA nameplate rating x 10) x .866

See page 124 this book which shows a detailed drawing of a 1 kVA volt, 208/32 volt, 2-winding single-phase transformer connected as an autotransformer to boost the voltage from 208 V to 240 volt. This 1 kVA transformer feeds a 7.5 kVA load and conforms to NEC 210-9.

Trade Competency Test 16

Name ______________________________ Date ______________

Section	Answer		
______	T	F	**1.** Seals must be installed in conduit runs passing between two adjacent Class I, Division 2 rooms.
______	T	F	**2.** Askarel-filled transformers may be used in Class II, Division 2 locations.
______	______		**3.** Locations where combustible dust is normally in the air are designated as ______. A. Class I, Division 2 — C. Class II, Division 2 B. Class II, Division 1 — D. Class III, Division 1
______	T	F	**4.** Locations where concentrations of explosive dust are likely to be encountered are designated as Class II, Division 1.
______	______		**5.** Surgical and other lighting fixtures above hazardous anesthetizing locations shall conform to Section ______.
______	T	F	**6.** Open squirrel-cage induction motors that have no sliding contacts may be used in Class I, Division 1 locations.
______	______		**7.** Sealing compound is used with mineral-insulated cable in a Class I location for the purpose of ______. A. preventing passage of gas or vapor — C. limiting the extent of a possible explosion B. excluding moisture — D. preventing the escape of insulating powder
______	T	F	**8.** Hazardous concentrations of gas or vapor are not normally present in Class I, Division 1 locations.
______	______		**9.** Mobile aircraft energizers shall carry a sign reading "WARNING—KEEP ______ FEET CLEAR OF AIRCRAFT ENGINES AND FUEL TANK AREAS."
______	______		**10.** In Class I, Division 1 locations, conduit seals shall be placed no farther than ______″ from spark-producing devices. A. 12 — C. 24 B. 18 — D. 30
______	______		**11.** In a conduit run leaving a Class I, Division 1 location, the seal may be placed on either side of the ______ within 10′ from the boundary.
______	______		**12.** Where gases and dust are present at the same time, both shall be considered when determining the operating ______ of equipment.
______	______		**13.** Equipment that has been approved for a Division 1 location shall be permitted in a Division 2 location of the same class and ______.
______	______		**14.** For limited flexibility of motor connections in a Class I, Division 2 location, flexible conduit with ______ fittings is permitted.

______ ______ **15.** In a night club, a lighting dimmer installed in an ungrounded conductor shall have overcurrent protection rated at no more than ______% of its own rating.

______ ______ **16.** Fused isolating switches for transformers not intended to interrupt current in a Class I, Division 2 location ______.

A. shall be explosionproof
B. may be installed in a general-use type enclosure
C. shall have interlocking devices
D. may not have doors or openings in hazardous area

______ ______ **17.** In Class I, Division 2 locations, concentrations of explosive gases are normally ______ in closed containers or systems.

______ T F **18.** General-use circuit breaker lighting panels may be used in Class I, Division 2 locations with some restrictions.

______ ______ **19.** A flexible cord may be used in a Class I location between a portable appliance and the ______ portion of its supply circuit.

______ ______ **20.** In Class I, Division 1 locations, pendant fixtures with rigid conduit stems over 12″ long shall be ______.

______ ______ **21.** Locations where combustible fibers are stored are designated as ______.

A. Class II, Division 2
B. Class III, Division 1
C. Class III, Division 2
D. nonhazardous

______ ______ **22.** In a Class II location, a horizontal run of conduit not less than ______′ long may be used as the sealing device.

______ ______ **23.** The circuit to a 1-phase motor on a gasoline dispensing pump shall have a(n) ______ or disconnecting means.

______ ______ **24.** In a parking garage where only limited maintenance work is done, an area up to 18″ is considered to be a(n) ______ area.

______ T F **25.** A location where cloth is woven is designated as Class III, Division 1.

______ ______ **26.** Flexible cords suitable for the type of service and approved for ______ usage may be used as pendants in a commercial garage.

______ ______ **27.** Underground conduit runs within 20′ of a gasoline dispensing location are considered to be in a Class I, Division ______ location.

______ ______ **28.** Class III locations are those where ______ fibers are present.

______ ______ **29.** Where electrodes penetrate an enclosure, ______ listed for the purpose shall be provided unless receptacles are installed.

______ ______ **30.** The secondary voltage of an isolating transformer used in an anesthetizing location may not exceed ______ V.

______ ______ **31.** Only ______ flexible conduit is permitted in Class I, Division 1 areas.

______ ______ **32.** Disappearing stage footlights shall have provisions for ______ disconnecting the current supply when the footlights are replaced in the recess designed for them.

______ ______ **33.** Incandescent lamps less than 8′ from the floor of a theatrical dressing room shall be equipped with ______ guards.

______ ______ **34.** In Class I, Division 1 locations, conduit seals are required adjacent to boxes containing splices if the conduit size is equal to or larger than ______″.

A. $\frac{3}{4}$
B. 1
C. $1\frac{1}{2}$
D. 2

______ ______ **35.** A switch controlling a plug receptacle in a theatrical dressing room shall be equipped with a(n) ______.

Trade Competency Test 17

Name ______________________ Date ______________

Section	Answer	
______	T F	**1.** Dust-tight receptacles for attachment plugs are acceptable in Class II, Division 1 locations.
______	T F	**2.** In general, explosionproof portable electric lamps may be used inside a paint spray booth during operation.
______	______	**3.** Where magnesium, aluminum, or aluminum-bronze powders may be present, ______ are not permitted to be installed.
______	T F	**4.** Portable appliances used inside an aircraft hangar need be approved only for Class I, Division 2 locations.
______	______	**5.** Class 2 and 3 circuits can occupy the same raceway as communication circuits if they are ______ as such.
______	______	**6.** For general wiring in Class I, Division 1 locations, ______ may be used. A. rigid metal conduit B. electrical metallic tubing C. flexible metal conduit D. all of the above
______	T F	**7.** In a commercial garage, a stockroom whose floor level is 18″ higher than that of the garage is normally considered a nonhazardous area.
______	______	**8.** Unit equipment to be used as emergency lighting can be cord-and-plug connected if the cord is not over ______′ long.
______	______	**9.** A permanently installed electric sign shall be controlled by a(n) ______ operable switch or circuit breaker.
______	______	**10.** Areas within 18″ horizontally of gas-dispensing pumps are considered Class I, Division 2 locations to the top of the ______. A. island B. curb C. concrete D. dispenser
______	______	**11.** Amplifiers and rectifiers shall not be installed where they are subjected to ______ temperature.
______	______	**12.** Markings for signs and outline lighting systems shall be ______ after installation.
______	______	**13.** Locations with dried, flammable paints that have ventilating equipment interlocked with the electrical equipment may be designated as ______. A. Class I, Division 2 B. Class II, Division 1 C. Class II, Division 2 D. unclassified
______	______	**14.** Portable lamps used in commercial garages shall be equipped with a hook and ______.
______	______	**15.** Neon tubing that is ______ to pedestrians must be protected from physical damage.
______	T F	**16.** Open spaces within 20′ horizontally of a gas-dispensing pump are designated as Class I, Division 1 locations.

17. Audible and visual signal devices shall be used, where practical, to give warning of _________ of emergency power.

18. If acceptable to the authority having jurisdiction, a separate _________ is suitable for use as an emergency power source.

19. Internal combustion engines used as prime movers shall have at least a(n) _________ hour supply of fuel.

T F 20. Open outside areas within 20′ horizontally from the fill pipe of an underground gasoline tank shall be considered as a Class I, Division 2 location to a height of 18″.

21. Two or more Class 1 circuits can be in the same box, cable, or raceway where all wires are insulated for the _________ voltage.

T F 22. Spray booths may be illuminated through suitable glass panels.

23. Power supply for Class 2 and 3 circuits shall not be connected in _________ unless they are listed.

24. Storerooms and similar areas adjacent to aircraft hangars, but effectively isolated, shall be designated as _________.
   A. Class I, Division 2
   B. Class II, Division 1
   C. Class II, Division 2
   D. unclassified

25. The output to Class 1 remote-control and signaling circuits is unlimited for systems rated at _________ V or less.

26. Wiring to lamps in a motion picture projector shall have insulation with high ambient temperature if exposed to over _________ °C.

27. Lampholders operating on systems rated less than 50 V are required to at least _________ W.

28. When a separate service is used as the emergency power source, it shall be electrically and physically _________ from the normal service.

29. In a Class II location where electrically conducting dust is present, flexible connections at motors may be made with _________.
   A. flexible metal conduit
   B. Type AC armored cable
   C. hard-usage cord
   D. none of the above

30. Branch circuits used for emergency lighting shall be _________ transferred upon failure of normal service.

31. OCPDs for feeders can be set at _________% of their conductor rating in buildings used for motion picture production.

32. Transfer equipment shall disconnect the conductors of one source before connecting the conductor of _________ source.

33. No electrical equipment, other than permanent lighting fixtures, shall be located in a(n) _________ with motion picture film.

34. In commercial garages, the floor area at a height of 18″ above grade is designated as _________.
   A. Class I, Division 1
   B. Class I, Division 2
   C. Class II, Division 1
   D. Class II, Division 2

35. A single switch can be used to disconnect all contactors on a location board that is not more than _________′ away.

Trade Competency Test 18

Name ______________________ Date ______________

Section	Answer	
______	T F	**1.** Resistance and reactor dimmers for stage lighting may be placed in the grounded conductor of a circuit.
______	______	**2.** If the flash point of a liquid is below 100°F in a gasoline dispensing station, the enforcing authority can classify the area where the liquid is not handled as ________.
______	T F	**3.** Branch circuits that supply only neon tubing installations shall be rated at 30 A or less.
______	______	**4.** A well-ventilated room where flammable anesthetics are stored in approved containers shall be considered a ________ area. A. Class I, Division 1 B. Class I, Division 2 C. Class II, Division 2 D. nonhazardous
______	T F	**5.** Storage batteries, motor-generator sets, or isolating transformers may be used to provide ungrounded circuits in anesthetizing locations.
______	______	**6.** Cranes operating over combustible fibers shall be supplied by a(n) ________ power source.
______	______	**7.** For calculating the load for the required branch-circuit for signs or outline lighting in commercial occupancies, ________ VA is used.
______	______	**8.** Conduits installed below the surface of a gasoline dispenser are considered to be in a(n) ________ location.
______	T F	**9.** A low-voltage isolating transformer supplying circuits within an anesthetizing location shall be grounded.
______	______	**10.** Pendant lampholders are not permitted in ________ rooms of theaters.
______	______	**11.** Each commercial building and occupancy with ground floor footage shall have at least one outside sign outlet rated at ________ A. A. 15 B. 20 C. 25 D. 30
______	______	**12.** Boxes and fittings used in Class III, Division 1 locations are required to be ________.
______	______	**13.** A feeder supplying power to stage lighting units with a load of 70,000 VA can have a demand factor of ________% applied.
______	T F	**14.** Lamps in cellulose nitrate film vaults may be installed in general-use fixtures provided they are equipped with suitable guards.
______	______	**15.** Autotransformer dimmers used in theaters shall exceed ________ V between conductors.
______	______	**16.** Dimmers used for theater lighting shall have protection set at ________% of their rating.

_________ _________ **17.** In motion picture studios, the ampacity of feeder conductors to the stage may be protected at a maximum value of _________%.

A. 200 C. 400
B. 250 D. 500

_________ _________ **18.** Cable tray runs shall be _________ before conductors are installed.

_________ T F **19.** Surgical fixtures more than 5′ above the floor of an operating room may be installed on a grounded circuit.

_________ _________ **20.** All circuits routed through a _________ pump shall be switched by a double-pole switch.

_________ _________ **21.** Lamps used in Class III, Division 2 locations shall be protected against the _________ of lint and fibers.

_________ _________ **22.** The maximum fill in wireways or gutters used for sound-recording installations is _________%.

A. 20 C. 58
B. 50 D. 75

_________ _________ **23.** All metal parts around a dispensing pump shall be _________.

_________ _________ **24.** Switches and controllers used in Class III locations shall be installed in _________ enclosures.

_________ _________ **25.** In general, live parts of electric signs and outline lighting other than lamps and neon tubing shall be _________.

_________ _________ **26.** The _________ is the first fitting after a conduit emerges from the earth or concrete in a dispenser pump.

_________ _________ **27.** A storage battery supplying emergency lighting and power shall maintain not less than 87½% of full voltage at total load for a period of at least _________ hour(s).

A. ¼ C. 1½
B. 1 D. 2

_________ _________ **28.** In a Class II location, a 50 kVA transformer containing askarel shall be located at least _________″ from combustible material.

_________ _________ **29.** Storeroom office and control rooms are considered _________ if proper ventilation is provided.

_________ _________ **30.** Portable lamps used in Class III locations shall be of the _________ type.

_________ _________ **31.** Motor-driven utilization equipment shall be dust-_________ if installed in a Class II location.

_________ _________ **32.** Dry-type transformers with open terminals that are installed in Class II locations are required to be installed in _________.

_________ _________ **33.** The primary voltage of an isolating transformer used in an anesthetizing location shall not exceed _________ V.

A. 150 C. 300
B. 250 D. 600

_________ _________ **34.** A 10′ or longer length of horizontal conduit can serve as a(n) _________ in Class II locations.

_________ _________ **35.** The entire floor area in a commercial garage can be classified as nonhazardous with ventilation providing _________ air changes per hour.

_________ T F **36.** The equipment grounding terminal bars of the normal and essential branch circuit panelboard shall be bonded together with an insulated No. 12, or larger, copper conductor.

Trade Competency Test 19

Name ______________________ Date ______________

Section	Answer	
______	T F	**1.** A plug receptacle exclusively for cleaning purposes may be tapped from the emergency circuit wires.
______	______	**2.** Conductors on poles, where not placed on racks or brackets, shall be separated by no less than ______″. A. 6 C. 18 B. 12 D. 24
______	T F	**3.** Power in a Class I remote-control and signal circuit is limited.
______	______	**4.** The color ______ can be used to identify intrinsically safe circuits.
______	T F	**5.** Metal raceways shall not be required where AC cable is used as a wiring method when the metal jacket is identified as a grounding means of a listed cable assembly.
______	______	**6.** If the normal power source is lost, the alternate power source shall be automatically connected within ______ seconds.
______	______	**7.** Intrinsically safe circuits have outputs of very ______ energies.
______	T F	**8.** Testing each level is required when equipment ground fault protection is first installed.
______	______	**9.** The rating of a lampholder on a circuit operating at less than 50 V shall be at least ______ VA. A. 220 C. 550 B. 330 D. 660
______	T F	**10.** A connection on the supply side of the service disconnect may be permitted as an emergency service.
______	______	**11.** On circuits of 600 V or less, overhead feeder spans up to 50′ in length shall have copper conductors no smaller than No. ______. A. 6 C. 10 B. 8 D. 12
______	______	**12.** Lighting fixtures installed at least ______′ above finished grade in commercial garages can be of the general purpose type.
______	T F	**13.** Business offices in health care facilities are not considered as patient care areas.
______	______	**14.** Electrical equipment that is approved to be installed in hazardous locations shall show ______, group, and operating temperature.
______	______	**15.** A pit in a commercial garage with at least ______ changes of air per hour is a Class I, Division 2 location where a dispenser is present.

_________ _________ **16.** In operating rooms using flammable anesthetics, the entire area is a Class I, Division 1 location up to a level of _________′ above the floor.

A. 4 C. 6
B. 5 D. 7

_________ _________ **17.** Surge arrestors installed in Class I, Division 1 locations are to be in _________ enclosures.

_________ _________ **18.** The alternate source of power shall have capacity to supply connected loads for a minimum of _________ hours.

A. 1.5 C. 2.5
B. 2 D. 3

_________ _________ **19.** Class I locations shall not extend beyond a(n) _________ partition floor.

_________ T F **20.** In patient care areas, the grounding terminals of all receptacles shall be connected to an insulated copper equipment grounding conductor.

_________ _________ **21.** Plug receptacles installed in Class I, Division I locations shall be _________ for the location.

_________ _________ **22.** Signaling apparatus used in Class II locations shall be in _________ enclosures.

_________ _________ **23.** Open-type generators shall not be used in Class I, Division 1 locations if they have _________ devices exposed.

_________ _________ **24.** Boxes and fittings used in Class II locations shall have close-fitting covers to contain _________.

_________ _________ **25.** In unattended self-service stations, emergency controls shall be located at least _________′ from dispensers.

_________ _________ **26.** Rubber cord used for flexible connections in a Class II location shall be of the _________ type.

_________ _________ **27.** The isolated circuit conductors for a 3-phase system supplying an anesthetizing location shall have the third conductor identified as _________.

A. yellow C. red
B. black D. blue

_________ _________ **28.** General-purpose enclosures can be used in _________ locations for transformers under certain rules.

_________ _________ **29.** A lighting fixture with an identification number of _________ has an operating temperature of 842°F.

_________ _________ **30.** Portable lamps shall be approved for Class II locations and marked for _________ lamp wattage.

_________ _________ **31.** No transformer or capacitor is permitted in a Class II location where _________ is present.

_________ _________ **32.** A Class I, Division 1 location is an area where flammable gases and vapors are _________ present.

_________ T F **33.** Telephone wires are permitted on the same crossarm with light or power wires if separated by at least 2′.

_________ _________ **34.** Explosionproof equipment contains an explosion without rupturing and _________ escaping gases.

_________ _________ **35.** Where a Class II, Group E motor operates without being subject to overload, its operating temperature is _________°C.

Final Test 1

Name ______________________ Date ______________________

Section	Answer	
______	T F	**1.** The smallest permissible hard-drawn copper service drop conductor is No. 10.
______	______	**2.** A No. ______ aluminum equipment grounding conductor is required for an automatic overcurrent protection device set at 200 A. A. 2 B. 4 C. 6 D. 8
______	T F	**3.** The grounding electrode conductor for a small, one-family dwelling shall be permitted to be spliced with listed means or exothermic weld.
______	______	**4.** The neutral conductor for a high-impedance grounded system shall be at least No. ______ Cu.
______	______	**5.** A 120/240 V feeder shall have a rating of at least ______ A to supply a 9 kVA noncontinuous load and a 12 kVA continuous load.
______	T F	**6.** Any available wall space in a dwelling shall be within 7′ of a plug receptacle.
______	T F	**7.** Switch enclosures, under certain conditions, may be used as junction boxes.
______	T F	**8.** Under certain conditions, a 20′ length of metal raceway need not be grounded.
______	______	**9.** When the initial computed load for a one-family dwelling is 10 kVA, the service disconnecting means rating shall not be less than ______ A. A. 50 B. 60 C. 70 D. 100
______	______	**10.** Conductors of dissimilar metals shall not be intermixed in a terminal or splicing connector where physical contact occurs, unless the device is ______ for the purpose and conditions of use. A. marked B. stamped C. intended D. identified
______	T F	**11.** The service disconnecting means may consist of not more than six switches or circuit breakers.
______	______	**12.** The service or feeder capacity necessary for four 10 kVA electric ranges is ______ kVA. A. 15 B. 17 C. 19 D. 21
______	______	**13.** When initially installed, flexible cords shall be used only in ______ lengths.
______	______	**14.** The clearance of 120/240 V service drop conductors over residential driveways not subject to truck traffic shall be no less than ______′.

__________ __________ **15.** The first 20,000 VA of computed load in a hotel shall be assessed at ________%.

__________ __________ **16.** Unless otherwise specified, conductors used to carry current shall be of ________.

A. copper
B. aluminum
C. copper-clad aluminum
D. neither A, B, nor C

__________ __________ **17.** The copper grounding electrode conductor for a 350 kcmil, 120/240 V service installation, if run to a waterpipe electrode, shall be no smaller than No. ________.

__________ __________ **18.** The smallest permissible driven ground rod electrode made of nonferrous materials is ________″ in diameter.

A. $\frac{3}{8}$
B. $\frac{1}{2}$
C. $\frac{5}{8}$
D. $\frac{3}{4}$

__________ T F **19.** It is permissible to install ten No. 12 conductors in a 4″ by $1\frac{1}{2}$″ square box that also contains one switch.

__________ __________ **20.** The minimum clearance from ground of service drop conductors operating at 600 V or less, which pass over public streets, alleys, roads, and driveways, on other than residential property, shall be ________′.

A. 10
B. 12
C. 15
D. 18

__________ __________ **21.** Under the optional calculation for a one-family dwelling, the central space heating load is assessed at ________%.

__________ __________ **22.** In general, type AC cable shall be supported at intervals not exceeding ________′.

__________ __________ **23.** Heating cables installed on drywall ceilings shall be secured at intervals not exceeding ________″.

A. 10
B. 12
C. 14
D. 16

__________ __________ **24.** The first 3000 VA of computed load in a multifamily dwelling shall be assessed at ________%.

__________ __________ **25.** Receptacles for a 20 A circuit shall be rated no less than ________ A.

__________ __________ **26.** The leads of a 120 V heating cable are marked with ________.

__________ __________ **27.** Except in dwellings, each ________′ or fraction of multioutlet assembly strip is usually considered as a load of 180 VA.

__________ T F **28.** The cross-sectional area of all conductors in any part of an metal auxiliary gutter shall not exceed 15% of the cross-sectional area of the gutter.

__________ __________ **29.** A 5 hp, 3-phase, 230 V squirrel-cage motor with no code letter is marked 40°C and has a nameplate rating of 15 A. The setting of an adjustable overload device shall not exceed ________ A.

A. 19
B. 20
C. 21
D. 25

__________ __________ **30.** The ampacity of No. 2 Type THWN-THHN, dual-rated copper conductors in rigid conduit and installed in a wet location is ________ A.

__________ T F **31.** A header used with precast cellular concrete floor raceways shall not be run at right angles to the cells.

__________ T F **32.** An AC general-use snap switch controlling an inductive load shall have an ampere rating equal to 150% of the load.

______ ______ **33.** Liquidtight flexible nonmetallic conduit is limited in length to ______′.

______ ______ **34.** Metal surface raceways used for signaling and lighting circuits shall have two separate compartments, each identified by contrasting colors of the ______ finish.

______ ______ **35.** Splices and taps in underfloor raceway installations shall be made only in ______.

______ ______ **36.** Nonmetallic underground conduit with conductors is permitted to be used ______.

A. for direct burial installations
B. in cinder fill
C. encased or embedded in concrete
D. A, B, or C

______ ______ **37.** Sheet metal auxiliary gutters, installed in wet locations, shall be ______ for the location.

______ ______ **38.** The motor ______ shall be in sight of the motor and the driven machinery location or shall be individually capable of being locked in the open position.

______ T F **39.** The maximum rating of a 250 V cartridge fuse is 400 A.

______ T F **40.** No. 18 copper wire may be used to ground a portable device connected to a 30 A circuit.

______ T F **41.** Conductors in underfloor raceway systems shall not be tapped at individual outlets.

______ ______ **42.** The minimum size copper grounding electrode conductor required for a 120/240 V service consisting of one 250 kcmil per phase is # ______. The grounding electrode is a ground rod and it is a sole connection to the electrode.

______ ______ **43.** The allowable ampacity of a conductor is 55 A. It is considered protected by a(n) ______ A fuse or circuit breaker.

______ T F **44.** Secondary circuits of wound-rotor induction motors require separate overload protection.

______ ______ **45.** Two or more services are permitted when the rating of the service is over ______ A.

A. 1500
B. 2000
C. 2500
D. 3000

______ T F **46.** Transformers are never required to have overcurrent protection devices in both primary and secondary circuits.

______ ______ **47.** A wood building, without water, is built on the earth without footings. If a 3000 A, 460 V service is installed and grounded by ground rods, the minimum size copper grounding electrode conductor is ______.

A. No. 3/0
B. No. 4
C. No. 6
D. 400 kcmil

______ ______ **48.** The conductor fill of an underfloor raceway shall not exceed ______% of the interior area of the raceway.

A. 20
B. 40
C. 75
D. 100

______ T F **49.** A motor branch circuit short-circuit device may sometimes be rated in excess of 300% of the full load motor current.

__________ T F **50.** The ampacity of the ungrounded conductors from the generator to the first overcurrent device shall not be less than 115% of the nameplate current of the generator.

__________ __________ **51.** Branch circuit conductors supplying a 50 hp, 3-phase, 460 V induction motor shall be rated for at least __________ A.

A. 65 C. 91
B. 82 D. 130

__________ T F **52.** A ½ hp pump motor controlled by a float switch requires running overcurrent protection.

__________ T F **53.** The primary windings of potential transformers installed indoors shall always be fused.

__________ __________ **54.** The largest size conductor permitted in an underfloor raceway is __________.

A. 250 kcmil C. 500 kcmil
B. No. 0 D. determined by design

__________ T F **55.** An 8 A, 115 V, 1-phase, sealed (hermetic) motor compressor shall show the locked rotor current on its nameplate.

__________ T F **56.** The minimum size of Type THW copper conductors supplying a 70 A, 480 V transformer arc welder used at 30% duty application is No. 8.

__________ __________ **57.** A lighting and power service has two No. 3/0 ungrounded copper conductors and a No. 1/0 neutral. A copper grounding electrode conductor attached to a waterpipe electrode shall not be smaller than No. __________.

A. 2 C. 6
B. 4 D. 8

__________ T F **58.** All receptacles on the property must be located a minimum of 10′ from the inside wall of a permanently installed swimming pool.

__________ __________ **59.** The maximum initial rating of branch circuit nontime-delay fuses for a 5 hp, 3-phase, 230 V squirrel-cage motor with no code letter is __________ A.

A. 40 C. 50
B. 45 D. 60

__________ T F **60.** The color coding for intrinsically safe conductors is light green.

__________ __________ **61.** A circuit breaker disconnecting means for a 50 hp, 3-phase, 460 V motor shall have an ampere rating of at least __________ A.

A. 63 C. 91
B. 75 D. 126

__________ __________ **62.** The maximum size of nontime-delay fuses for branch circuit overcurrent protection for a 50 hp, 3-phase, 460 V motor is __________ A.

A. 90 C. 250
B. 200 D. 300

__________ __________ **63.** Duct heaters must include a fan circuit interlock which ensures that the fan circuit is __________ when any heater circuit is energized.

A. locked out C. energized
B. locked in D. neither A, B, nor C

__________ __________ **64.** A 5 hp, 3-phase, 230 V squirrel-cage motor with no code letter is started at full voltage. The minimum ampacity of the branch circuit conductor shall be __________ A.

A. 15.2 C. 19
B. 18 D. 20

Final Test 2

Name ______________________ Date ______________

Section	Answer	
________	T F	**1.** When the power for a fire pump is provided by a tap ahead of the service disconnecting means, the tap is permitted to be made within the service disconnecting means compartment.
________	________	**2.** A demand factor of ________% may be applied to the neutral feeder load for electric ranges. A. 40 C. 70 B. 60 D. 80
________	T F	**3.** Flexible metal conduit used to provide flexibility at equipment shall be equipped with an EGC.
________	T F	**4.** A tap, for an individual lampholder or fixture more than 12″ long, may not be smaller than the branch circuit conductors.
________	T F	**5.** Under certain conditions, the neutral feeder conductor is permitted to be smaller than the ungrounded conductors.
________	________	**6.** In general, wiring for emergency systems shall be kept entirely ________ of all other wiring and equipment and shall not occupy the same raceway, box, cable or cabinet with other wiring.
________	T F	**7.** Service entrance conductors, in general, shall not be spliced.
________	________	**8.** The ampacity of a No. 3/0 Type THW copper conductor at a room temperature of 50°C is ________ A. A. 125 C. 175 B. 150 D. 200
________	________	**9.** The rating of conductors supplying power to a capacitor shall be at least one third of the motor circuit conductors and ________% of the FLC rating of the capacitor.
________	________	**10.** Conduit bodies shall not contain splices, taps, or devices unless they are marked by the manufacturer to indicate their ________ capacity.
________	________	**11.** For electrical equipment rated over ________ A or more than 6′ wide, that contains overcurrent devices, there shall be at least one entrance 24″ wide and 6½′ high at each end of the work space.
________	T F	**12.** Embedded electric space heating cables shall not be spliced.
________	T F	**13.** Nonflammable, fluid-insulated transformers installed indoors and rated at 15,000 V shall be installed in vaults.
________	________	**14.** A transformer with a primary FLC rating of less than 2 A shall be protected at ________% of its primary current.
________	________	**15.** The ampacity of conductors supplying a 20 A oscillator induction heating unit shall not be less than ________ A.

__________ __________ **16.** The primary overcurrent protection device, when properly selected, may protect the secondary conductors only if the secondary of the transformer is 1-phase and feeds a(n) _________-wire circuit.

__________ __________ **17.** The feeder load for three electric ranges (13 kVA, 18 kVA, and 20 kVA) is _________ kVA.

A. 15 C. 19
B. 17.5 D. 21

__________ __________ **18.** Where thermal cutouts are used for motor overload protection for a 3-phase motor, a total of _________ thermal cutouts is required.

__________ __________ **19.** Where explosionproof enclosures cannot be used in Class I, Division 1 locations due to grouping, _________, and pressurized enclosures can be used.

__________ __________ **20.** Construction site receptacles mounted on a(n) _________ kW or less generator do not require GFCI protection.

A. 5 C. 15
B. 10 D. 20

__________ __________ **21.** To supply a feeder for night lights in an apartment, the required size of Type THW copper conductors for a continuous 100 A load shall not be smaller than No. _________.

A. 0 C. 2
B. 1 D. 3

__________ T F **22.** The screw shell of a fuseholder shall not be connected to the line side of the circuit.

__________ __________ **23.** Inside locations of spray booths are classified as a Class I, _________ location.

__________ __________ **24.** The general lighting load for an office building is calculated to be _________ VA per sq ft.

A. 2 C. 3.5
B. 2.5 D. 5

__________ __________ **25.** Transformers shall not be installed in locations where _________ dust is present.

__________ __________ **26.** In general, raceways shall not be used as a means of support for other _________, cables, or nonelectric equipment.

__________ __________ **27.** The maximum rating of a cord-and-plug connected appliance used on a 20 A branch circuit shall be _________ A.

A. 10 C. 16
B. 12 D. 20

__________ __________ **28.** The area within _________′ horizontally form aircraft fuel tanks is classified as a Class I, Division 2 location.

__________ __________ **29.** The electrical systems in aircraft shall be _________ when they are stored in hangars.

__________ __________ **30.** PVC can be used to supply power to a dispenser pump if buried at least _________′ in the earth.

__________ __________ **31.** Under certain conditions, an unbroken run of conduit can pass through a Class I, Division 1 location without _________ being provided.

__________ __________ **32.** Approved sealing _________ shall be used to seal fittings in a conduit run.

______ ______ **33.** Receptacles in marinas and boatyards that provide shore power to boats shall be rated not less than 20 A and shall be of the locking and ________ types.

______ ______ **34.** Nonmedical X-ray, high-voltage ________ cables shall be grounded.

______ ______ **35.** Spas and hot tubs with associated electrical components shall be ________-protected.

______ ______ **36.** The ampacity of conductors supplying an oscillator-type, induction-heating unit need not exceed its ________ rating.

______ ______ **37.** The full load current of a 5 hp, 3-phase, 208 V induction motor is ________ A.

______ ______ **38.** The radius of bends in mineral-insulated cable with a ¾″ external diameter shall be not less than ________ times the cable diameter.

______ ______ **39.** All exposed noncurrent-carrying metal parts of information technology equipment shall be ________.

______ ______ **40.** Horsepower rated switches are not required as a disconnecting means for 300 V stationary motors of ________ hp or less.

______ ______ **41.** In general, the controller disconnect shall be in sight of the ________.

______ T F **42.** Open wiring on insulators is permitted to be used for branch circuit temporary wiring on construction sites.

______ ______ **43.** Open transformers installed indoors and rated less than 112.5 kVA shall have a clearance of at least ________″ from combustible material.

______ ______ **44.** A psychiatric hospital is used exclusively for such care on a 24-hour basis for ________ or more inpatients.

______ ______ **45.** The isolated circuit conductors for a 2-wire system supplying an anesthetizing location shall be identified by using ________ conductors.

A. orange and brown
B. black and red
C. purple and pink
D. none of the above

______ ______ **46.** Flexible cords and cables shall not be used as a substitute for the ________ wiring of a structure.

______ ______ **47.** Intrinsic Safety Wiring methods shall be marked and identified every ________′.

______ ______ **48.** Unshielded lead-in conductors of amateur transmitting stations shall clear the building surface that is wired over by a distance of at least ________″.

A. 1
B. 2
C. 3
D. 4

______ ______ **49.** General purpose enclosures can be used in Class I, Division 2 locations if the contacts are immersed in ________.

______ ______ **50.** A basic parking garage is not considered a(n) ________ location.

______ ______ **51.** Flexible cords feeding 120 V equipment within a hazardous anesthetizing location shall be of the ________ type.

______ ______ **52.** A 10′ or longer length of horizontal conduit can serve as a(n) ________ in Class II locations.

______ ______ **53.** The entire floor area in a commercial garage can be classified as nonhazardous with ventilation providing ________ air changes per hour.

______ ______ **54.** Ten 10 kVA ovens located in an apartment complex can have a demand factor of ________ kVA.

__________ __________ **55.** In general, deck boxes for swimming pools must be located at least 4′ from the inside wall of the pool and they shall be located not less than 4″ from the pool deck or 8″ from the maximum level of the pool water, whichever provides the _________ elevation.

__________ __________ **56.** With no overcurrent protection in the secondary, the primary overcurrent device for a 600 V dry-type transformer with a rated primary current of 100 A shall not exceed _________ A.

A. 125
B. 150
C. 250
D. 600

__________ __________ **57.** The use of a continuous white or natural gray covering on a conductor or termination is reserved for the _________ conductor.

__________ __________ **58.** A bathroom is an area with a(n) _________ and one or more of the following: toilet, tub, or shower.

__________ __________ **59.** Receptacles installed over kitchen countertops 12″ or wider shall be not more than _________″ apart.

__________ __________ **60.** A service with phase conductors rated at 300 kcmil per phase shall have a minimum grounded conductor of No. _________ copper.

__________ T F **61.** Two to six disconnecting means can serve to disconnect the service equipment of a multifamily dwelling.

__________ __________ **62.** In commercial occupancies a maximum of _________ receptacle outlets may be installed on a 20 A OCPD.

__________ T F **63.** The lowest point of attachment for a service drop to a building is 12′.

__________ __________ **64.** THHN copper conductors in a circuit routed through an ambient temperature of 121°F shall be derated _________%.

__________ __________ **65.** A No. 6 copper conductor rated at 60°C terminated to an OCPD has an ampacity of _________ A.

__________ __________ **66.** The smallest size EGC permitted in cable trays is No. _________ or larger.

__________ T F **67.** The ungrounded conductors of a 3-pole OCPD shall be disconnected simultaneously.

__________ __________ **68.** The feeder circuit from the service equipment to a mobile home shall have a(n) _________ EGC.

__________ __________ **69.** A 6′ fence with 1′ of barbed wire on top is equivalent to a(n) _________′ fence for guarding equipment of over 600V.

__________ __________ **70.** A(n) _________ VA rating is required for each linear foot of show window.

__________ __________ **71.** A control transformer with a primary FLC of 1 A can be protected with a(n) _________ A fuse.

__________ __________ **72.** A disconnecting means located within _________′ and within sight can serve as the disconnecting switch for a motor.

__________ __________ **73.** A 4 $^{11}/_{16}$″ by 1¼″ square box has a(n) _________ cu in. rating.

ENCLOSURES – ELECTRICAL EQUIPMENT (1000 VOLTS MAXIMUM) NATIONAL ELECTRICAL MANUFACTURERS ASSOCIATION (NEMA)

INTRODUCTION

An enclosure is a surrounding case constructed to provide a degree of protection to personnel against incidental contact with the enclosed equipment and to provide a degree of protection to the enclosed equipment against specified environmental conditons.

A brief description of the more common types of enclosures used by the electrical industry relating to their environmental capabilities follows. Refer to the appropriate sections of this standards publication for more information regarding applications, features, and design tests.

NEMA Standard 1-10-1979.

NEC Art. 100 Basic definition of an enclosure is a case or housing of apparatus to prevent personnel from accidentally contacting energized parts, or to protect the equipment from physical damage.

DEFINITIONS PERTAINING TO NONHAZARDOUS LOCATIONS

Type 1 Enclosure

Type 1 enclosures are intended for indoor use primarily to provide a degree of protection against contact with the enclosed equipment.

NEMA Standard 1-10-1979.

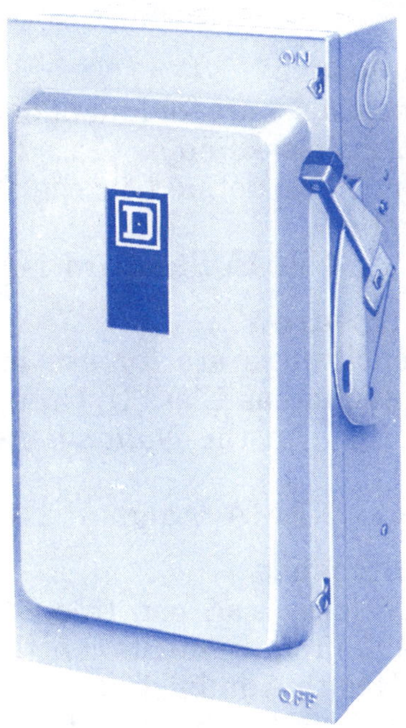

NEMA TYPE I ENCLOSURE

(Courtesy of Square D Company)

Type 2 Enclosure

Type 2 enclosures are intended for indoor use primarily to provide a degree of protection against limited amounts of falling water and dirt.

NEMA Standard 1-10-1979.

Type 3 Enclosure

Type 3 enclosures are intended for outdoor use primarily to provide a degree of protection against windblown dust, rain, sleet, and external ice formation.

NEMA Standard 1-10-1979.

Type 3R Enclosure

Type 3R enclosures are intended for outdoor use primarily to provide a degree of protection against falling rain, sleet, and external ice formation.

NEMA Standard 1-10-1979.

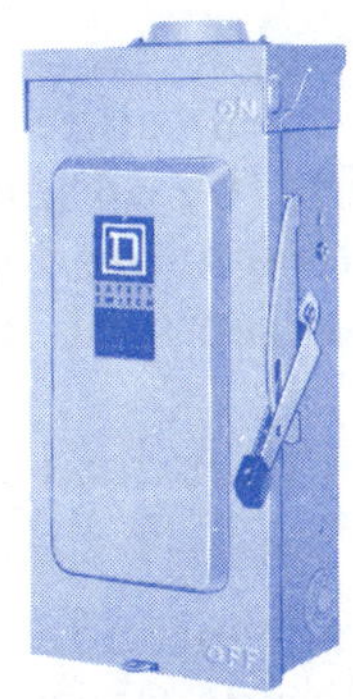

NEMA TYPE 3R ENCLOSURE

(Courtesy of Square D Company)

Type 3S Enclosure

Type 3S enclosures are intended for outdoor use primarily to provide a degree of protection against windblown dust, rain, sleet, and provide for operation of external mechanisms when ice laden.

NEMA Standard 1-10-1979.

Type 4 Enclosure

Type 4 enclosures are intended for indoor or outdoor use primarily to provide a degree of protection against windblown dust and rain, splashing water, and hose-directed water.

NEMA Standard 1-10-1979.

Type 4X Enclosure

Type 4X enclosures are intended for indoor or outdoor use primarily to provide a degree of protection against corrosion, windblown dust and rain, splashing water, and hose-directed water.

NEMA Standard 1-10-1979.

ENCLOSURES – ELECTRICAL EQUIPMENT (1000 VOLTS MAXIMUM)
NATIONAL ELECTRICAL MANUFACTURERS ASSOCIATION (NEMA) – Continued

Type 5 Enclosure

Type 5 enclosures are intended for indoor use primarily to provide a degree of protection against dust and falling dirt.

NEMA Standard 1-10-1979.

Type 6 Enclosure

Type 6 enclosures are intended for indoor or outdoor use primarily to provide a degree of protection against the entry of water during occasional temporary submersion at a limited depth.

NEMA Standard 1-10-1979.

Type 6P Enclosure

Type 6P enclosures are intended for indoor or outdoor use primarily to provide a degree of protection against the entry of water during prolonged submersion at a limited depth.

NEMA Standard 1-10-1979.

Type 11 Enclosure

Type 11 enclosures are intended for indoor use primarily to provide, by oil immersion, a degree of protection to enclosed equipment against the corrisive effects of liquids and gases.

NEMA Standard 1-10-1979.

Type 12 Enclosure

Type 12 enclosures are intended for indoor use primarily to provide a degree of protection against dust, falling dirt, and dripping noncorrosive liquids.

NEMA Standard 1-10-1979.

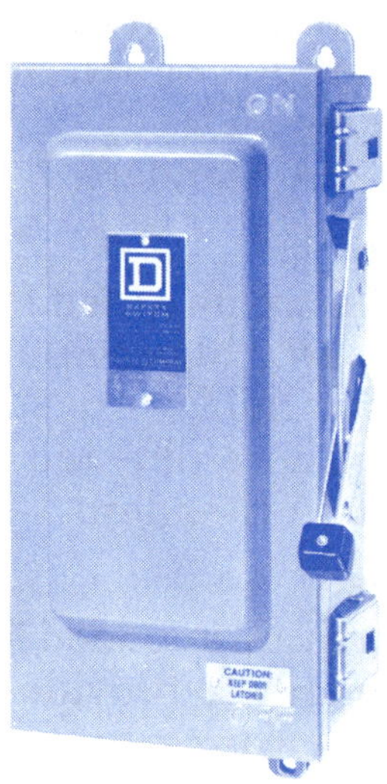

NEMA TYPE 12 ENCLOSURE

(Courtesy of Square D Company)

Type 12K Enclosure

Type 12K enclosures with knockouts are intended for indoor use primarily to provide a degree of protection against dust, falling dirt, and dripping noncorrosive liquids other than at knockouts.

NEMA Standard 1-10-1979.

Type 13 Enclosure

Type 13 enclosures are intended for indoor use primarily to provide a degree of protection against dust, spraying of water, oil, and noncorrosive coolant.

NEMA Standard 1-10-1979.

DEFINITIONS PERTAINING TO HAZARDOUS (CLASSIFIED) LOCATIONS

Type 7 Enclosure

Type 7 enclosures are for use indoors in locations classified as Class I, Groups A, B, C, or D, as defined in the *National Electrical Code*.

NEMA Standard 1-10-1979.

NEMA TYPE 7 & 9 ENCLOSURE

(Courtesy of Square D Company)

Type 8 Enclosure

Type 8 enclosures are for indoor or outdoor use in locations classified as Class I, Groups A, B, C, or D, as defined in the *National Electrical Code*.

NEMA Standard 1-10-1979.

Type 9 Enclosure

Type 9 enclosures are for use in indoor locations classified as Class II, Groups E, F, or G, as defined in the *National Electrical Code*.

NEMA Standard 1-10-1979.

Type 10 Enclosure

Type 10 enclosures are constructed to meet the applicable requirements of the Mine Safety and Health Administration.

NEMA Standard 1-10-1979.

Electrical Symbols Commonly Used on Blueprints

GENERAL OUTLETS

Ceiling Wall

- Outlet.
- Blanked Outlet.
- Drop Cord.
- Electrical Outlet; for use only when circle used alone might be confused with columns, plumbing symbols, etc.
- Fan Outlet.
- Junction Box.
- Lamp Holder.
- Lamp Holder with Pull Switch.
- Pull Switch.
- Outlet for Vapor Discharge Lamp.
- Exit Light Outlet.
- Clock Outlet. (Specify Voltage.)

CONVENIENCE OUTLETS

- Duplex Convenience Outlet.
- Convenience Outlet other than Duplex. 1=Single, 3=Triplex, etc.
- Weatherproof Convenience Outlet.
- Range Outlet.
- Switch and Convenience Outlet.
- Radio and Convenience Outlet.
- Special Purpose Outlet. (Des. in Spec.)
- Floor Outlet.

SWITCH OUTLETS

S	Single Pole Switch.
S_2	Double Pole Switch.
S_3	Three-Way Switch.
S_4	Four-Way Switch.
S_D	Automatic Door Switch.
S_E	Electrolier Switch.
S_K	Key Operated Switch.
S_P	Switch and Pilot Lamp.
S_{CB}	Circuit Breaker.
S_{WCB}	Weatherproof Circuit Breaker.
S_{MC}	Momentary Contact Switch.
S_{RC}	Remote Control Switch.
S_{WP}	Weatherproof Switch.
S_F	Fused Switch.
S_{WF}	Weatherproof Fused Switch.

SPECIAL OUTLETS

O a,b,c,etc
⊖ a,b,c,etc
S a,b,c,etc

Any Standard Symbol as given above with the addition of a lower case subscript letter may be used to designate some special variation of Standard Equipment of particular interest in a specific set of Architectural Plans.

When used, they must be listed in the Key of Symbols on each drawing and if necessary further described in the specifications.

Courtesy of American National Standards Institute New York, N.Y.

PANELS, CIRCUITS, AND MISCELLANEOUS

- Lighting Panel.
- Power Panel.
- Branch Circuit; Concealed in Ceiling or Wall.
- Branch Circuit; Concealed in Floor.
- Branch Circuit; Exposed.
- Home Run to Panel Board. Indicate number of Circuits by number of arrows.
  Note: Any circuit without further designation indicates a two-wire circuit. For a greater number of wires indicate as follows —///— (3 wires) —////— (4 wires), etc.
- Feeders. Note: Use heavy lines and designate by number corresponding to listing in Feeder Schedule.
- Underfloor Duct and Junction Box. Triple System. For double or single systems, eliminate 1 or 2 lines. This symbol equally adaptable to auxiliary system layouts.
- Generator.
- Motor.
- Instrument.
- Power Transformer. (Or draw to scale.)
- Controller.
- Isolating Switch.

AUXILIARY SYSTEMS

- Push Button.
- Buzzer.
- Bell.
- Annunciator.
- Outside Telephone.
- Interconnecting Telephone.
- Telephone Switchboard.
- Bell Ringing Transformer.
- D — Electric Door Opener.
- F — Fire Alarm Bell.
- F — Fire Alarm Station.
- City Fire Alarm Station.
- FA — Fire Alarm Central Station.
- FS — Automatic Fire Alarm Device.
- W — Watchman's Station.
- W — Watchman's Central Station.
- H — Horn.
- N — Nurse's Signal Plug.
- M — Maid's Signal Plug.
- R — Radio Outlet.
- SC — Signal Central Station.
- Interconnection Box.
- Battery.
- Auxiliary System Circuits.
  Note: Any line without further designation indicates a 2-Wire System. For a greater number of wires designate with numerals in manner similar to —-—12-No. 18W-¾″ C., or designate by number corresponding to listing in Schedule.
- □ a,b,c — Special Auxiliary Outlets. Subscript letters refer to notes on plans or detailed description in specs.

ELECTRICAL SYMBOLS

ELECTRICAL SYMBOLS: Normally a list of Electrical Symbols is submitted with each set of electrical blueprints. Most of the material and equipment on a blueprint can be readily identified by using the standard ANSI Electrical Symbols, but in some cases where special material or equipment is used, such as Automatic Fire Alarm Systems, Smoke Detectors or receptacles for special equipment, it becomes necessary for the engineer to concoct a symbol which will serve to identify these special items. The following is a typical application of Electrical Symbols which would be submitted with a blueprint.

- INCANDESCENT OR H.I.D. FIXTURE
- INCANDESCENT OR H.I.D. WALL BRACKET
- FLUORESCENT FIXTURE
- EXIT LIGHT
- TYPICAL OUTLET DESIGNATION: FIXTURE TYPE (14), CIRCUIT (2), SWITCHING (b)
- JUNCTION BOX
- CLOCK OUTLET ....7′-6″
- DELTA
- PHASE

- DUPLEX RECEPTACLE ....1′-0″
- DUPLEX RECEPTACLE SPLIT WIRED ....1′-0″
- 120/208, 1ϕ COMBINATION RECEPTACLE ..1′-0″
- FLOOR OUTLET (ATTACH ADDITIONAL SYMBOL FOR RECEPT. TEL., ETC.)
- SPECIAL OUTLET—AS NOTED
- 240V, 3W, 4P RECEPTACLE—AMPS AS NOTED
- 240V, 2W, 3P RECEPTACLE—AMPS AS NOTED
- 120V, 2W 3P, 30-AMP RECEPTACLE
- GROUND
- FUSE
- CIRCUIT BREAKER

- S — SINGLE POLE SWITCH ....4′-0″
- S_2 — DOUBLE POLE SWITCH ....4′-0″
- S_3 — THREE-WAY SWITCH ....4′-0″
- S_4 — FOUR-WAY SWITCH ....4′-0″

- S_D — DIMMER SWITCH ....4′-0″
- S_K — KEY OPERATED SWITCH ....4′-0″
- S_P — SWITCH AND PILOT LIGHT ....4′-0″
- S_{MC} — MOMENTARY CONTACT SWITCH ....4′-0″

- BRANCH CIRCUIT PANEL
- DISTRIBUTION PANEL
- EXPOSED CONDUIT
- CONDUIT CONCEALED IN CEILING OR WALLS
- CONDUIT CONCEALED IN OR UNDER FLOOR
- SM — SURFACE METAL RACEWAY

- CONDUIT UP
- CONDUIT DOWN
  CROSSLINES INDICATE NUMBER OF WIRES—SEE SPECIFICATIONS
- UNDERFLOOR DUCT & JUNCTION BOX, NUMBER OF LINES INDICATES NUMBER OF DUCTS

- M-3 — MOTOR—NUMBER, IF ANY, REFERS TO MOTOR SCHEDULE
- S_T — THERMAL SWITCH
- MOTOR STARTER
- SAFETY SWITCH
- AQUASTAT ....5′-0″

- THERMOSTAT ....5′-0″
- HUMIDISTAT
- HOA — CONTROL DEVICE—LETTERS INDICATE SPECIAL FUNCTION SUCH AS:
  HOA—HAND OFF AUTO
  PBPL—PUSH BUTTON, PILOT LIGHT
- REMOTE CONTROL SWITCH OR CONTACTOR

ALL MOUNTING HEIGHTS ARE AS LISTED ABOVE UNLESS NOTED OTHERWISE ON PLANS.

APPENDIX

Symbol	Description
T	TRANSFORMER
	PUSH BUTTON
	BUZZER
	BELL
H	CHIME
	TELEPHONE OUTLET ....1′-0″
	INTERCOM OUTLET
S	SPEAKER
T	TELEVISION OUTLET
M	MICROPHONE OUTLET

FIRE ALARM SYMBOLS

Symbol	Description
F	FIRE ALARM SENDING STATION ....5′-0″
F	FIRE ALARM HORN ....7′-6″
F	FIRE ALARM BELL ....7′-6″
A	AUTOMATIC FIRE ALARM STATION
A H	AUTOMATIC FIRE ALARM STATION—HIGH TEMP.
ELR	FIRE ALARM END OF LINE RESISTOR
FA	FIRE ALARM PANEL
F H	FIRE ALARM PRE-SIGNAL CHIME
A S	AUTOMATIC FIRE ALARM STATION—IONIZATION SMOKE DETECTOR
F D	MAGNETIC DOOR HOLDER

ALL MOUNTING HEIGHTS ARE AS LISTED ABOVE UNLESS NOTED OTHERWISE ON PLANS.

ABBREVIATIONS

A or AMP	AMPERE
AC	ALTERNATING CURRENT
APPROX	APPROXIMATE
AWG	AMERICAN WIRE GAGE
BLDG	BUILDING
C or CND	CONDUIT
CB	CIRCUIT BREAKER
CAB	CABINET
CIR or CKT	CIRCUIT
CLG	CEILING
CO	COMPANY
COMPR or CPRSR	COMPRESSOR
CONN	CONNECTED OR CONNECTION
CONTR	CONTRACTOR
DC	DIRECT CURRENT
DISTR	DISTRIBUTION
DT	DUST TIGHT
EMT	ELECTRICAL METALLIC TUBING
EP	EXPLOSION PROOF
EXH	EXHAUST
EXIST	EXISTING
FNSH	FINISH
FL	FLOOR
FLA	FULL LOAD AMPS
FPN	FINE PRINT NOTE
FUT	FUTURE
GFCI	GROUND FAULT CIRCUIT INTERRUPTER
GRND or GND	GROUND
HOA	HAND-OFF-AUTO
HTG	HEATING
IMC	INTERMEDIATE METAL CONDUIT
JB	JUNCTION BOX
LOC	LOCATION
LT	LIGHT
LTG	LIGHTING
LTS	LIGHTS
MAN or MNL	MANUAL
MCC	MOTOR CONTROL CENTER
MECH	MECHANICAL
MIN	MINIMUM
MTR or MOT	MOTOR
MS	MOTOR SWITCH
MTD	MOUNTED
NEUT	NEUTRAL
NO	NUMBER
OC	OVERCURRENT
OUT	OUTLET
PB	PUSH BUTTON
PF	POWER FACTOR
PL	PILOT LIGHT
PNL	PANEL
PR	PAIR
PVC	POLYVINYL CHLORIDE
RECEPT or RCPT	RECEPTACLE
RM	ROOM
RT	RAINTIGHT
SE	SERVICE ENTRANCE
SHT or SH	SHEET
SP	SPARE
STR or START	STARTER
SW	SWITCH
TEL	TELEPHONE
TRANS or XFMR	TRANSFORMER
TERM	TERMINAL
V	VOLTAGE
VD	VOLTAGE DROP
VT	VAPOR TIGHT
WP or WTRPRF	WEATHERPROOF
W	WATT

Shown below are different types of bases for incandescent lamps. The bayonet, candelabra, and intermediate base are used on the small-sized (miniature) lamps. The medium base, used on general-service lamps of 300 watts or less, is the most common type. The mogul base is used on sizes of 300 watts and up. The admedium is slightly larger in diameter than the medium and is used on some of the mercury Mazda lamps. The three-contact base is used with a three-lite type of lamp. The disc base is used on lumiline lamps. The medium and mogul prefocused bases are used on certain types of concentrated-filament lamps, such as those for picture projection and aviation service where it is desirable to have the light source accurately located. The medium bipin base is for fluorescent lamps. The medium bipost base is made in 500, 750, and 1,000 watt lamp sizes for use principally with indirect fixtures, where it allows a better radiation of heat than is obtainable with the mogul-base type. For the large size lamps of 1,500 watts and for floodlights, the mogul bipost is the standard.

BASES FOR INCANDESCENT LAMPS

Listed below are the dimensions of six common bases for incandescent lamps.

	Depth	Diameter
1. Mogul	1⅝	1½
2. Admedium	1⅛	1⅛
3. Medium	1$\frac{5}{16}$	1
4. Intermediate	$\frac{25}{32}$	$\frac{21}{32}$
5. Candelabra	⅝	⅝
6. Miniature	$\frac{15}{32}$	⅜

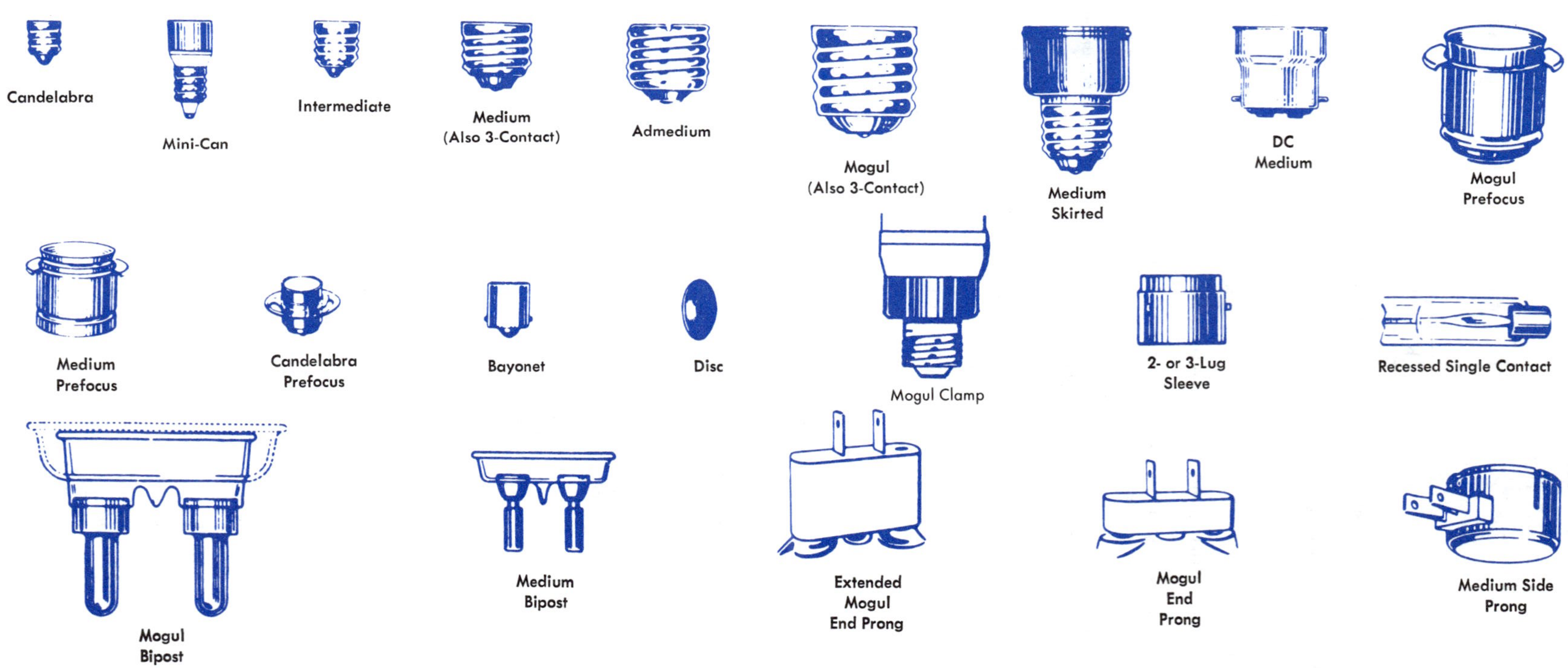

NEC 210-24 Branch circuits rated 30, 40 and 50 amps having two or more outlets may supply fixed lighting units with heavy-duty (mogul) lampholders in other than dwelling occupancies.

NEC 210-6 The voltage on branch circuits supplying lampholders of 15-amp or less rating shall not exceed 150 volts. (See Exceptions.)

NEC 410-47 The identified conductor, where used, shall be connected to the screw-shell of a lampholder.

NEC 410-53 An incandescent lamp for general use on lighting branch circuits shall not be equipped with a medium base when rated over 300 watts, nor with a mogul base when rated over 1,500 watts. Above 1,500 watts, special approved bases or other devices shall be used.

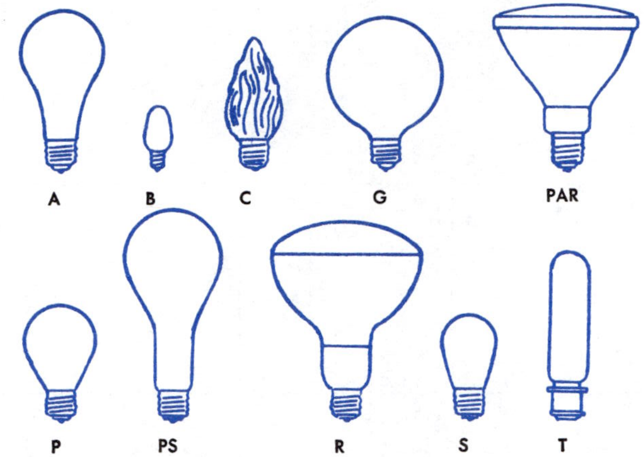

BULB DESIGNATIONS OF INCANDESCENT LAMPS

Shape of Bulb	Designating Letter	Shape of Bulb	Designating Letter
Standard Line	A	Pear Shaped	P or PS
Cone Shaped	D	Reflector	R
Flame Shaped	F	Straight Side	S
Globular	G	Tubular	T
Parabolic	PAR		

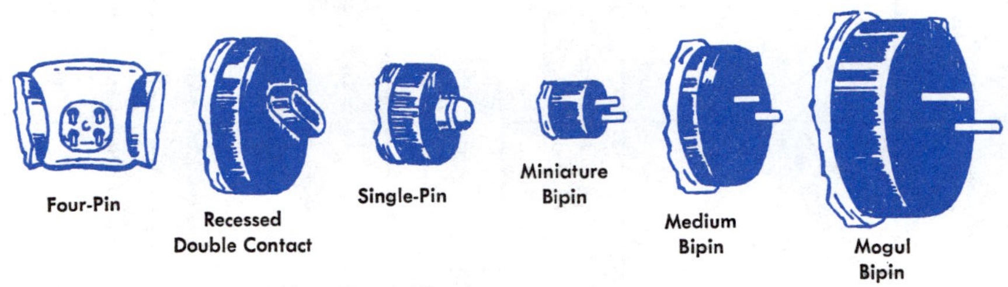

BASES FOR FLUORESCENT LAMPS

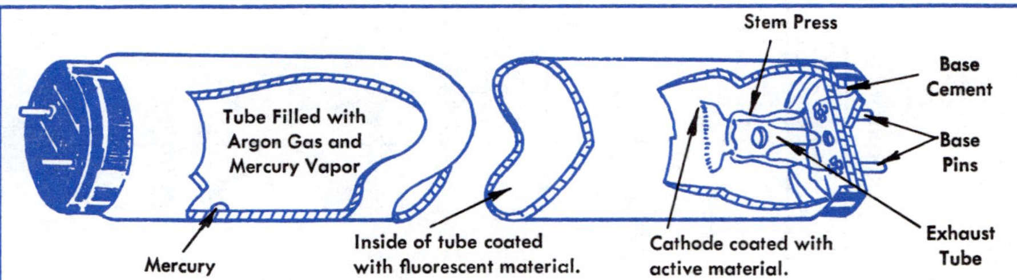

FLUORESCENT LAMP CONSTRUCTION

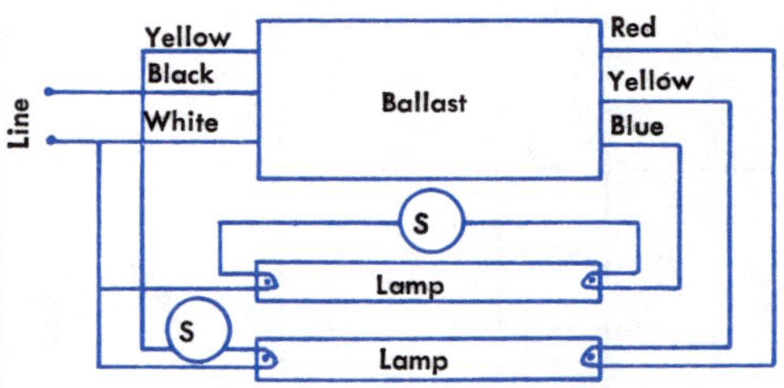

BASIC PREHEAT (SWITCH-START) CIRCUIT

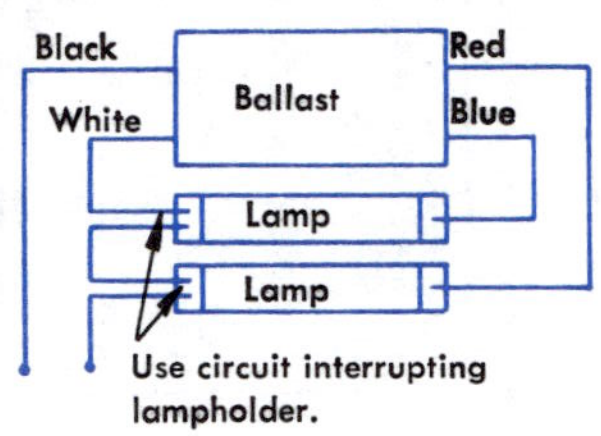

LEAD-LAG CIRCUIT

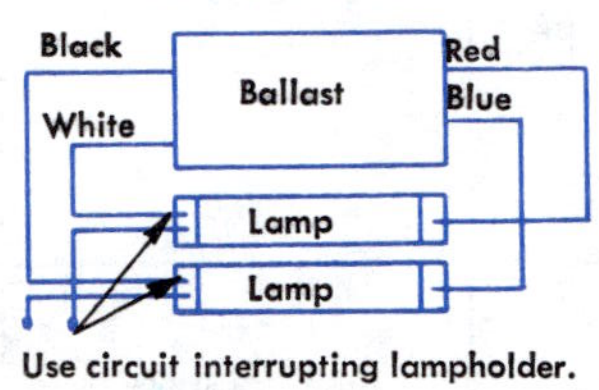

SERIES-SEQUENCE CIRCUIT

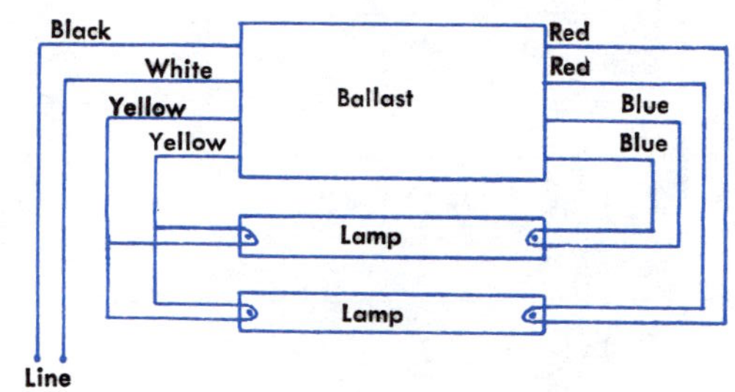

RAPID-START (AND TRIGGER-START) CIRCUIT

CASES FOR FLUORESCENT LAMP BALLASTS

NEC 410-31 Fixtures shall not be used as a raceway for circuit conductors unless they meet requirements for approved raceways, except a single branch circuit supplying fixtures if fixtures are specifically approved. (See Exceptions.) Branch-circuit conductors within 3 inches of a ballast within ballast compartment shall be rated at not less than 90°C.

CONFIGURATIONS FOR GENERAL-PURPOSE NONLOCKING PLUGS AND RECEPTACLES

			15 AMPERE		20 AMPERE		30 AMPERE		50 AMPERE		60 AMPERE	
			RECEPTACLE	PLUG	RECEPTACLE	PLUG	RECEPTACLE	PLUG	RECEPTACLE	PLUG	RECEPTACLE	PLUG
2-POLE 2-WIRE	1	125 V	1-15R	1-15P								
	2	250 V		2-15P	2-20R	2-20P	2-30R	2-30P				
	3	277 V	(RESERVED FOR FUTURE CONFIGURATIONS)									
	4	600 V	(RESERVED FOR FUTURE CONFIGURATIONS)									
2-POLE 3-WIRE GROUNDING	5	125 V	5-15R	5-15P	5-20R	5-20P	5-30R	5-30P	5-50R	5-50P		
	6	250 V	6-15R	6-15P	6-20R	6-20P	6-30R	6-30P	6-50R	6-50P		
	7	277 V AC	7-15R	7-15P	7-20R	7-20P	7-30R	7-30P	7-50R	7-50P		
	24	347 V AC	24-15R	24-15P	24-20R	24-20P	24-30R	24-30P	24-50R	24-50P		
	8	480 V AC	(RESERVED FOR FUTURE CONFIGURATIONS)									
	9	600 V AC	(RESERVED FOR FUTURE CONFIGURATIONS)									
3-POLE 3-WIRE	10	125/250 V			10-20R	10-20P	10-30R	10-30P	10-50R	10-50P		
	11	3 ø 250 V	11-15R	11-15P	11-20R	11-20P	11-30R	11-30P	11-50R	11-50P		
	12	3 ø 480 V	(RESERVED FOR FUTURE CONFIGURATIONS)									
	13	3 ø 600 V	(RESERVED FOR FUTURE CONFIGURATIONS)									
3-POLE 4-WIRE GROUNDING	14	125/250 V	14-15R	14-15P	14-20R	14-20P	14-30R	14-30P	14-50R	14-50P	14-60R	14-60P
	15	3 ø 250 V	15-15R	15-15P	15-20R	15-20P	15-30R	15-30P	15-50R	15-50P	15-60R	15-60P
	16	3 ø 480 V	(RESERVED FOR FUTURE CONFIGURATIONS)									
	17	3 ø 600 V	(RESERVED FOR FUTURE CONFIGURATIONS)									
4-POLE 4-WIRE	18	3 ø 208Y/120 V	18-15R	18-15P	18-20R	18-20P	18-30R	18-30P	18-50R	18-50P	18-60R	18-60P
	19	3 ø 480Y/277 V	(RESERVED FOR FUTURE CONFIGURATIONS)									
	20	3 ø 600Y/347 V	(RESERVED FOR FUTURE CONFIGURATIONS)									
4-POLE 5-WIRE GROUNDING	21	3 ø 208Y/120 V	(RESERVED FOR FUTURE CONFIGURATIONS)									
	22	3 ø 480Y/277 V	(RESERVED FOR FUTURE CONFIGURATIONS)									
	23	3 ø 600Y/347 V	(RESERVED FOR FUTURE CONFIGURATIONS)									

Courtesy National Electrical Manufacturers Association (NEMA)

CONFIGURATIONS FOR LOCKING TYPE PLUGS AND RECEPTACLES

			15 AMPERE RECEPTACLE	15 AMPERE PLUG	20 AMPERE RECEPTACLE	20 AMPERE PLUG	30 AMPERE RECEPTACLE	30 AMPERE PLUG
2-POLE 2-WIRE	125 V	L1	L1-15R	L1-15P				
	250 V	L2			L2-20R	L2-20P		
	277 V AC	L3	F	U	T	U	R	E
	600 V	L4	F	U	T	U	R	E
2-POLE 3 WIRE GROUNDING	125 V	L5	L5-15R	L5-15P	L5-20R	L5-20P	L5-30R	L5-30P
	250 V	L6	L6-15R	L6-15P	L6-20R	L6-20P	L6-30R	L6-30P
	277 V AC	L7	L7-15R	L7-15P	L7-20R	L7-20P	L7-30R	L7-30P
	480 V AC	L8			L8-20R	L8-20P	L8-30R	L8-30P
	600 V AC	L9			L9-20R	L9-20P	L9-30R	L9-30P
3-POLE 3-WIRE	125/250 V	L10			L10-20R	L10-20P	L10-30R	L10-30P
	3Ø 250 V	L11	L11-15R	L11-15P	L11-20R	L11-20P	L11-30R	L11-30P
	3Ø 480 V	L12			L12-20R	L12-20P	L12-30R	L12-30P
	3Ø 600 V	L13					L13-30R	L13-30P
3-POLE 4-WIRE GROUNDING	125/250 V	L14			L14-20R	L14-20P	L14-30R	L14-30P
	3Ø 250 V	L15			L15-20R	L15-20P	L15-30R	L15-30P
	3Ø 480 V	L16			L16-20R	L16-20P	L16-30R	L16-30P
	3Ø 600 V	L17					L17-30R	L17-30P
4-POLE 4-WIRE	3Ø 208Y/120 V	L18			L18-20R	L18-20P	L18-30R	L18-30P
	3Ø 480Y/277 V	L19			L19-20R	L19-20P	L19-30R	L19-30P
	3Ø 600Y/347 V	L20			L20-20R	L20-20P	L20-30R	L20-30P
4-POLE 5-WIRE GROUNDING	3Ø 208Y/120 V	L21			L21-20R	L21-20P	L21-30R	L21-30P
	3Ø 480Y/277 V	L22			L22-20R	L22-20P	L22-30R	L22-30P
	3Ø 600Y/347 V	L23			L23-20R	L23-20P	L23-30R	L23-30P

CONFIGURATIONS FOR LOCKING TYPE PLUGS AND RECEPTACLES

			50 AMPERE RECEPTACLE	50 AMPERE PLUG	60 AMPERE RECEPTACLE	60 AMPERE PLUG
2 POLE 3 WIRE GROUNDING	125 V	L5	L5-50R	L5-50P	L5-60R	L5-60P
	250 V	L6	L6-50R	L6-50P	L6-60R	L6-60P
	277 V AC	L7	L7-50R	L7-50P	L7-60R	L7-60P
	480 V AC	L8	L8-50R	L8-50P	L8-60R	L8-60P
	600 V AC	L9	L9-50R	L9-50P	L9-60R	L9-60P
3 POLE 4 WIRE GROUNDING	125/250 V	L14	L14-50R	L14-50P	L14-60R	L14-60P
	3Ø 250 V	L15	L15-50R	L15-50P	L15-60R	L15-60P
	3Ø 480 V	L16	L16-50R	L16-50P	L16-60R	L16-60P
	3Ø 600 V	L17	L17-50R	L17-50P	L17-60R	L17-60P
4 POLE 5 WIRE GROUNDING	3Ø 208Y/120V	L21	L21-50R	L21-50P	L21-60R	L21-60P
	3Ø 480Y/277V	L22	L22-50R	L22-50P	L22-60R	L22-60P
	3Ø 600Y/347V	L23	L23-50R	L23-50P	L23-60R	L23-60P

Courtesy National Electrical Manufacturers Association (NEMA)

ALTERNATING CURRENT (AC) CIRCUITS

ALTERNATING CURRENT OR VOLTAGE CHANGES WITH TIME, RISING FROM 0 TO MAXIMUM VALUE IN ONE DIRECTION, DECREASING TO 0, THEN RISING TO THE SAME MAXIMUM VALUE IN THE OPPOSITE DIRECTION, AGAIN DECREASING TO 0, THEN REPEATING THESE VALUES AT EQUAL INTERVALS OF TIME. IN THE FOREGOING, THE NUMBER OF CYCLES PER SECOND IS CALLED FREQUENCY (F) OR HERTZ (HZ). IN AC CIRCUITS THE EFFECTIVE VALUES OF CURRENT OR VOLTAGE ARE ALWAYS USED UNLESS OTHERWISE SPECIFIED.

EFFECTIVE VALUE = .707 x MAXIMUM VALUE

MISCELLANEOUS FORMULAS

To find CM (circular mil area) of a conductor, square the diameter.
CM = Diameter in Inches x 1,000 (Also see Chapter 9, Table 8--NEC)

To find CM of a bus bar:

$$CM = \frac{(\text{width in inches} \times 1000) \times (\text{thickness in inches} \times 1000)}{.7854}$$

CM = (width in inches x 1000) x (thickness in inches x 1000) x 1.2732
To find the sq mil area when the CM is known: sq mil = CM x .7854

To find the CM area when the sq mil is known:

$$CM = \frac{\text{sq mil}}{.7854}$$

Temperature Conversions
C = Centigrade degrees F = Fahrenheit degrees
(Also see Note 13 to Ampacity Tables in NEC.)

$$F = \left(\frac{9}{5} \times C\right) + 32 \qquad C = \frac{5}{9} \times (F - 32)$$

ELECTRICAL FORMULAS

DC CIRCUITS--BASIC OHM'S LAW

E = VOLTS R = RESISTANCE I = AMPS OR CURRENT

$$E = IR \quad \text{or} \quad I = \frac{E}{R} \quad \text{or} \quad R = \frac{E}{I}$$

Resistance in Series Circuit
R_t = Total circuit resistance (ohms)
R_1, R_2, etc. = Each resistance in circuit
$R_t = R_1 + R_2 + R_3$ etc.

Resistance in Parallel Circuit

$$\frac{1}{R_t} = \frac{1}{R_1} + \frac{1}{R_2} + \frac{1}{R_3} \text{ etc.}$$

Amps in a Series Circuit
Same amps flow through all parts of circuit.

Amps in a Parallel Circuit
$I_t = I_1 + I_2 + I_3$ etc.

Volts in a Parallel Circuit
Applied volts are the same in all parts of circuit.

Volts in a Series Circuit
Applied volts equal to total volt drop for each load.
$E_t = E_1 + E_2 + E_3$ etc.

Electrical Power (Watts = W)

$$W = EI = I^2R = \frac{E^2}{R}$$

To find I, E and R: $I = \frac{W}{E} = \sqrt{\frac{W}{R}}$; $E = \frac{W}{I} = \sqrt{RW}$; $R = \frac{E^2}{W} = \frac{W}{I^2}$

Convert Mechanical Power to Electrical Power
HP = Horsepower = 746 watts

$HP = \frac{\text{Watts}}{746}$, or Watts x .00134

$HP = \frac{\text{kilowatts}}{.746}$ or kilowatts x 1.34

Efficiency (Ratio of energy output to energy input)

$$Eff = \frac{\text{output}}{\text{input}} \times 100\%$$

$$\text{output} = \frac{\text{input} \times \text{eff}}{100\%}$$

Torque (Expressed in pound-feet or pounds of force at a certain radius.)
Torque (T) and horsepower (HP)
RPM = Speed (revolutions per minute)

$$HP = \frac{6.28 \times RPM \times T}{33{,}000}; \qquad HP = \frac{T \times RPM}{5250}$$

$$T = \frac{33{,}000 \times HP}{6.28 \times RPM}$$

Motors--To find speed, frequency and poles of AC motors:
RPM = Revolutions per minute
F = Frequency in Hertz (cycles per second)
P = Number of field poles

$$RPM = \frac{F \times 120}{P}; \qquad F = \frac{RPM \times P}{120}; \qquad P = \frac{F \times 120}{RPM}$$

Impedance:
Impedance in an AC circuit is similar to resistance in a DC circuit in that both are the total opposition to the flow or current. In AC circuits, Ohm's law could read: "The current is directly proportional to the applied voltage and inversely proportional to the impedance."
Impedance could be defined as the total resultant ohmic combination of resistance and reactance.
Impedance is the vector sum of resistance, capacitive reactance and inductive reactance.
Z = Impedance
Xc = Capacitive Reactance
X_L = Inductive reactance
R = Resistance
F = Hertz (cycles per second)
C = Capacitance (farads)
L = Inductance (henrys)
π = 3.14 (Greek letter pi)

$$Z = \sqrt{R^2 + (X_L - X_C)^2} = 2\ FL;\ X_C = \frac{1}{2\pi FC}$$

AC CIRCUITS--OHM'S LAW

$$E = IZ \qquad I = \frac{E}{Z} \qquad Z = \frac{E}{I}$$

Power Factor:
The power factor of an AC circuit is the ratio between the true power in watts and the apparent power in volt-amperes. Power factor is normally expressed in percent.

PF is actually the cosine of the phase angle that the current leads or lags the voltage in an AC circuit.

PF = Power Factor

$$PF = \frac{\text{True Power}}{\text{Apparent Power}} = \frac{\text{Watts}}{\text{Volt-Amps}} = \frac{kW}{KVA} = \frac{R}{Z}$$

Single-Phase AC Circuits

$$kVA = \frac{E_I}{1{,}000} \quad \frac{kW}{PF} \qquad kW = kVA \times PF$$

$$I = \frac{W}{E \times PF} \qquad E = \frac{W}{I \times PF} \qquad PF = \frac{W}{E_I}$$

Three-Phase Circuits (Balanced Star or Wye)
In (neutral current) = 0
I_l (line current) = I_p (phase current)
E_l (line voltage = 1.73 × E_p (phase voltage)

$$E_p \text{ (phase voltage)} = \frac{E_l \text{ (line voltage)}}{1.73} = E_l \times 0.577$$

Formula for current in the neutral of a three-phase, unbalanced Star or Wye at unity power factor:

In = Neutral current
Ia = Phase A current
Ib = Phase B current
Ic = Phase C current

$$In = \sqrt{Ia^2 + Ib^2 + Ic^2 - IaIb - IbIc - IcIa}$$

Three-Phase Circuits (Balanced Delta)

$$I_l = 1.73 \times I_p$$

$$I_p = \frac{I_l}{1.73} = I_l \times 0.577$$

$$E_p = E_l$$

Watts (Power) in Three-Phase Circuits
W = watts (true power)
VA = volt-amps (apparent power)
E = line volts
I = line amps
W = 1.73 x E x I x PF
VA = 1.73 x E x I

$$E = \frac{W}{1.73 \times I \times PF} = \frac{0.577 \times W}{I \times PF}$$

$$I = \frac{W}{1.73 \times E \times PF} = \frac{0.577 \times W}{E \times PF}$$

$$PF = \frac{W}{1.73 \times E \times I} = \frac{0.577 \times W}{E \times I}$$

Watt (Power) Loss in AC or DC circuits.

$$W = I^2R \qquad I = \sqrt{\frac{W}{R}} \qquad R = \frac{W}{I^2}$$

TYPICAL FAULT CURRENT FORM

DEPARTMENT OF INSPECTIONS
City of Minneapolis

To ______________________________ Date ______________
(Electrical Contractor)

______________________________ Electrical Permit No. ______________
(Address)

The following information is requested to determine whether or not the electrical

equipment installed at ______________________________
(street address) (Name of occupant)

Electrical Permit No. ______________, meets the requirements of the

National Electrical Code as it pertains to available short-circuit current.

Electrical Inspector ______________________ Phone ______________

Trans. size ________ Imp ________ Sec Volt ________ Phase ________ 3-4 Wire ________

Service Conductors Size ________ Copper ________ Aluminum ________ Length ________

Available Fault current at line side of service disconnect ______________

Type, size and fault current interrupting rating of overcurrent device used at main

panel or panels ______________________________

____

Available fault current* at power panel #1 ______________

Available fault current* at power panel #2 ______________

Available fault current* at lighting panel #1 ______________

Available fault current* at lighting panel #2 ______________

Available fault current* at lighting panel #3 ______________

Available fault current* at lighting panel #4 ______________

Available fault current* at lighting panel #5 ______________

Available fault current* at emergency panel ______________

Use back of form for data on additional panels. Show one-line diagram of service and all related equipment on back of form also.

*If panels are protected by fuses, indicate the fault current interrupting rating and the let-thru current of fuses and also include fuse type and size. For circuit breakers indicate the fault current interrupting rating and the let-thru current, voltage, type and size.

(See NEC-110-10 "Circuit Impedence and Other Characteristics" for specific code requirements.) (Also see NEC 110-9, 230-65 and 240-1)

Electrical Contractor: ______________________ Date: ______________

Address: ______________________ Phone ______________

Form No. 32483

(Courtesy of Department of Inspections – Minneapolis, Minnesota)

Index

H

I

J-K

L

M

N

O

U

V

W

X